Kala Azar in South Asia

Eisei Noiri · T.K. Jha
Editors

Kala Azar in South Asia

Current Status and Sustainable Challenges

Second Edition

Springer

Editors
Eisei Noiri
The University of Tokyo Hospital
Bunkyō, Tokyo
Japan

T.K. Jha
Kalazar Research Center
Muzaffarpur
India

ISBN 978-3-319-43611-1   ISBN 978-3-319-47101-3   (eBook)
DOI 10.1007/978-3-319-47101-3

Library of Congress Control Number: 2016953299

1st edition: © Springer Science+Business Media B.V. 2011
2nd edition: © Springer International Publishing 2016
This work is subject to copyright. All rights are reserved by the Publisher, whether the whole or part of the material is concerned, specifically the rights of translation, reprinting, reuse of illustrations, recitation, broadcasting, reproduction on microfilms or in any other physical way, and transmission or information storage and retrieval, electronic adaptation, computer software, or by similar or dissimilar methodology now known or hereafter developed.
The use of general descriptive names, registered names, trademarks, service marks, etc. in this publication does not imply, even in the absence of a specific statement, that such names are exempt from the relevant protective laws and regulations and therefore free for general use.
The publisher, the authors and the editors are safe to assume that the advice and information in this book are believed to be true and accurate at the date of publication. Neither the publisher nor the authors or the editors give a warranty, express or implied, with respect to the material contained herein or for any errors or omissions that may have been made.

Printed on acid-free paper

This Springer imprint is published by Springer Nature
The registered company is Springer International Publishing AG
The registered company address is: Gewerbestrasse 11, 6330 Cham, Switzerland

Preface

> *There are no whole truths; all truths are half-truths. It is trying to treat them as whole truths that plays the devil.*
>
> Alfred North Whitehead

Neglected tropical diseases (NTDs) are a group of diseases with a devastating impact on people in their prime or younger in endemic areas. Some illnesses are abrupt in onset and cured relatively easily. Others occur in chronic or subacute fashion and prove rather difficult to remedy. The impacts are partly linked to socioeconomic factors; risks of losing a daily job due to treatments, insufficient drug efficacy further increasing hospitalization necessity or stay duration, medical expenses, prevalence of fake drugs, folk remedies, and so on. Because these patients often belong to the lowest economic class, they do not have a stable place to live and move frequently, and may not retain the same cell phone number, prognostic surveillance in endemic areas is extremely difficult to conduct.

If a disease has strong infectivity and is life threatening, policy makers must cope with it immediately and completely. NTDs are not at that threat level and have been neglected historically. However, the social burden is high because the victims, often children and the prime generation in each country, become encumbrances. Visceral leishmaniasis (VL), also known as kala-azar, is a part of the spectrum of NTDs and combines all the earlier mentioned traits. Many stakeholders have contributed from the beginning of the twenty-first century in the Indian subcontinent to reducing the incidence of VL; this goal formed the basis of a memorandum of understanding shared through WHO with ministers in the countries concerned. SATREPS, funded by JICA and JST, is one of the first projects stating the necessity of simple and reliable diagnostic procedures in endemic areas. Ours has been the only project targeting NTDs from Japan until 2015.

This scope was introduced in the first edition. The present edition further explores the practical applicability of methods of disease control that have been lacking in the NTD field, and data collection is still ongoing. All of the contributing authors are experts and have devoted part of their careers, in cooperation with other stakeholders, to VL disease control for people living in endemic areas. I hope this volume will tack the sail for disease eradication and allow it to catch the wind and make way to a speedy conclusion.

Tokyo, Japan Eisei Noiri
March 2016

Contents

Part I   Introduction Into Kala-Azar

1 **Pathology and Mechanism of Disease in Kala-Azar**   3
   Shyamal Paul and Monira Pervin

2 **Containing Post Kala-Azar Dermal Leishmaniasis (PKDL): Pre-requisite for Sustainable Elimination of Visceral Leishmaniasis (VL) from South Asia**   7
   Poonam Salotra, Himanshu Kaushal and V. Ramesh

3 **Characteristics of Patients Visiting Suruya Kanta Kala-Azar Research Center in Mymensingh, Bangladesh; Data from a Patient Registry System**   23
   Masao Iwagami, Bumpei Tojo and Eisei Noiri

Part II   Therapeutic Strategy to Deal with Emergence of Drug Resistance

4 **Epidemiology of Drug-Resistant Kala-Azar in India and Neighboring Countries**   33
   T.K. Jha

5 **A Therapeutic Strategy for Treating Visceral Leishmaniasis in Regions with Drug Resistance**   53
   Shyam Sundar and Dipti Agarwal

6 **Treatment of Post-kala-azar Dermal Leishmaniasis**   67
   V. Ramesh and Prashant Verma

7 **Combination Therapy for Leishmaniases**   79
   Farrokh Modabber

8 **Vaccine Development for Leishmaniasis**   89
   Yasuyuki Goto

9 Siccanin Is a Novel Selective Inhibitor of Trypanosomatid Complex II (Succinate-Ubiquinone Reductase) and a Potent Broad-Spectrum Anti-trypanosomatid Drug Candidate 101
Nozomu Nihashi, Daniel Ken Inaoka, Chiaki Tsuge, Emmanuel Oluwadare Balogun, Yasutaka Osada, Yasuyuki Goto, Yoshitsugu Matsumoto, Takeshi Nara, Tatsushi Mogi, Shigeharu Harada and Kita Kiyoshi

Part III Diagnostic Strategy Enhancing Kala-Azar Elimination Program

10 Challenges in the Diagnosis of Visceral Leishmaniasis on the Indian Subcontinent. 125
Suman Rijal, François Chappuis, Jorge Alvar and Marleen Boelaert

11 Changes of *Leishmania* Antigens in Kala-Azar Patients' Urine After Treatment 135
Sharmina Deloer, Sohel Mohammad Samad, Hidekazu Takagi, Chatanun Eamudomkarn, Kazi M. Jamil, Eisei Noiri and Makoto Itoh

12 Potentiality of Urinary L-FABP Tests to Kala-Azar Disease Management. 141
Eisei Noiri, Yoshifumi Hamasaki, Bumpei Tojo, Kazi M. Jamil, Kent Doi and Takeshi Sugaya

13 Antibody Capture Direct Agglutination Test (abcDAT) for Diagnosis of Visceral Leishmaniasis with Urine 161
Fumiaki Nagaoka, Hidekazu Takagi, Eisei Noiri and Makoto Itoh

14 Loop-Mediated Isothermal Amplification (LAMP): Molecular Diagnosis for the Field Survey of Visceral Leishmaniasis 167
Hidekazu Takagi, Makoto Itoh and Eisei Noiri

15 Applicability of Multiplex Real-Time PCR to Visceral Leishmaniasis. 173
Yoshifumi Hamasaki, Hirofumi Aruga, Chizu Sanjoba, Hidekazu Takagi, Shyamal Paul, Yoshitsugu Matsumoto and Eisei Noiri

Part IV Pathogenesis of Kala-Azar and PKDL

16 Immunoglobulins in the Pathophysiology of Visceral Leishmaniasis. 187
Satoko Omachi, Yoshitsugu Matsumoto and Yasuyuki Goto

Contents

Part V Knowledge and Practice for Vector Control in Kala-Azar

17 Geographical Distribution and Ecological Aspect of Sand Fly Species in Bangladesh.................................. 199
Yusuf Özbel, Chizu Sanjoba and Yoshitsugu Matsumoto

18 The Efficacy of Long Lasting Insecticidal Nets for Leishmaniasis in Asia.. 211
Chizu Sanjoba, Yusuf Özbel and Yoshitsugu Matsumoto

Part VI New Challenges in Kala-Azar Elimination Programme

19 Environmental Change and Kala-Azar with Particular Reference to Bangladesh...................................... 223
Ashraf Dewan, Masahiro Hashizume, Md. Masudur Rahman, Abu Yousuf Md. Abdullah, Robert J. Corner, Md. Rakibul Islam Shogib and Md. Faruk Hossain

20 Drug Safety Monitoring for Liposomal Amphotericin B......... 249
Eisei Noiri and Kosuke Minami

21 High Resolution Mapping of Kala-Azar Hot Spots Using GPS Logger and Urinary Antibody Measurements............. 257
Bumpei Tojo, Makoto Itoh, Mohammad Sohel Samad, Emi Ogasawara, Dinesh Mondal, Rashidul Haque and Eisei Noiri

22 Geography and Reality of Kala-Azar Endemic in Bangladesh, Analysis Using GIS and Urine-Based Mass Screening........... 271
Bumpei Tojo, Makoto Itoh, Shyamal Kumar Paul, Mohammad Sohel Samad, Emi Ogasawara, Fumiaki Nagaoka, Dinesh Mondal, Rashidul Haque and Eisei Noiri

23 Animal Models of Visceral Leishmaniasis and Applicability to Disease Control.. 287
Yasutaka Osada, Satoko Omachi, Chizu Sanjoba and Yoshitsugu Matsumoto

24 Pharmacovigilance on Therapeutic Protocols for Visceral Leishmaniasis.. 297
Eisei Noiri, Bumpei Tojo, Yoshifumi Hamasaki, Masao Iwagami, Takeshi Sugaya, Michiyo Harada, Progga Nath, Ariful Basher, Dinesh Mondal, Rashidul Haque, Fasifur Rahman and Shyamal Paul

Index 307

Contributors

Abu Yousuf Md. Abdullah Department of Geography & Environment, University of Dhaka, Dhaka, Bangladesh

Dipti Agarwal Department of Paediatrics, Sarojini Naidu Medical College, Agra, India

Jorge Alvar Leishmaniasis Program, Drugs for Neglected Diseases initiative, Geneva, Switzerland

Hirofumi Aruga Research and Development, Thermo Fisher Scientific, Tokyo, Japan

Emmanuel Oluwadare Balogun Department of Biomedical Chemistry, Graduate School of Medicine, The University of Tokyo, Tokyo, Japan; Department of Biochemistry, Ahmadu Bello University, Zaria, Nigeria

Ariful Basher Surya Kanta Kala-azar Research Center, Mymensingh, Bangladesh

Marleen Boelaert Epidemiology and Disease Control Unit, Department of Public Health, Institute of Tropical Medicine, Antwerp, Belgium

François Chappuis Head of Division of International and Humanitarian Medicine, Geneva University Hospitals, Geneva 14, Switzerland

Robert J. Corner Department of Spatial Sciences, Curtin University, Perth, Australia

Sharmina Deloer Department of Microbiology and Immunology, Aichi Medical University School of Medicine, Nagakute, Japan; Department of Infection and Immunology, Aichi Medical University, Aichi, Nagakute, Japan

Ashraf Dewan Department of Spatial Sciences, Curtin University, Perth, Australia

Kent Doi Nephrology 107 Lab, The University of Tokyo Hospital, Tokyo, Japan

Chatanun Eamudomkarn Department of Microbiology and Immunology, Aichi Medical University School of Medicine, Nagakute, Japan; Department of Parasitology, Faculty of Medicine, Khon Kaen University, Khon Kaen, 40000, Thailand; Department of Infection and Immunology, Aichi Medical University, Aichi, Nagakute, Japan

Yasuyuki Goto Laboratory of Molecular Immunology, Department of Animal Resource Sciences, Graduate School of Agricultural and Life Sciences, The University of Tokyo, Tokyo, Japan

Yoshifumi Hamasaki Total Renal Care Medicine, Nephrology 107 Lab, The University of Tokyo Hospital, Tokyo, Japan; 22nd Century Medical and Research Center, The University of Tokyo Hospital, Tokyo, Japan; Department of Hemodialysis and Apheresis, The University of Tokyo Hospital, Tokyo, Japan; Department of Nephrology and Endocrinology, The University of Tokyo Hospital, Tokyo, Japan

Rashidul Haque Department of Hemodialysis and Apheresis, The University of Tokyo Hospital, Bunkyo-ku, Tokyo, Japan; Infectious Diseases, ICDDR, B, Dhaka, Bangladesh

Michiyo Harada Department of Hemodialysis and Apheresis, The University of Tokyo Hospital, Tokyo, Japan; Department of Nephrology and Endocrinology, The University of Tokyo Hospital, Tokyo, Japan

Shigeharu Harada Department of Applied Biology, Kyoto Institute of Technology, Kyoto, Japan

Masahiro Hashizume Institute of Tropical Medicine, Nagasaki University, Nagasaki, Japan

Md. Faruk Hossain Department of Geography & Environment, University of Dhaka, Dhaka, Bangladesh

Daniel Ken Inaoka Department of Biomedical Chemistry, Graduate School of Medicine, The University of Tokyo, Tokyo, Japan

Makoto Itoh Department of Microbiology and Immunology, Aichi Medical University School of Medicine, Nagakute, Aichi, Japan; Department of Infection and Immunology, Aichi Medical University, Aichi, Nagakute, Japan; Department of Hemodialysis and Apheresis, The University of Tokyo Hospital, Bunkyo-ku, Tokyo, Japan

Masao Iwagami Faculty of Epidemiology and Population Health, London School of Hygiene and Topical Medicine, London, UK; Department of Hemodialysis and Apheresis, The University of Tokyo Hospital, Tokyo, Japan; Department of Nephrology and Endocrinology, The University of Tokyo Hospital, Tokyo, Japan; Department of Hemodialysis and Apheresis, The University of Tokyo Hospital, Tokyo, Japan

Contributors

Kazi M. Jamil Clinical Science Division, International Center for Diarrhoeal Disease Research, Bangladesh, (ICDDR,B), Dhaka, Bangladesh; International Centre for Diarrhoeal Disease Research, Dhaka, Bangladesh

T.K. Jha Kalazar Research Centre Brahmpura, Muzaffarpur, Bihar, India

Himanshu Kaushal National Institute of Pathology (ICMR), New Delhi, India

Kita Kiyoshi Department of Biomedical Chemistry, Graduate School of Medicine, The University of Tokyo, Tokyo, Japan; School of Tropical Medicine and Global Health, Nagasaki University, Nagasaki, Japan

Yoshitsugu Matsumoto Laboratory of Molecular Immunology, Graduate School of Agricultural and Life Sciences, The University of Tokyo, Tokyo, Japan; Graduate School of Agricultural and Life Sciences, The University of Tokyo, Tokyo, Japan; Laboratory of Molecular Immunology, Graduate School of Agricultural and Life Sciences, The University of Tokyo, Bunkyo-ku, Tokyo, Japan; Department of Molecular Immunology, School of Agricultural and Life Sciences, University of Tokyo, Tokyo, Japan; Faculty of Medicine, Department of Parasitology, Ege University, İzmir, Turkey; Laboratory of Molecular Immunology, Department of Animal Resource Sciences, Graduate School of Agricultural and Life Sciences, The University of Tokyo, Bunkyo-ku, Tokyo, Japan

Kosuke Minami International Research Center for Materials Nanoarchitectonics, National Institute for Materials Science, Tsukuba, Japan

Farrokh Modabber Honorary Faculty, Center for Research and Training on Skin Diseases and Leprosy (CRTSDL), Tehran University of Medical Sciences, Tehran, Iran

Tatsushi Mogi Department of Biomedical Chemistry, Graduate School of Medicine, The University of Tokyo, Tokyo, Japan; President's Office, Medical & Biological Laboratories Co., Ltd., Nagoya, Japan

Dinesh Mondal Nutrition and Clinical Services, ICDDR, B, Dhaka, Bangladesh; Department of Hemodialysis and Apheresis, The University of Tokyo Hospital, Bunkyo-ku, Tokyo, Japan; Department of Hemodialysis and Apheresis, The University of Tokyo Hospital, Tokyo, Japan

Fumiaki Nagaoka Department of Microbiology and Immunology, Aichi Medical University School of Medicine, Nagakute, Aichi, Japan; Department of Hemodialysis and Apheresis, The University of Tokyo Hospital, Tokyo, Japan

Takeshi Nara Department of Molecular and Cellular Parasitology, Juntendo University, Tokyo, Japan

Progga Nath Surya Kanta Kala-azar Research Center, Mymensingh, Bangladesh

Nozomu Nihashi Department of Biomedical Chemistry, Graduate School of Medicine, The University of Tokyo, Tokyo, Japan

Eisei Noiri Department of Hemodialysis and Apheresis, The University of Tokyo Hospital, Tokyo, Japan; Hemodialysis and Apheresis, Nephrology 107 Lab, The University of Tokyo Hospital, Tokyo, Japan; Department of Nephrology & Endocrinology, Department of Hemodialysis & Apheresis, University Hospital, The University of Tokyo, Tokyo, Japan; Hemodialysis and Apheresis, Nephrology 107 Lab, The University of Tokyo Hospital, Tokyo, Japan; Department of Hemodialysis and Apheresis, The University of Tokyo Hospital, Bunkyo-ku, Tokyo, Japan; Department of Hemodialysis and Apheresis, The University of Tokyo Hospital, Tokyo, Japan; Department of Nephrology and Endocrinology, The University of Tokyo Hospital, Tokyo, Japan

Emi Ogasawara Department of Hemodialysis and Apheresis, The University of Tokyo Hospital, Bunkyo-ku, Tokyo, Japan; Department of Hemodialysis and Apheresis, The University of Tokyo Hospital, Tokyo, Japan

Satoko Omachi Laboratory of Molecular Immunology, Graduate School of Agricultural and Life Sciences, The University of Tokyo, Bunkyo-ku, Tokyo, Japan; Laboratory of Molecular Immunology, Department of Animal Resource Sciences, Graduate School of Agricultural and Life Sciences, The University of Tokyo, Bunkyo-ku, Tokyo, Japan

Yasutaka Osada Laboratory of Molecular Immunology, Department of Animal Resource Sciences, Graduate School of Agricultural and Life Sciences, The University of Tokyo, Bunkyo-ku, Tokyo, Japan

Yusuf Özbel Faculty of Medicine, Department of Parasitology, Ege University, Bornova, İzmir, Turkey

Shyamal Paul Department of Microbiology, Mymensingh Medical College, Mymensingh, Bangladesh

Shyamal Kumar Paul Department of Hemodialysis and Apheresis, The University of Tokyo Hospital, Tokyo, Japan

Monira Pervin Department of Virology, Dhaka Medical College, Dhaka, Bangladesh

Fasifur Rahman Armed Forces Medical College, Dhaka, Bangladesh

Md. Masudur Rahman Department of Geography & Environment, University of Dhaka, Dhaka, Bangladesh

V. Ramesh Department of Dermatology & STD, Safdarjung Hospital and Vardhman Mahavir Medical College, New Delhi, India; Department of Dermatology & STD, Vardhman Mahavir Medical College & Safdarjang Hospital, New Delhi, India

Suman Rijal Drugs for Neglected Diseases Initiative, India Regional Office, New Delhi, India

Poonam Salotra National Institute of Pathology (ICMR), New Delhi, India

Mohammad Sohel Samad Department of Hemodialysis and Apheresis, The University of Tokyo Hospital, Bunkyo-ku, Tokyo, Japan; Department of Hemodialysis and Apheresis, The University of Tokyo Hospital, Tokyo, Japan

Sohel Mohammad Samad Department of Microbiology and Immunology, Aichi Medical University School of Medicine, Nagakute, Japan; Department of Infection and Immunology, Aichi Medical University, Aichi, Nagakute, Japan

Chizu Sanjoba Graduate School of Agricultural and Life Sciences, The University of Tokyo, Tokyo, Japan; Department of Molecular Immunology, School of Agricultural and Life Sciences, University of Tokyo, Tokyo, Japan; Faculty of Medicine, Department of Parasitology, Ege University, İzmir, Turkey; Laboratory of Molecular Immunology, Department of Animal Resource Sciences, Graduate School of Agricultural and Life Sciences, The University of Tokyo, Bunkyo-ku, Tokyo, Japan

Md. Rakibul Islam Shogib Department of Geography & Environment, University of Dhaka, Dhaka, Bangladesh

Takeshi Sugaya CMIC Holdings Co., Ltd., Tokyo, Japan

Shyam Sundar Department of Medicine, Institute of Medical Sciences, Banaras Hindu University, Varanasi, India

Hidekazu Takagi Department of Infection and Immunology, Aichi Medical University, Aichi, Nagakute, Japan; Department of Microbiology and Immunology, Aichi Medical University School of Medicine, Nagakute, Aichi, Japan; Department of Parasitology, Aichi Medical University School of Medicine, Nagakute, Aichi, Japan

Bumpei Tojo Department of Hemodialysis and Apheresis, The University of Tokyo Hospital, Tokyo, Japan; Hemodialysis and Apheresis, Nephrology 107 Lab, The University of Tokyo Hospital, Tokyo, Japan; Department of Hemodialysis and Apheresis, The University of Tokyo Hospital, Bunkyo-ku, Tokyo, Japan; Department of Hemodialysis and Apheresis, The University of Tokyo Hospital, Tokyo, Japan; Department of Nephrology and Endocrinology, The University of Tokyo Hospital, Tokyo, Japan

Chiaki Tsuge Department of Biomedical Chemistry, Graduate School of Medicine, The University of Tokyo, Tokyo, Japan

Prashant Verma Department of Dermatology & STD, Vardhman Mahavir Medical College & Safdarjang Hospital, New Delhi, India

Abbreviations

ABC	ATP binding cassette
abcDAT	Antibody capture direct agglutination test
ACD	Active case detection
AIDS	Acquired immune deficiency syndrome
AKI	Acute kidney injury
ALT	Alanine aminotransferase
AMPH-B	Amphotericin B
Ampho	Amphotericin-B deoxycholate
ART	Anti retroviral therapy treatment
AST	Aspartate aminotransferase
ATP	Adenosine triphosphate
AUC	Area under curve
AVNIR-2	Advanced Visible and Near Infrared Radiometer-II
BAFF	B-cell activating factor
BCC	Behavior change communication
BCG	Bacille Calmette- Guerin
BMI	Body mass index
BUN	Blood urea nitrogen
CA	Circulating antigen
CCP	Cyclic citrullinated peptides
CDDP	*cis*-platinum
CI	Confidence interval
Cin	Inulin clearance
CKD	Chronic kidney disease
CL	Cutaneous leishmaniasis
COX2	Cyclooxygenase 2
CPB	Cysteine protease B
CPB	Cardio-pulmonary bypass surgery
CRP	C-reacting protein
DAT	Direct agglutination test

DCIP	2,4-dichlorophenolindophenol
DDM	n-dodecyl-D-maltoside
DDT	Dicophane (dichlorodiphenyltrichloroethane)
DGHS	Director General of Health Services
DND*i*	The Drugs for Neglected Diseases *initiative*
DOT	Directly observed treatment
DSPG	Distearoyl phosphatidylglycerol
DTH	Delayed type hypersensitivity
ELISA	Enzyme-linked immunosorbent assay
ENSO	El Nino/Southern Oscillation
EVM	Environmental management
FABP1	Fatty acid binding protein 1
FAD	Flavin adenine dinucleotide
FFA	Free fatty-acids
FGS	Focal glomerular sclerosis
FGT	Formol gel test
FPF	False positive fraction
G3P	Glycerol 3-phosphate
G3PDH	Glycerol 3-phosphate dehydrogenase
GIS	Geographical Information System
GM-CSF	Granulocyte macrophage -colony stimulating factor
GPS	Geographical Positioning System
HHE	4-hydroxy-2-*hexenal*
HIF-1	Hypoxia inducible factor-1
HIV	Human immunodeficiency virus
HNE	4-Hydroxy-2,3-nonenal
HNF	Hepatocyte nuclear factor
HQNO	2-heptyl-4-hydroxyquinoline 1-oxide
hrCNE	High resolution clear native electrophoresis
HRE	Hypoxia responsive element
ICT	Immunochromatography
IEC	Information, education, communication
IFN-γ	Interferon-γ
IFN	Interferon
IFNy	Interferon gamma
Ig	Immunoglobulin
IL	Interleukin
iNOS	Inducible nitric oxide synthase
IPCC	Intergovernmental Panel on Climate Change
IQR	Interquartile range
IRS	Indoor residual spraying
ITN	Insecticide-treated nets
KA	Kala-azar
KAMRC	Kala-azar Medical Research Center
KAtex	Latex agglutination test

Abbreviations

kDNA	Kinetoplast DNA
kDNA	Kinetoplastid
KIM-1	Kidney inflammatory molecule-1
L-AMB	Liposomal amphotericin B
LAMB	Liposomal amphotericin B
LAMB	Liposomal formulations
LAMP	Loop-mediated isothermal amplification
LC-MS/MS	Liquid chromatography-mass spectrometry/ tandem mass spectrometry
LDU	Leishman Donovan units
L-FABP	L-type fatty acid binding protein, equivalent to FABP1
LLIN	Long-lasting insecticidal net
LnPCR	Nested PCR
LtSQR	SQR from *L. tarentolae*
MCL	Mucocutaneous leishmaniasis
MDRD	Modification of diet in renal disease study group
MKD	Milligram per kilogram of body weight daily
MMT	3-(4, 5-dimethyl-2-thiazolyl)-2, 5-diphenyl-2H-tetrazolium bromide
NADH	Nicotinamide adenine dinucleotide
NAG	N-acetyl-β-D-glucosaminidase
NDH2	Quinone reductase
NEHC	Non endemic healthy controls
NGAL	Neutrophil gelatinase-associated lipocalin
NO	Nitric oxide
NTDs	Neglected tropical diseases
PBS	Phosphate buffered saline
PCR	Polymerase chain reaction
PEG	Polyethylene glycol
PKDL	Post kala-azar-dermal leishmaniasis
PMS	Phenazinium methylsulfate
PMX-F	Polymyxin B-immobilized fiber
PPAR	Peroxisome proliferator activated receptor
RA	Rheumatoid arthritis
RBC	Red blood cells
RDT	Rapid diagnostic test
REC	Reticulo endothelial cell
RF	Rheumatoid factor
rKRP42	Recombinant kinesin related protein 42
RMRI	Rajendra Memorial Research Institute
RNA	Ribonucleic acid
RNase P	Ribonuclease P
ROC	Receiver operation curve
ROOH	Hydroperoxide radicals
RPPH1	Ribonuclease P RNA component H1

RT	Room temperature
SAG	Sodium antimony gluconate
Sb^V	Pentavalent antimonial
SCr	serum creatinine
SD	standard deviation
SDH	succinate dehydrogenase
SDS-PAGE	sodium dodecyl sulfate-poly-acrylamide gel
SKKRC	Surya Kanta Kala-azar Research Center
SKKRC	Surya Kanto VL (Kala-Azal) Research Center
SLE	Systemic lupus erythematosus
SML	Sucrose monolaurate
SQR	Succinate-ubiquinone oxidoreductase, equivalent to succinate-ubiquinone reductase
SSG	Sodium stibogluconate
TAO	Trypanosoma alternative oxidase
TCA	Tricarboxylic acid
TcSQR	SQR from *T. cruzi*
TEM	Transmission electron microscope
TGF	Transforming growth factor
Th1	2T helper cell type 1, 2
TPF	True positive fraction
TTFA	Thenoyltrifluoroacetone
UHC	Upazila Health Complex
UQ_1	Ubiquinone 1
VL	Visceral leishmaniasis
VLEP	VL Elimination Program
WHO	World Health Organization
WHOPES	WHO Pesticide Evaluation Scheme
$\alpha 1MG$	α_1-microglobulin
$\beta 2MG$	β_2-microglobulin

Part I
Introduction Into Kala-Azar

Chapter 1
Pathology and Mechanism of Disease in Kala-Azar

Shyamal Paul and Monira Pervin

Abstract Visceral leishmaniasis (VL) or kala-azar is a neglected tropical disease caused by the intracellular protozoan *Leishmania donovani* and is transmitted by an infected female sand fly vector. The sand fly inoculates the promastigote form of the parasite to the human host through the dermis during a blood meal. This extracellular form is rapidly taken up, principally by the host macrophage, where it undergoes metamorphosis to an amastigote. After rupture of the macrophage, the amastigote circulates in the blood and other body tissues. These amastigotes are again taken up by another sand fly during its next blood meal and again transform in the fly gut to its infective promastigote form to infect the next victim. After a variable incubation period there develops a spectrum of clinical manifestations like fever, weight loss, anaemia, and splenomegaly. These clinical manifestations are associated with immune suppression (CMI) leading to parasite survival accompanied by induction of IL-10 and/or IL-4 in tissues, hypergammaglobulinemia, and an increased amount of IL-4 in the blood of VL patients.

Keywords Visceral leishmaniasis · Kala-azar · Pathogenesis

1.1 Life Cycle of *Leishmania Donovani*

Leishmania donovani is a species of intracellular parasitic protozoan belonging to the genus *Leishmania* that causes the disease kala-azar. The natural transmission in humans is carried out by the bite of an infected sand fly of the genus *Phlebotomus*, which transmits the promastigote form of the parasite to the susceptible human host. The parasite has 2 stages in its life cycle: 1. the amastigote form occurring in humans, and 2. the promastigote form occurring in the sand fly.

S. Paul (✉)
Department of Microbiology, Mymensingh Medical College, Mymensingh, Bangladesh
e-mail: drshyamal10@yahoo.com

1.1.1 The Amastigote Form Occurring in Humans

In humans the promastigote form is injected by the sand fly using its proboscis through the skin during a blood meal. Many promastigotes are taken up through phagocytosis by mononuclear phagocytes, and inside these cells they undergo spontaneous transformation into oval-shaped amastigote forms called *L. donovani* bodies. While residing inside the cells the amastigotes multiply by binary fission. Multiplication continues until the host cell becomes packed with the parasites and eventually ruptures. In a fully packed cell there can be 50–200 amastigotes that are released into tissue spaces upon rupture. Each individual amastigote is then capable of invading fresh cells; thus entire systems are progressively infected. A number of free amastigotes then enter the blood stream where many are phagocytosed by macrophages and this process is repeated. These free and phagocytosed amastigotes in the peripheral blood are then taken up by blood-feeding sand flies.

1.1.2 The Promastigote Form Occurring in Sand Flies

Amastigotes ingested by blood feeding sand flies undergo further development only in the digestive tract of the female sand fly. They enter the mid gut of the sand fly and undergo structural modification into flagellated promastigotes, becoming larger and elongated. They multiply rapidly by binary fission in the gut epithelium. They then migrate back towards the anterior part of the digestive system: the pharynx and buccal cavity. This process is known as anterior station development. A heavy infection of the pharynx can be observed within 6–9 days of the sand fly's infective blood meal. The promastigotes become infective only at this time, and the stage is called the metacyclic stage. The metacyclic promastigotes then enter the hollow proboscis where they accumulate and completely block the esophagus; the sand flies then regurgitate promastigotes from their buccal cavity into the puncture wound of the host skin when they bite again, which invariably results in infection [1, 2] Fig. 1.1.

1.2 Pathogenesis and Pathology of Visceral Leishmaniasis

Visceral leishmaniasis (VL), also known as kala-azar, is caused by obligate intracellular protozoan parasites of the *L. donovani* species complex (i.e., *L. donovani, L. infantum,* and *L. chagasi*), and is transmitted by sandflies (*Phlebotomus* species). It exists as extracellular flagellated promastigotes in the gut of female sand flies where it multiplies and develops extracellularly, after which it is transmitted during blood meal into the mammalian host. After inoculation in the dermis of the host, the parasites attract macrophages, which serve as the host for parasite multiplication.

1 Pathology and Mechanism of Disease in Kala-Azar

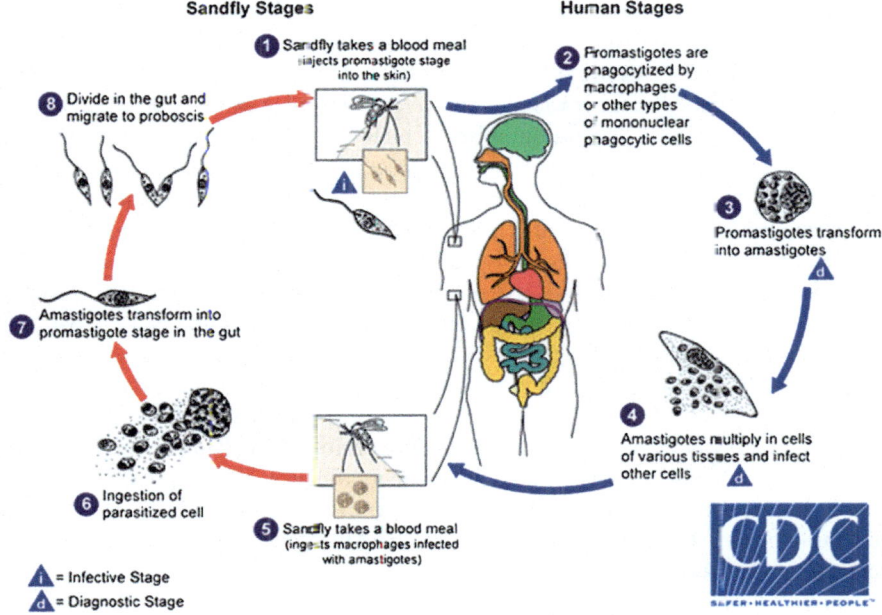

Fig. 1.1 Life cycle of *Leishmania donovani* obtained from CDC web site http://www.dpd.cdc.gov/dpdx/HTML/Leishmaniasis.htm

Promastigotes are rapidly phagocytosed by dermal macrophages through a complement receptor-3 dependent mechanism and eventually metamorphose to amastigotes by promastigote surface glycoprotein gp63. The amastigotes settle in acidic parasitophorous vacuoles and avoid hydrolysis with the help of another promastigote surface molecule: lipophosphoglycan that prevents the formation of a phagolysosome.

In VL the incubation period commonly varies from 2 months to 1 year, but it could be as short as 10 days or as long as 2 years or more. The infection may be asymptomatic or may lead to full-blown kala-azar, depending upon the host immunological response. Patients with VL present symptoms and signs of low-grade recurrent fever and malaise, followed by progressive wasting, anaemia, and hepatosplenomegaly. If untreated, VL proves fatal within 2–3 years. Common histological features are the early accumulation of mononuclear phagocytic cells in the invaded tissues leading to hyperplasia of reticuloendothelial cells (REC) of the organs involved. The pathology of VL is subjugated by the specific suppression of cell-mediated immunity; this permits the dissemination and uncontrolled multiplication of the parasite within the macrophage leading to cellular rupture and resulting in various complications. The REC hyperplasia that follows infection with *L. donovani* affects the spleen, liver, mucosa of small intestine, bone marrow, and lymph nodes. The skin may also become infected as a consequence of kala-azar, a condition known as post kala-azar leishmaniasis.

Haematopoiesis is initially normal but later becomes depressed. The life span of leukocytes and erythrocytes is reduced due to spleenomegaly and stasis of blood in the spleen sinusoids, causing granulocytopenia and anaemia. Liver function is normal and few hepatocytes are invaded; however, prothrombin production is decreased. Thrombocytopenia and the prothrombin depletion may result in severe mucosal haemorrhage. Intercurrent infections are frequent, especially pneumonia, dysentery, and tuberculosis, and these are the common causes of death in the advanced stage.

During progressive infection, Th2-type CD4+ T cells proliferate and secrete interleukin(IL)-4, resulting in polyclonal B-cell activation with nonspecific immunoglobulin, especially IgG and specific immunoglobulin of different isotypes and subtypes of IgG. These have no protective role against this intracellular parasite and may contribute to disease progression and immune-complex diseases including immune-mediated vasculitis and glomerulonephritis. Patients who have recovered from VL are protected, but reactivation can occur. In VL patients, the inability to control *L. donovani* infection is associated with a profound T cell unresponsiveness to *L. donovani* antigens due to cellular anergy. Inappropriate antigen presentation and communication between the antigen-presenting cells and T cells, as well as the induction of IL-10 and IL-10 producing FOXP3-Treg cells, may be associated with the anergy [1, 3–6].

References

1. Chappuis F, Sundar S, Hailu A, Ghalib H, Rijal S, Peeling R, J Alvar, Boelaert M. Visceral leishmaniasis: what are the needs for diagnosis, treatment and control? Nat Rev Microbiol. 2007;5:873–82.
2. Chatterjee KD. Parasitology. 13th ed. Chapter II. New Delhi: CBS Publishers & Distributors Pvt. Ltd.; 2011. p. 64–89.
3. Asad MD, Ali N. Dynamicity of immune regulation during visceral leishmaniasis. Proc Indian Natn Sci Acad. 2014;809(2):247–67.
4. Mutiso J, Macharia J, Gicheru M, Ozwar H. Immunology of leishmaniasis. Sci Parasitol. 2013;14(2):51–61.
5. Malla N, Mahajan R. Pathophysiology of visceral leishmaniasis—some recent concepts. Indian J Med Res. 2006;123:267–74.
6. Shrma U, Singh S. Immunology of leishmaniasis. Indian J Exp, Biol. 2009;47:412–23.

Chapter 2
Containing Post Kala-Azar Dermal Leishmaniasis (PKDL): Pre-requisite for Sustainable Elimination of Visceral Leishmaniasis (VL) from South Asia

Poonam Salotra, Himanshu Kaushal and V. Ramesh

Abstract Post Kala-Azar Dermal Leishmaniasis (PKDL) is a chronic dermal manifestation which appears in a small proportion of cases following cure from visceral leishmaniasis (VL) episode, and occasionally in patients with no history of VL. The global prevalence of PKDL is not well studied and the available data are based only on estimates. As per the available reports, the incidence of PKDL varies considerably within endemic countries. PKDL diagnosis remains a challenge more because serology does not have much relevance while parasitological and molecular diagnostic tests show either low sensitivity or are difficult to decentralize in the field. The available treatment options are costly, lengthy and frequently toxic. It is believed that PKDL has a multi-factorial and complex origin combining host and parasite factors and perhaps the treatment rendered in VL treatment. PKDL patients harbor *Leishmania* parasites in the skin, therefore, are considered a durable reservoir of infection that may propagate VL transmission, especially during inter-epidemic periods. Hence, PKDL poses a serious threat to the success of VL elimination program in South Asia and calls for combined and coordinated efforts towards its surveillance and management in India, Nepal, and Bangladesh. In a nutshell, containing PKDL is a must for sustainable elimination of VL from South Asia where VL transmission is anthroponotic.

Keywords Bangladesh · India · Kala-azar · Leishmania · Nepal · Post kala-azar dermal leishmaniasis · Visceral leishmaniasis elimination program · Visceral leishmaniasis

---

P. Salotra (✉)
National Institute of Pathology (ICMR), New Delhi, India
e-mail: poonamsalotra@hotmail.com

© Springer International Publishing 2016
E. Noiri and T.K. Jha (eds.), *Kala Azar in South Asia*,
DOI 10.1007/978-3-319-47101-3_2

2.1 Epidemiology

Post kala-azar dermal leishmaniasis (PKDL) is well known to occur in regions endemic for *Leishmania donovani,* with variable incidence in affected countries. Information on global epidemiology of PKDL is fragmentary since only limited studies are available. The highest incidence of PKDL has been recorded from Sudan in East Africa, and in Bangladesh in the Indian subcontinent. More importantly, its distribution predominantly reflects the distribution of visceral leishmaniasis (VL) or kala-azar (KA) caused by *L. donovani*. The factors that predispose apparently cured VL patients to PKDL in an endemic zone are not well understood but appear to be related to age at the time of contracting VL and incomplete treatment course [1]. PKDL is reported to occur not only after treatment of VL with sodium stibogluconate (SSG) [2], but also following treatment with miltefosine, amphotericin B [3] and paromomycin [4]. Longitudinal follow-up studies would help to find out the relation between the type of drug used for VL treatment and the likelihood of patients developing PKDL thereafter. Furthermore, approximately 20 % PKDL cases do not report a history of VL [5] which poses another intriguing question in precise understanding of PKDL pathogenesis.

The first population-based study of PKDL incidence was carried out in Bangladesh [6]. It combined a house-to-house survey (2007–2008) and a retrospective study that included 22,699 people from 4,553 households. In total, 813 cases of VL were detected with 79 PKDL cases (9.7 %) among them. It is noteworthy that 8 of the 79 PKDL cases (10.1 %) reported no history of VL. The median duration between VL episode and PKDL was 22 months. More specifically, 20 % of PKDL cases appeared within 6 months, 40 % between 6–24 months, and 40 % after 24 months. Uncommon events were also reported such as PKDL occurring concomitantly with VL or a VL relapse and spontaneous cure. Subsequent studies suggested that the epidemiology varies widely, even among districts of Bangladesh and showed a prevalence of PKDL of 6–16/10,000 population [7, 8].

In India, PKDL appears in up to 10 % of VL treated cases and usually occurs within 5 years after treatment [9]. In a highly endemic area of Bihar, a three stage house-to-house survey of 2020 households' demonstrated local prevalence of confirmed PKDL of 4.4/10,000 and 7.8 if probable cases were also considered [10].

In Nepal, the incidence of VL is much lower compared to Bangladesh or India. Most of the cases are recorded from 12 endemic districts residing in 25 % of the country's population. In the 1990s, there was an upsurge in the number of VL cases, but the incidence has declined since 2006. The prevalence of PKDL has been estimated to be 2.3 % of patients with history of VL [11]. The prevalence of PKDL may be higher in areas adjacent to Indian border, where SSG resistance has been identified. In 2010, a survey found that the median onset of PKDL following VL episode was 23 months [11]. Patients who received inadequate treatment (<20 injections of SSG) were 11 times more likely to develop PKDL than those who completed full treatment course.

2.2 Clinical Features

PKDL patients are healthy except for their skin rash, therefore, they may not always visit health centres promptly. Unlike VL cases, they do not have fever, splenomegaly or weight loss, and the physical examination looks usually normal. However, severe form of PKDL when present on face can cause significant social and clinical discomfort. Since the initial diagnosis is based on clinical characteristics, the differential diagnosis is very important.

In the Indian subcontinent, the polymorphic form of PKDL is the most frequently encountered: hypopigmented macules, indurated erythematous plaques, papules and nodules are seen in varying proportions [1]. The typical form is that of lesions clustered around the chin and mouth, with discrete to no lesions in the rest of the face or body (Fig. 2.1). At times the lesions may be present on the sides of the cheeks and the ears, leading to potential for confusion with leprosy. Over time, these lesions can coalesce and give rise to large tumor-like plaques which are well circumscribed. Disease progression is nearly always from the face to the rest of the body, including the feet. Approximately 20 % of PKDL cases display mucosal lesions affecting the glans penis and oral cavity. If the outer lips are taken into account, this figure rises considerably since lesions are frequently present in this area. Impact on the mucosa without involvement of the skin is extremely rare and has been observed only occasionally in the highly endemic area of Bihar [12].

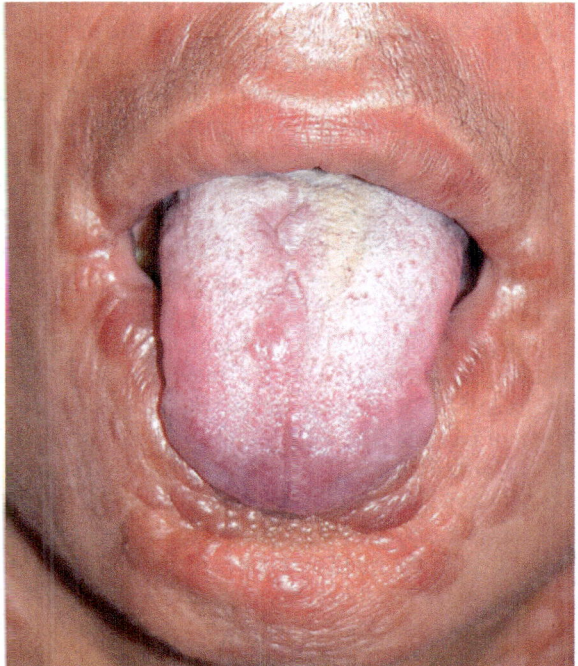

Fig. 2.1 Erythematous nodules in the peri-oral area and tongue

Fig. 2.2 Large, irregular and coalescent hypopigmented macules on chest and upper limbs studded with discrete papules and nodules

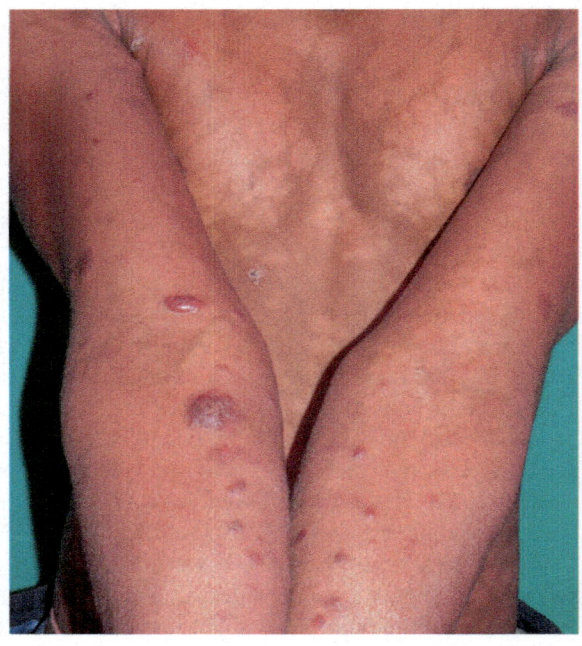

Similarly, ocular lesions in the cornea and sclera were rarely seen in the distant past. The unusual forms of the disease are those that are often missed and unsuspected. Of these, the macular form of PKDL is the most important. It is more likely to be mistaken for vitiligo rather than leprosy since the degree of pigment loss can be more than that seen in leprosy. Distribution of macular PKDL may be localized, generalized or extensive; such lesions lack a photosensitive evolution and at times the face may be minimally or not affected [13, 14]. Remarkably, hypopigmentation may affect the entire skin surface leaving few areas of normal or hyperpigmented skin, particularly at the flexures (Fig. 2.2). Others have also observed a similar incidence of macular PKDL in their studies [15]. The other presentation of PKDL is the fibroid variety with plaques on the dorsa of fingers and toes, with an appearance like knuckle pads.

African PKDL, mainly concentrated in Sudan, has been increasingly reported and has been reviewed [16, 17]. In contrast to PKDL in Indian subcontinent, which occurs in 5–15 % of those treated for VL [1], Sudanese PKDL appears in as high as 50 % of patients following VL [16]. The distribution of lesions in Sudan is similar to that of Indian form but differs in the ulceration that can occur in severe cases. Lesions typically start on the face as papules, spreading to other parts of the body. The lesions may remain confined to the face or may spread to other parts of the body: first, to the trunk and upper limbs, and subsequently to rest of the body parts. Lesions are generally symmetrical and not itchy. The size of papules may increase, and turn into nodules or plaques, or a combination of these; alternatively, the lesion may be macular. Maculopapular lesions are common; a micropapular form resembling measles may be seen [1]. Unlike PKDL in the Indian subcontinent, three grades of PKDL severity have been described in Sudan [1]. An interesting study in Sudan showed that the severity of facial involvement was related to ultraviolet B radiation in sunlight that appears to modify the immune system, promoting lesion development [18]. Mucosal involvement in African cases also differs from Indian PKDL in that lesions have been more frequently noted in the eyes or restricted to ocular mucosa [19], giving rise to a number of terms like post-kala-azar ocular or mucosal leishmaniasis. These could well be subsumed under the term PKDL which also covers mucosal lesions, avoiding further confusion.

PKDL is relatively common in HIV-positive individuals and often has atypical presentations, such as large nodular lesions. The typical clinical distribution i.e. the spread from face to other parts of the body is not always followed. The majority of these patients present florid disease with various descriptions including macules, disseminated miliary papules, papulo-erythematous eruption, nodules and plaques, or a mixed picture [20–22]. PKDL in co-infected patients have been reported to be caused by both *L. donovani* and *Leishmania infantum/Leishmania chagasi*; in most cases parasites were easily demonstrated [21–23].

2.3 Diagnosis

The variability of clinical presentations, poor awareness of this otherwise asymptomatic dermatosis, the migration of patients to non-endemic areas and lack of laboratory facilities to confirm diagnosis remain the major impediments in early recognition of PKDL [24]. The diagnosis of PKDL was initially based only on the typical clinical features (macules, papules, and nodules). The most frequent differential diagnoses are leprosy, vitiligo, sarcoidosis and secondary syphilis. However two diseases deserve special mention: neurofibromatosis and leprosy, which can be differentiated by the ulnar nerve thickness and the loss of skin sensitivity on hypopigmented patches. The diagnosis of PKDL is usually supported by the epidemiological background: most cases report a previous VL episode, usually in a focus of anthroponotic transmission, and/or origin from an area endemic for VL. The laboratory methods of diagnosis are immunological and parasitological. Serological methods used for VL diagnosis, such as the direct agglutination test (DAT), rapid rK39 immuno-chromatographic strip test, and ELISA based on recombinant antigens have been applied successfully to PKDL diagnosis [24, 25]. However, a positive antibody test may be the result of the previous VL episode rather than current PKDL. Nevertheless, serology can be helpful when other diseases (for example, leprosy) are considered in the differential diagnosis, or if a history of VL is uncertain. Antigen detection in tissues, blood, and urine is also of some help [26, 27]. However, only parasitological methods are confirmatory. The slit skin smear and culture are the standard methods but lack high sensitivity. In all types of PKDL, the finding of organisms is a bedside aid to diagnosis which the nodular type of lesion is most likely to show *Leishmania* amastigotes. The likelihood of finding amastigotes diminishes in less indurated type of skin lesions and is the lowest in the macular variety. Molecular methods based on nucleic acid detection, using gene amplification techniques like PCR, provide a reliable means of species-specific diagnosis of the disease. The methods are more sensitive than immunohistochemical or serological methods. Gene amplification is carried out by targeting multicopy sequences like ribosomal RNA genes, kinetoplastid DNA (kDNA), miniexon derived RNA genes or genomic repeats [24, 25]. A kDNA based PCR assay developed in India detected the parasite in 45 out of 48 PKDL patients with 93.8 % sensitivity [28]. Different PCR assays are now available: real-time PCR, nested PCR, LAMP PCR, etc. that can detect *Leishmania* parasites in blood or on slit skin aspirates [29]. The LAMP assay in diagnosis of PKDL was demonstrated to be rapid and reliable, with sensitivity of 96.8 % and specificity of 98.5 % [30]. LAMP PCR has several advantages: an easier to run kit is available, no sophisticated equipment needed, the required temperature for DNA amplification is lower (62–65 °C), and the test duration is shorter (1 h only). In addition, the reading is easy (turbidity and color change). It is as sensitive as nested PCR, which is claimed to be more sensitive than traditional PCR. It is also claimed to be a good test of cure as it becomes negative 2 weeks after treatment. However, it is difficult to decentralize the test further than district level. Parasitological confirmation using minimally invasive skin slit aspirate sample demonstrated to have equally reliable results as with tissue biopsy [29] and therefore should be encouraged, as the procedure will reduce discomfort and permanent scarring and thus motivate more patients to come forward for timely diagnosis and monitoring after treatment.

The main limitation is that most of the molecular tests are unavailable or not performed in local institutions. The diagnosis of PKDL at primary health centres when *Leishmania* amastigotes are not demonstrable often rests on the clinical picture and the response to anti-leishmanial therapy. In Africa, PKDL frequently occurs during or soon after the treatment of kala-azar, making the diagnosis relatively easy and with less likelihood of confusion with other dermatoses, unlike the case of PKDL from Indian subcontinent.

2.4 Histopathology

Histopathological examination in PKDL shows several changes that often occur in combination. The epidermal changes include hyperkeratosis, parakeratosis, follicular plugging, focal acanthosis or, rarely, atrophy. Dermis shows mixed inflammatory infiltrate consisting of histiocytes, lymphocytes, and plasma cells [1].

A recent study from India with 88 biopsies from patients of PKDL showed that the dermal infiltrates were arranged predominantly in 3 patterns reflecting the clinical type of lesion; superficial and perivascular in macules, perivascular and perifollicular in some, and the third pattern being diffuse infiltration in those with indurated lesions; *Leishmania* amastigotes were seen in approximately 30 % of the cases and were better seen in biopsies from mucosal lesions [23]. Anecdotal report from Spain has also shown many parasites in biopsy from the mucosal lesion as compared to none from the skin of the same patient [31].

In African cases, lymphocytes are the predominant cells followed by histiocytes and some plasma cells. Epithelioid cell granulomas were seen in about 20 % of cases and scattered epithelioid cells in about half [17]. Compact granulomas were seen more commonly in nodules than in papules [32]. Neuritis, an unusual finding, has been reported in four of 15 cases and has to be differentiated from leprosy [17]. Parasites are seen in 17–20 % of cases on routinely stained sections; this sensitivity increases to 88 % when a specific monoclonal antibody was used to stain *L. donovani* [33].

2.5 Treatment

Treatment of PKDL is an important component of the kala-azar eradication programme. The pentavalent antimonials which had been at the helm of therapy now stand excluded because of the duration of therapy, toxicity and increasing antimony resistance. Miltefosine and amphotericin B are the current treatments of choice. They can be used separately or combined depending on the severity. Studies for combination therapy are far from complete. The choice of drugs, dose and duration vary from that used in Sudanese PKDL and has been discussed in a separate chapter.

2.6 Pathogenesis

The precise understanding of PKDL pathogenesis is still obscure and, importantly, the immunopathobiology varies between Sudanese and Asian PKDL. PKDL is considered to be immunological triggered, with host, parasite, and drug all perhaps contributing to the pathogenesis.

- Host characteristics are mainly immunological and seem crucial for PKDL development. The host's role is suspected, based on the: (1) high incidence of PKDL in immunosuppressed people (i.e. transplant recipients and patients with HIV/AIDS, tuberculosis, malaria, measles), (2) efficacy of therapeutic vaccines with good and extended immunogenicity, (3) clinical healing associated with a conversion of Leishmanin Skin Test (LST) from negative to positive [34].
- Parasite characteristics are still not well understood. To date, no parasite strain has been associated with one single entity (VL or PKDL) in endemic areas [35, 36]. Molecular studies on genetic typing revealed monomorphism between *L. donovani* isolates from VL and PKDL cases in India [37]. It is clear that PKDL incidence is much higher in *L. donovani* foci than in *L. infantum*. However, there

is still an unanswered question of whether the higher incidence is related to the parasite and/or to the epidemiology. PKDL's higher incidence could be due to the anthroponotic (human-to-human) transmission in *L. donovani* foci compared to the zoonotic (animal-to-human) transmission in the *L. infantum*.

- Drug characteristics may also be important to PKDL development. However, approximately 20 % PKDL cases appear without any previous VL episode [5, 38] and, consequently, prior to receiving any drug treatment. Much of the PKDL reported in the past has been observed after treatment of VL with pentavalent antimonials. The introduction of newer drugs like amphotericin B and miltefosine has led researchers to investigate the relationship between the efficacy of a given drug and the subsequent occurrence of PKDL. Having analyzed cases seen over 35 years from a high endemic area, the authors concluded that the incidence of PKDL declined after the introduction of amphotericin B for VL in areas with high refractoriness to antimonials [39, 40]. It is now established that any drug can lead to PKDL but in variable proportions. PKDL cases after VL treatment with SSG are reported more frequently than with amphotericin B treatment; however this may be because SSG has been the most commonly and extensively used drug over many decades. In recent times PKDL cases have been reported after using miltefosine, amphotericin B and paromomycin for VL treatment [3, 4, 41]. PKDL could be the result of an immunological attack on *Leishmania* parasites which have survived in the skin despite chemotherapy. It is probably not the drug itself which leads to PKDL but the type of cellular immune response induced by the drug (cytokine profile) and the level of *Leishmania* parasite burden remaining in the body after treatment. $Sb5^+$ has a specific influence on the immune response through its effects on cell signaling, cytokines and immune complex induced levels of granulocyte macrophage colony stimulating factor (GM-CSF) [42]. $Sb5^+$ and amphotericin B have contrasting effects on IL-10 and TGF beta in PKDL patients [43]. Only prospective clinical trials on VL patients treated with different drugs, ($Sb5^+$, amphotericin B, miltefosine, paromomycin, and AmBisome) and longitudinal follow up will allow clear and final conclusions on the comparative rate of PKDL induced by each drug [39].

In South Asian PKDL, one of the important effector cell implicated in pathogenesis is $CD3^+CD8^+$ T cells that have been found both in lesions and circulation [44, 45]. The observation of high mRNA expression of FoxP3, CTLA-4, and CD25 at lesion site suggested involvement of Treg cells [46, 47]. Besides, an increased IL-10-expressing $CD3^+CD8^+$ T lymphocytes has been observed in circulation which gets restored to normal post treatment [45]. The cytokine profile at the lesion site show enhanced expression of IL-10, TGF-β, IFN-γ, and TNF-α in both South Asian and Sudanese PKDL. However, the expression of IFN-γR and TNFR1 was lower in Indian PKDL which was restored following treatment [48, 49]. Similar observations were made in Sudanese PKDL, a genetic polymorphism was found in IFN-γR [50]. In Sudanese PKDL, expression of IL-10 was considered as an important predictor for onset of PKDL, particularly following VL [17].

Furthermore, high levels of IL-17, its transcription factor ROR-γt, and IL-22, both in lesions and circulation indicated involvement of Th17 cells in PKDL pathogenesis [51]. Overall, the studies on immunobiology of PKDL suggest that it is a systemic disease rather than localized one.

2.7 Challenges for Sustainable Elimination of VL

In South East Asia, the transmission of VL is known to be anthroponotic; therefore PKDL has a major role as a reservoir of infection, especially during inter-epidemic period [52]. The experimental laboratory work carried out in India, in 1992, demonstrated the infectivity of PKDL cases. One hundred and four laboratory-bred *P. argentipes* sandflies were fed by xenodiagnosis on nodular PKDL cases. Sixty survived and 32 (53 %) were infected as evidence by the presence of *Leishmania* promastigotes in the mid-gut. It was concluded that Indian nodular PKDL cases were highly infective for the sandfly *P. argentipes* and it was extrapolated that PKDL cases can play a role as inter-epidemic reservoir [53]. A few years earlier, in 1988, another study concluded that the presence of as few as 0.5 % durably infectious PKDL patients during an epidemic may cause VL to become endemic [54].

PKDL poses a second challenge: the perception of PKDL by the patients themselves. They consider PKDL as a chronic, cosmetic, non-fatal disease with a stigma only if nodular. Hence the motivation for early diagnosis and treatment is low. The delay between the onset of clinical manifestations and the treatment is usually long and leads to an increased risk of VL transmission to people living in the same household or neighborhood. Another challenge is of PKDL diagnosis: The diagnosis of PKDL is intriguing and the disease is often misdiagnosed especially at primary health centres and/or private clinics, primarily as leprosy, a co-endemic dermatosis with high prevalence in the same areas. In addition, the treatment is long, costly, and frequently toxic. Increasing incidence of drug resistance is another major impediment to deal with to achieve VL elimination.

The presence of asymptomatic VL cases in the endemic areas could be an important issue towards VL eliminination. A majority of the *Leishmania* infected human population do not develop into full blown VL cases, and are considered asymptomatic [55–57] and these cases could play an important role in maintaining transmission dynamics of *Leishmania* infection [58]. However, the actual estimate of asymptomatic cases in endemic area is difficult to assess. A few studies have reported the presence of asymptomatic cases in high endemic areas of VL in Bihar in the range of 10–34 % [55, 59, 60] and the conversion rate to symptomatic VL was 17.85 per 1000 persons [59]. These "asymptomatic carriers" could prove an important impediment towards VL eliminination program.

2.8 Methods and Strategies to Control PKDL

In 2005, India, Nepal and Bangladesh signed a Memorandum of Understanding to work regionally towards the elimination of VL by 2015; recently extended up to 2017. In order to increase the chances of success of the ongoing VL elimination program in the Indian subcontinent, PKDL has to be addressed more precisely and seriously. In this regard, WHO consultative meeting on the management and control of PKDL made several recommendations [1]. A series of control measures and research activities need to be undertaken quickly and simultaneously:

(I) **Control measures**

(I.1) Monitoring of PKDL combining both passive and active case detection (ACD) should be undertaken routinely for a better estimate of PKDL prevalence. Point of care diagnostic testing, adaptable to community-based ACD will enhance surveillance.

(I.2) PKDL surveillance should be part of the national surveillance system for leishmaniasis or the national communicable diseases surveillance system.

(I.3) Ensure complete and successful treatment of VL.

(I.4) Early diagnosis and prompt treatment of PKDL together with an improved referral system.

(I.5) Greater awareness among the communities based on IEC and BCC campaigns is urgently needed for improved reporting and acceptability of treatment.

(I.6) Capacity building through institutional strengthening and training at all levels (especially health volunteers, lab technicians and physicians) involving both governmental and private sectors.

(I.7) Distribution of Long Lasting Nets (LLNs) to limit the infectivity of untreated PKDL cases.

(I.8) Vector controls measures such as Indoor residual spraying of houses and animal shelters should be undertaken regularly.

(II) **Research priorities**

(II.1) Identification of more effective, safe, short-course, affordable, accessible and acceptable treatment regimens for PKDL.

(II.2) Identification of new treatment regimens (combinations) for VL to prevent the appearance of PKDL.

(II.3) Post VL-treatment longitudinal follow-up to detect the appearance of PKDL and an assessment of the drug-specific incidence rates.

(II.4) Evaluate the characteristics of medicines and their capacity to penetrate the skin; modify medicines to target the skin.

(II.5) Identify immunological markers predictive of PKDL to monitor the immune response after treatment of VL and follow-up for development of PKDL.

(II.6) Identify serum markers to monitor progression towards cure, and define the end-point of treatment.

Relation of VL to PKDL

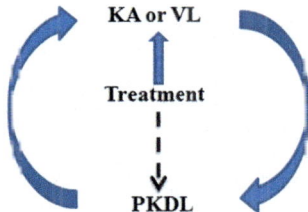

Fig. 2.3 Weak link (*broken lines*) in kala-azar control

In an increasingly global world, it is important for practitioners of tropical medicine to be familiar with PKDL as it can be seen in immigrants residing in the West several years after the episode of kala-azar [61] or in those from non-endemic countries like Japan who return after a period of time in India [62]. The importance of detecting and treating PKDL cases in VL control program is vital (Fig. 2.3). Actions have to be regularly monitored or PKDL cases will remain a major impediment for the VL elimination program. The pace and enthusiasm along with heightened political and financial commitments should be maintained in implementing the above recommendations.

References

1. World Health Organization. Post-kala-azar dermal leishmaniasis: a manual for case management and control. Rep. a WHO consulative Meet. 2012:2–3.
2. Croft SL, Sundar S, Fairlamb AH. Drug resistance in leishmaniasis. Clin Microbiol Rev. 2006;19(1):111–26. doi:10.1128/CMR.19.1.111-126.2006.
3. Kumar D, Ramesh V, Verma S, Ramam M, Salotra P. Post-kala-azar dermal leishmaniasis (PKDL) developing after treatment of visceral leishmaniasis with amphotericin B and miltefosine. Ann Trop Med Parasitol. 2009;103(8):727–30. doi:10.1179/000349809X12554106963438.
4. Pandey K, Das VNR, Singh D, et al. Post-kala-azar dermal leishmaniasis in a patient treated with injectable paromomycin for visceral leishmaniasis in India. J Clin Microbiol. 2012;50(4):1478–9. doi:10.1128/JCM.05966-11.
5. Desjeux P, Ghosh RS, Dhalaria P, Strub-Wourgaft N, Zijlstra EE. Report of the post kala-azar dermal leishmaniasis (PKDL) consortium meeting, New Delhi, India, 27–29 June 2012. Parasit Vectors. 2013;6(1):196. doi:10.1186/1756-3305-6-196.
6. Rahman KM, Islam S, Rahman MW, et al. Increasing incidence of post-kala-azar dermal leishmaniasis in a population-based study in Bangladesh. Clin Infect Dis. 2010;50(1):73–6. doi:10.1086/648727.
7. Mondal D, Nasrin KN, Huda MM, et al. Enhanced case detection and improved diagnosis of PKDL in a kala-azar-endemic area of Bangladesh. PLoS Negl Trop Dis. 2010;4(10):e832. doi:10.1371/journal.pntd.0000832.

8. Islam S, Kenah E, Bhuiyan MAA, et al. Clinical and immunological aspects of post-kala-azar dermal leishmaniasis in Bangladesh. Am J Trop Med Hyg. 2013;89(2):345–53. doi:10.4269/ajtmh.12-0711.
9. Ramesh V, Mukherjee A. Post-kala-azar dermal leishmaniasis. Int J Dermatol. 1995;34(2):85–91.
10. Singh RP, Picado A, Alam S, et al. Post-kala-azar dermal leishmaniasis in visceral leishmaniasis-endemic communities in Bihar, India. Trop Med Int Heal. 2012;17(11):1345–8. doi:10.1111/j.1365-3156.2012.03067.x.
11. Uranw S, Ostyn B, Rijal A, et al. Post-kala-azar dermal leishmaniasis in nepal: a retrospective cohort study (2000–2010). PLoS Negl Trop Dis. 2011;5(12):e1433. doi:10.1371/journal.pntd.0001433.
12. Singh RP. Observation on dermal leishmanoid in Bihar. Indian J Dermatol. 1968;13(3):59–63.
13. Ramesh V, Singh N. A clinical and histopathological study of macular type of post-kala-azar dermal leishmaniasis. Trop Doct. 1999;29(4):205–7.
14. Ramesh V, Ramam M, Singh R, Salotra P. Hypopigmented post-kala-azar dermal leishmaniasis. Int J Dermatol. 2008;47(4):414–6. doi:10.1111/j.1365-4632.2008.03621.x.
15. Saha S, Mazumdar T, Anam K, et al. Leishmania promastigote membrane antigen-based enzyme-linked immunosorbent assay and immunoblotting for differential diagnosis of Indian post-kala-azar dermal leishmaniasis. J Clin Microbiol. 2005;43(3):1269–77. doi:10.1128/JCM.43.3.1269-1277.2005.
16. Musa AM, Khalil EAG, Raheem MA, et al. The natural history of Sudanese post-kala-azar dermal leishmaniasis: clinical, immunological and prognostic features. Ann Trop Med Parasitol. 2002;96(8):765–72. doi:10.1179/000349802125002211.
17. Zijlstra EE, Musa AM, Khalil EAG. el-Hassan IM, el-Hassan AM. Post-kala-azar dermal leishmaniasis. Lancet Infect Dis. 2003;3(2):87–98.
18. Ismail A, Khalil EAG, Musa AM, et al. The pathogenesis of post kala-azar dermal leishmaniasis from the field to the molecule: does ultraviolet light (UVB) radiation play a role? Med Hypotheses. 2006;66(5):993–9. doi:10.1016/j.mehy.2005.03.035.
19. El Hassan AM, Khalil EA, el Sheikh EA, Zijlstra EE, Osman A, Ibrahim ME. Post kala-azar ocular leishmaniasis. Trans R Soc Trop Med Hyg. 1998;92(2):177–9.
20. Rihl M, Stoll M, Ulbricht K, Bange F-C, Schmidt R-E. Successful treatment of post-kala-azar dermal leishmaniasis (PKDL) in a HIV infected patient with multiple relapsing leishmaniasis from Western Europe. J Infect. 2006;53(1):e25–7. doi:10.1016/j.jinf.2005.09.015.
21. Stark D, Pett S, Marriott D, Harkness J. Post-kala-azar dermal leishmaniasis due to leishmania infantum in a human immunodeficiency virus type 1-infected patient. J Clin Microbiol. 2006;44(3):1178–80. doi:10.1128/JCM.44.3.1178-1180.2006.
22. Antinori S, Longhi E, Bestetti G, et al. Post-kala-azar dermal leishmaniasis as an immune reconstitution inflammatory syndrome in a patient with acquired immune deficiency syndrome. Br J Dermatol. 2007;157(5):1032–6. doi:10.1111/j.1365-2133.2007.08157.x.
23. Singh A, Ramesh V, Ramam M. Histopathological characteristics of post kala-azar dermal leishmaniasis: a series of 88 patients. Indian J Dermatol Venereol Leprol. 2015;81(1):29–34. doi:10.4103/0378-6323.148562.
24. Salotra P, Singh R. Challenges in the diagnosis of post kala-azar dermal leishmaniasis. Indian J Med Res. 2006;123(3):295–310.
25. Salotra P, Sreenivas G, Nasim AA, Subba Raju BV, Ramesh V. Evaluation of enzyme-linked immunosorbent assay for diagnosis of post-kala-azar dermal leishmaniasis with crude or recombinant k39 antigen. Clin Diagn Lab Immunol. 2002;9(2):370–3.
26. Riera C, Fisa R, Lopez P, et al. Evaluation of a latex agglutination test (KAtex) for detection of Leishmania antigen in urine of patients with HIV-Leishmania coinfection: value in diagnosis and post-treatment follow-up. Eur J Clin Microbiol Infect Dis. 2004;23(12):899–904. doi:10.1007/s10096-004-1249-7.
27. Singh R, Subba Raju BV, Jain RK, Salotra P. Potential of direct agglutination test based on promastigote and amastigote antigens for serodiagnosis of post-kala-azar dermal

leishmaniasis. Clin Diagn Lab Immunol. 2005;12(10):1191–4. doi:10.1128/CDLI.12.10. 1191-1194.2005.
28. Salotra P, Sreenivas G, Pogue GP, et al. Development of a species-specific PCR assay for detection of leishmania donovani in clinical samples from patients with kala-azar and post-kala-azar dermal leishmaniasis. J Clin Microbiol. 2001;39(3):849–54. doi:10.1128/JCM. 39.3.849-854.2001.
29. Verma S, Bhandari V, Avishek K, Ramesh V, Salotra P. Reliable diagnosis of post-kala-azar dermal leishmaniasis (PKDL) using slit aspirate specimen to avoid invasive sampling procedures. Trop Med Int Health. 2013;18(3):268–75. doi:10.1111/tmi.12047.
30. Verma S, Avishek K, Sharma V, Negi NS, Ramesh V, Salotra P. Application of loop-mediated isothermal amplification assay for the sensitive and rapid diagnosis of visceral leishmaniasis and post-kala-azar dermal leishmaniasis. Diagn Microbiol Infect Dis. 2013;75 (4):390–5. doi:10.1016/j.diagmicrobio.2013.01.011.
31. Roustan G, Jiménez JA, Gutiérrez-Solar B, Gallego JL, Alvar J, Patrón M. Post-kala-azar dermal leishmaniasis with mucosal involvement in a kidney transplant recipient: treatment with liposomal amphotericin B. Br J Dermatol. 1998;138(3):526–8.
32. El Hassan AM, Ghalib HW, Zijlstra EE, et al. Post kala-azar dermal leishmaniasis in the Sudan: clinical features, pathology and treatment. Trans R Soc Trop Med Hyg. 1992;86 (3):245–8.
33. Ismail A, Kharazmi A, Permin H, el Hassan AM. Detection and characterization of leishmania in tissues of patients with post kala-azar dermal leishmaniasis using a specific monoclonal antibody. Trans R Soc Trop Med Hyg. 1997;91(3):283–5.
34. Musa AM, Khalil EAG, Mahgoub FAE, et al. Immunochemotherapy of persistent post-kala-azar dermal leishmaniasis: a novel approach to treatment. Trans R Soc Trop Med Hyg. 2008;102(1):58–63. doi:10.1016/j.trstmh.2007.08.006.
35. Dey A, Singh S. Genetic heterogeneity among visceral and post-kala-azar dermal leishmaniasis strains from eastern India. Infect Genet Evol. 2007;7(2):219–22. doi:10.1016/j.meegid.2006.09.001.
36. Subba Raju BV, Singh R, Sreenivas G, Singh S, Salotra P. Genetic fingerprinting and identification of differentially expressed genes in isolates of Leishmania donovani from Indian patients of post-kala-azar dermal leishmaniasis. Parasitology. 2008;135(Pt 1):23–32. doi:10. 1017/S0031182007003484.
37. Subba Raju BV, Gurumurthy S, Kuhls K, Bhandari V, Schnonian G, Salotra P. Genetic typing reveals monomorphism between antimony sensitive and resistant leishmania donovani isolates from visceral leishmaniasis or post kala-azar dermal leishmaniasis cases in India. Parasitol Res. 2012;111(4):1559–68. doi:10.1007/s00436-012-2996-5.
38. Das VNR, Ranjan A, Pandey K, et al. Short report: clinical epidemiologic profile of a cohort of post-kala-azar dermal leishmaniasis patients in Bihar, India. Am J Trop Med Hyg. 2012;86 (6):959–61. doi:10.4269/ajtmh.2012.11-0467.
39. Croft SL. PKDL—a drug related phenomenon? Indian J Med Res. 2008;128(1):10–1.
40. Thakur CP, Kumar A, Mitra G, et al. Impact of amphotericin-B in the treatment of kala-azar on the incidence of PKDL in Bihar, India. Indian J Med Res. 2008;128(1):38–44.
41. Das VNR, Pandey K, Verma N, et al. Short report: development of post-kala-azar dermal leishmaniasis (PKDL) in miltefosine-treated visceral leishmaniasis. Am J Trop Med Hyg. 2009;80(3):336–8.
42. Elshafie AI, Ahlin E, Mathssor L, ElGhazali G, Rönnelid J. Circulating immune complexes (IC) and IC-induced levels of GM-CSF are increased in sudanese patients with acute visceral Leishmania donovani infection undergoing sodium stibogluconate treatment: implications for disease pathogenesis. J Immunol. 2007;178(8):5383–9.

43. Saha S, Mondal S, Ravindran R, et al. IL-10- and TGF-beta-mediated susceptibility in kala-azar and post-kala-azar dermal leishmaniasis: the significance of amphotericin B in the control of Leishmania donovani infection in India. J Immunol. 2007;179(8):5592–603.
44. Rathi SK, Pandhi RK, Chopra P, Khanna N. Lesional T-cell subset in post-kala-azar dermal leishmaniasis. Int J Dermatol. 2005;44(1):12–3. doi:10.1111/j.1365-4632.2004.01579.x.
45. Ganguly S, Das NK, Panja M, et al. Increased levels of interleukin-10 and IgG3 are hallmarks of Indian post-kala-azar dermal leishmaniasis. J Infect Dis. 2008;197(12):1762–71. doi:10.1086/588387.
46. Ganguly S, Mukhopadhyay D, Das NK, et al. Enhanced lesional Foxp3 expression and peripheral anergic lymphocytes indicate a role for regulatory T cells in Indian post-kala-azar dermal leishmaniasis. J Invest Dermatol. 2010;130(4):1013–22. doi:10.1038/jid.2009.393.
47. Katara GK, Ansari NA, Verma S, Ramesh V, Salotra P. Foxp3 and IL-10 expression correlates with parasite burden in lesional tissues of post kala azar dermal leishmaniasis (PKDL) patients. PLoS Negl Trop Dis. 2011;5(5):e1171. doi:10.1371/journal.pntd.0001171.
48. Ansari NA, Ramesh V, Salotra P. Interferon (IFN)-gamma, tumor necrosis factor-alpha, interleukin-6, and IFN-gamma receptor 1 are the major immunological determinants associated with post-kala azar dermal leishmaniasis. J Infect Dis. 2006;194(7):958–65. doi:10.1086/506624.
49. Ansari NA, Katara GK, Ramesh V, Salotra P. Evidence for involvement of TNFR1 and TIMPs in pathogenesis of post-kala-azar dermal leishmaniasis. Clin Exp Immunol. 2008;154 (3):391–8. doi:10.1111/j.1365-2249.2008.03761.x.
50. Salih MA, Ibrahim ME, Blackwell JM, et al. IFNG and IFNGR1 gene polymorphisms and susceptibility to post-kala-azar dermal leishmaniasis in Sudan. Genes Immunol. 2007;8 (1):75–8. doi:10.1038/sj.gene.6364353.
51. Katara GK, Ansari NA, Singh A, Ramesh V, Salotra P. Evidence for involvement of Th17 type responses in post kala azar dermal leishmaniasis (PKDL). PLoS Negl Trop Dis. 2012;6 (6):e1703. doi:10.1371/journal.pntd.0001703.
52. World Health Organization. Control of the leishmaniases. World Health Organ. Tech Rep Ser. 2010;(949):xii–xiii, 1–186, back cover.
53. Addy M, Nandy A. Ten years of kala-azar in west Bengal, Part I. Did post-kala-azar dermal leishmaniasis initiate the outbreak in 24-Parganas? Bull. World Health Organ. 1992;70 (3):341–6.
54. Dye C, Wolpert DM. Earthquakes, influenza and cycles of Indian kala-azar. Trans R Soc Trop Med Hyg. 1988;82(6):843–50.
55. Das VNR, Siddiqui NA., Verma RB, et al. Asymptomatic infection of visceral leishmaniasis in hyperendemic areas of Vaishali district, Bihar, India: A challenge to kala-azar elimination programmes. Trans R Soc Trop Med Hyg. 2011;105(11):661–666. doi:10.1016/j.trstmh.2011.08.005.
56. Ostyn B, Gidwani K, Khanal B, et al. Incidence of symptomatic and asymptomatic leishmania donovani infections in high-endemic foci in India and Nepal: a prospective study. PLoS Negl Trop Dis. 2011;5(10). doi:10.1371/journal.pntd.0001284.
57. Stauch A, Sarkar RR, Picado A, et al. Visceral leishmaniasis in the indian subcontinent: modelling epidemiology and control. PLoS Negl Trop Dis. 2011;5(11). doi:10.1371/journal.pntd.0001405.
58. Sharma MC, Gupta AK, Das VNR, et al. Leishmania donovani in blood smears of asymptomatic persons. Acta Trop. 2000;76(2):195–6. doi:10.1016/S0001-706X(00)00068-1.
59. Topno RK, Das VNR, Ranjan A, et al. Asymptomatic infection with visceral leishmaniasis in a disease-endemic area in Bihar, India. Am J Trop Med Hyg. 2010;83:502–6. doi:10.4269/ajtmh.2010.09-0345.
60. Sudarshan M, Singh T, Singh AK, et al. Quantitative PCR in epidemiology for early detection of visceral leishmaniasis cases in India. PLoS Negl Trop Dis. 2014;8(12):e3366. doi:10.1371/journal.pntd.0003366.

61. Munro DD, Du Vivier A, Jopling WH. Post kala-azar dermal leishmaniasis. Br J Dermatol. 1972;87(4):374–8.
62. Ono H, Ghoreishi M, Yokozeki H, Katayama I, Nishioka K. A case of post-kala-azar dermal leishmaniasis. J Dermatol. 1998;25(2):118–20.

Chapter 3
Characteristics of Patients Visiting Suruya Kanta Kala-Azar Research Center in Mymensingh, Bangladesh; Data from a Patient Registry System

Masao Iwagami, Bumpei Tojo and Eisei Noiri

Abstract Mymensingh district is the most affected area of visceral leishmaniasis (VL) in Bangladesh. Surya Kanta Kala-azar Research Center (SKKRC) was established in 2012 to provide care for patients with VL and post-kala-azar dermal leishmaniasis (PKDL) and to advance research on associated conditions. We also started a patient registry system to provide short- and long-term follow up for patients given diagnoses of and treated for VL. This chapter demonstrates the baseline characteristics of patients with VL receiving treatment in SKKRC for the first year, from 1st April 2013 to 31st March 2014. We registered 181 patients, consisting of 68 patients with incipient VL and 113 patients with non-incipient VL. Seasonality (more cases in winter and fewer in summer) was observed, mainly for incipient VL. Compared with those with incipient VL, patients with non-incipient VL were more likely to have a family history of PKDL and splenomegaly at physical examination. During the study period, most patients were treated by liposomal amphotericin B (5 mg/day × 3 days). In conclusion, our patient registry system has been successfully established and we will be able to examine the incidence, associated factors, and short- and long-term outcomes of VL treated in SKKRC.

Keywords Epidemiology · Cohort · Mymensingh · Registry

M. Iwagami (✉)
Faculty of Epidemiology and Population Health, London School of Hygiene and Topical Medicine, London, UK; Department of Hemodialysis and Apheresis, The University of Tokyo Hospital, Tokyo, Japan
e-mail: iwagami-tky@umin.ac.jp

3.1 Introduction

Although estimating the true incidence of visceral leishmaniasis (VL) and cutaneous leishmaniasis (CL) may be difficult because of under-reporting bias, a previous systematic review estimated from available data that approximately 0.2–0.4 million VL and 0.7–1.2 million CL cases occur annually in the world in 2007–2011 [1]. Among these VL cases, more than 90 % were expected to occur in the following 6 countries: India, Bangladesh, Sudan, South Sudan, Ethiopia, and Brazil. In Bangladesh, between 1994 and 2013, 109,266 VL cases were reported in 37 endemic districts [2]. Of note, the Mymensingh district was reported to be the most affected area, accounting for nearly half of the VL cases in Bangladesh. Moreover, when comparing the statistics between the periods of 1998–2005 and 2006–2013, the number of reported VL cases has not changed in Mymensignh, while other areas saw large reductions in the number of VL cases [2].

To promote patient care and research in this most endemic area in Bangladesh, Surya Kanta Kala-azar Research Center (SKKRC) was established in December 2012, as part of the Surya Kanta Hospital in the Mymensingh district [3]. Since then, the SKKRC has been providing care for patients with VL who are coming from inside and outside of Mymensingh. Following this, we established at the end of March 2013 a case registration system named 'Epidemiology of Kala-Azar (Visceral Leishmaniasis) and Post-Kala-Azar Dermal Leishmaniasis in Bangladesh—A Prospective Cohort Study—' (UMIN Clinical Trials Registry, number UMIN000011426 [4]). This cohort aims to register all the patients who were given a diagnosis of VL in the SKKRC, and examine short- and long-term outcomes and associated factors after treatment. In this chapter, we will demonstrate the baseline characteristics of patients with VL visiting SKKRC during the first year from 1st April 2013 to 31st March 2014.

3.2 Methods

All patients who visited SKKRC and received a diagnosis of VL agreed to participate in the registration system. Medical staff in SKKRC recorded patient medical histories, and conducted physical examinations and blood and urinary sampling before treatment. During the study period, a final diagnosis of incipient VL was determined with a positive result of an rK39 *Leishmania* antigen test, while non-incipient VL was diagnosed based on spleen biopsy findings and polymerase chain reaction (PCR) assays by treatment physicians.

The number of patients was counted each month to examine seasonality of the incidence of VL, differentiating patients with incipient (primary) and non-incipient VL. Based on patients' addresses, the distribution of patients was plotted on a map of Mymensingh and the surrounding districts. Continuous variables such as age and height were presented as median values [interquartile range (IQR)], while categorical variables were presented as the number of patients and proportion. Baseline characteristics were compared between incipient and non-incipient VL cases. Statistical analysis included t-test or Wilcoxon rank sum test for continuous variables, and chi-square test or Fisher's exact test for categorical variables, as appropriate. A p value of less than 0.05 was considered statistically significant. All statistical analyses were conducted using Stata 14 software (Stata Corp, Texas, US).

3.3 Results

During the first year period, 181 patients received a diagnosis with VL and were treated in SKKRC; 68 patients with incipient VL and 113 patients with non-incipient VL. Figure 3.1 illustrates the number of patients by month. Incipient VL appears to have seasonality, with higher incidence during winter and lower incidence during summer. However, non-incipient VL does not seem to have a large seasonality except for the highest number occurring in March 2014. Figure 3.2 demonstrates the geographical distribution of VL cases. The majority of patients (nearly 60 %) came from the Mymensingh district, followed by fewer patients from Tangail, Jamalpur, Gazipur, Netrokona, and Kishorgonji. In the Mymensingh district, almost all cases were from south of the (old) Brahmaputra River.

Non-incipient VL cases consisted of 21 patients receiving previous treatment for VL in the past 1 year, 84 patients receiving VL treatment ≥ 1 year before, and 8 patients with unknown detailed history (Table 3.1). Of note, 27 % of previous VL

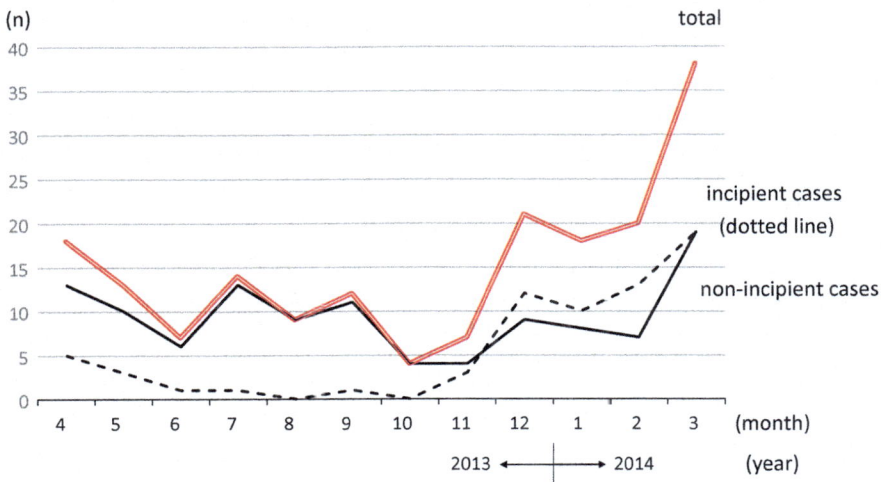

Fig. 3.1 Number of patients with visceral leishmaniasis visiting Surya Kanta Kala-azar Research Center by month

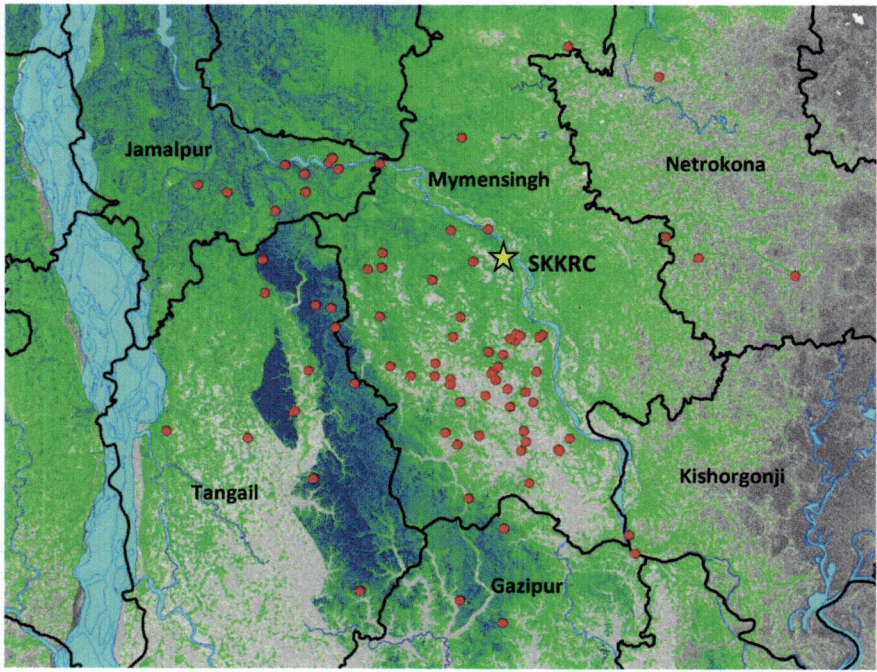

Fig. 3.2 Geographical distribution of patients with visceral leishmaniasis visiting Surya Kanta Kala-azar Research Center (SKKRC)

Table 3.1 Previous treatment history of 113 patients with non-incipient visceral leishmaniasis

Treatment regimen	Last episode of visceral leishmaniasis		
	<1 year ($n = 21$)	≥1 year ($n = 84$)	Not remembered/recorded ($n = 8$)
Miltefosine	12 (57 %)	38 (45 %)	2 (25 %)
Sodium stibogluconate	0	23 (27 %)	0
Paromomycin	1 (5 %)	2 (2 %)	0
Liposomal amphotericin B	6 (29 %)	15 (18 %)	3 (38 %)
Intravenous drug (not specifically remembered/recorded)	0	0	1 (13 %)
Combination of liposomal amphotericin B and miltefosine	1 (5 %)	1 (1 %)	0
Not remembered/recorded	1 (5 %)	5 (6 %)	2 (25 %)

patients (≥1 year) received sodium stibogluconate (SSG), while recently treated VL patients did not. Table 3.2 compares the baseline characteristics of patients with incipient VL and non-incipient VL. Age was not very different, with the majority of

Table 3.2 Baseline characteristics of patients with visceral leishmaniasis before treatment

	Incipient VL patients (n = 68)	Non-incipient VL patients (n = 113)	P value
Age (years)*	30 [14.5–45]	25 [12–39]	0.151
Sex (male)	41 (60 %)	78 (69 %)	0.231
Patient history			
Family history of VL (yes)	24 (35 %)	34 (30 %)	0.128
Family history of PKDL (yes)	0 (0 %)	5 (4 %)	0.017
Chief complaint: Fever / Others (abdominal pain, vomiting, etc.)	63 (93 %) / 5 (7 %)	109 (96 %) / 4 (4 %)	0.253
Duration of fever until visit (days)*	45 [30–120]	30 [20–60]	0.201
Weight loss	62 (91 %)	102 (90 %)	0.834
Anorexia	62 (91 %)	103 (91 %)	0.996
Diarrhea	2 (3 %)	6 (5 %)	0.385
Vomiting	10 (15 %)	6 (5 %)	0.031
Nasal bleeding	1 (1 %)	1 (1 %)	0.429
Gum bleeding	0 (0 %)	0 (0 %)	n/a
Physical examination			
Height (cm)*	153 [145–165]	155 [140–162]	0.941
Weight (kg)*	40.5 [34–50]	42 [30–50]	0.855
Body mass index (kg/m^2)*	17.3 [15.1–18.8]	17.4 [14.7–20.0]	0.595
Pulse (/min)*	84 [80–90]	85 [80–90]	0.525
Systolic blood pressure (mm Hg)*	110 [100–120]	110 [100–120]	0.362
Diastolic blood pressure (mm Hg)*	70 [60–80]	70 [70–80]	0.179
Body temperature (Fahrenheit)*	99.0 [99.0–100.0]	99.0 [98.4–100.0]	0.626
Pallor	64 (94 %)	108 (96 %)	0.555
Jaundice	10 (15 %)	24 (21 %)	0.238
Abdominal swelling	2 (3 %)	9 (8 %)	0.158
Ascites	1 (1 %)	2 (2 %)	0.594
Splenomegaly	33 (49 %)	91 (81 %)	<0.001
Hepatomegaly	0 (0 %)	3 (3 %)	0.226

PKDL = post-kala-azar dermal leishmaniasis, VL = visceral leishmaniasis, * median [interquartile range]

patients being in their 10s, 20s, and 30s. Sex was also similar, with the male proportion being around 60–70 %. Patients with non-incipient VL were more likely to have a family history of post-kala-azar dermal leishmaniasis (PKDL). Most

Table 3.3 Current treatment for patients with incipient and non-incipient visceral leishmaniasis

	Incipient visceral leishmaniasis ($n = 68$)	Non-incipient visceral leishmaniasis ($n = 113$)
Liposomal amphotericin B		
Dose: 5 mg/day × 3 days	58 (85 %)	99 (88 %)
Dose: 5 mg/day × 6 days	1 (1 %)	3 (3 %)
Dose: 10 mg/day × 1 day	9 (13 %)	1 (1 %)
Non-liposomal amphotericin B		
Dose: 1 mg/day × 15 days	0	4 (4 %)
Miltefosine		
Dose: 100 mg/day × 28 days	0	4 (4 %)
Dose: 40 mg/day × 28 days	0	1 (1 %)
Dose: 10 mg/day × 84 days	0	1 (1 %)

patients in both groups visited the hospital with fever as their chief complaint. Weight loss and anorexia were also involved for 90 % of the patients. The proportion of those reporting vomiting was significantly higher in patients with incipient VL. Body mass index (BMI) was obviously low in both groups, with the median slightly over 17 and the upper limit of the IQR 20 or lower. At physical examination, most patients were judged to have pallor. Splenomegaly was significantly more frequent in patients with non-incipient VL than that seen in patients with incipient VL.

As shown in Table 3.3, more than 90 % of patients were treated by liposomal amphotericin B, mostly with the dose of 5 mg/day for 3 days. A small number of patients, many of whom had received a course of liposomal amphotericin B before, received other treatments such as non-liposomal amphotericin B and miltefosine.

3.4 Discussion

We established a patient registry system in a hospital in the Mymensingh district, the most endemic area for VL in Bangladesh. To achieve the elimination target of VL (an annual incidence of less than 1 per 10,000 people as proposed by World Health Organization [5]) SKKRC would play an important role in caring for patients with VL and PKDL and reporting the incidence, associated factors, and short- and long-term outcomes. In particular, because humans are considered the only reservoir host, successful treatment of VL is essential to avoid spreading the disease through infected people. In our cohort, approximately one third of patients had a family history of VL, and there were several patients with a family history of PKDL in the group of non-incipient patients.

Characteristics of patients with VL were similar to those of other patient cohorts established in endemic areas in Bangladesh. For example, in a cohort of a Fulbaria clinic, Mymensingh, the study population was under-nourished with a mean BMI of around 17.5 kg/m² [6], and in another cohort of the Muktagacha upazila hospital in Mymensingh, systolic/diastolic blood pressure was close to 110/70 mm Hg [7]. However, our cohort included a larger proportion of non-incipient VL patients (113/181) than did the Muktagacha upazila hospital cohort (68/300). Patients with non-incipient VL had different characteristics from those with incipient VL in that seasonality of the incidence was smaller and a family history of PKDL was reported in some patients. Moreover, the proportion of splenomegaly, which was suggested to be a risk factor for relapse among patients with primary VL [6], was considerably higher in patients with non-incipient VL. How to treat and follow up these patients would be ever more important in endemic areas, including Mymensingh.

During the study period, most patients were treated by liposomal amphotericin B, mostly with the dose of 5 mg/day for 3 days. The study in the Fulbaria clinic, Mymensingh, reported that a regimen of 5 mg/day for 3 days for primary VL led to a >95 % cure rate and <1 % serious adverse events [6]. Meanwhile, in the Muktagacha upazila hospital, the regimen of 10 mg/day for 1 day was also successful with a >95 % cure rate and no serious adverse events [7]. Our cohort also included a small number of patients treated by 10 mg/day for 1 day. For non-incipient visceral leishmaniasis, non-liposomal amphotericin B and miltefosine were also treatment options. The associations between the treatment regimen and short- and long-term outcomes, including drug side effects, treatment failure, recurrence, and the incidence of PKDL, will be examined in our established cohort. We will also examine the association between baseline characteristics (including blood and urine test results that were not shown in this chapter) and outcomes.

3.5 Conclusion

In conclusion, a new patient registry system has been successfully established in SKKRC in the Mymensingh district to follow up treatment of patients with VL. Using this cohort, we will be able to examine the incidence, associated factors, and short- and long-term outcomes of patients with VL. This will ultimately contribute to the eradication of VL and PKDL in the region, as well as in the world.

References

1. Alvar J, Vélez ID, Bern C, et al. Leishmaniasis worldwide and global estimates of its incidence. PLoS One. 2012;7:e35671.
2. Chowdhury R, Mondal D, Chowdhury V, et al. How far are we from visceral leishmaniasis elimination in Bangladesh? An assessment of epidemiological surveillance data. PLoS Negl Trop Dis. 2014;8:e3020.
3. SATREPS project kala-azar research center officially opens in Bangladesh—Japan-Bangladesh collaboration contributes to elimination of visceral leishmaniasis. Japan Science and Technology Agency. http://www.jst.go.jp/report/2012/130121_e.html. Accessed 1 Oct 2015.

4. Epidemiology of Kala-azar (visceral leishmaniasis) and Post-kala-azar dermal leishmaniasis in Bangladesh—a prospective cohort study—', UMIN clinical trials registry. https://upload.umin.ac.jp/cgi-open-bin/ctr/ctr.cgi?function=brows&action=brows&type=summary&recptno=R000013374&language=J. Accessed 1 Oct 2015.
5. Regional strategic framework for elimination of kala-azar from the South-East Asia Region (2011–2015). WHO regional office for South-East Asia. http://apps.searo.who.int/pds_docs/B4870.pdf?ua=1. Accessed 1 Oct 2015.
6. Lucero E, Collin SM, Gomes S, et al. Effectiveness and safety of short course liposomal amphotericin B (AmBisome) as first line treatment for visceral leishmaniasis in Bangladesh. PLoS Negl Trop Dis. 2015;9:e0003699.
7. Mondal D, Alvar J, Hasnain MG, et al. Efficacy and safety of single-dose liposomal amphotericin B for visceral leishmaniasis in a rural public hospital in Bangladesh: a feasibility study. Lancet Glob Health. 2014;2:e51–7.

Part II
Therapeutic Strategy to Deal with Emergence of Drug Resistance

Part II
The hydrochemistry to deal with
transport of trace metal ions

Chapter 4
Epidemiology of Drug-Resistant Kala-Azar in India and Neighboring Countries

T.K. Jha

Abstract Sodium stibogluconate (SSG), a pentavalent antimonial compound, had been the first-line treatment for visceral leishmaniasis (VL) in the Indian subcontinent for more than 70 years, with a cure rate of over 90 %. However, by the year 2000, even a stronger regimen of this treatment failed to cure as many as 60 % of the cases. Among the drugs administered orally, miltefosine was found highly effective in the treatment of VL in different phases of trials in India, with a cure rate of 94, or 82–87 % in outpatients. In the VL elimination program, miltefosine has been used as a first-line drug in India, Nepal, and Bangladesh since 2008. Miltefosine has a long half-life and has been found to be teratogenic in females of child-bearing age. Paromomycin, administered intramuscularly for 21 days, was found to have a cure rate of 94 % in different phases of trials in India. It has been registered in India since August 2006, and is a potential anti-leishmanial drug that is being used in combination therapy. Amphotericin B, or its liposomal compound (AmBisome), remains the most potent drug with a cure rate of 98–100 % in all resistant cases of VL. Since 2015, single dose AmBisome (10 MKD) is being used in the elimination program. To counter the problem of drug resistance, particularly in patients of VL co-infected with HIV, AmBisome in a total dose of 40 MKD in a divided dosage, or a combination therapy of 2 potent drugs, or alternatively, 2 courses of AmBisome, is recommended.

Keywords Drug resistance unresponsiveness · Kala-azar · Visceral leishmaniasis · Indian subcontinent

T.K. Jha (✉)
Kalazar Research Centre Brahmpura, Muzaffarpur 842003, Bihar, India
e-mail: dr_tkjha@hotmail.com

© Springer International Publishing 2016
E. Noiri and T.K. Jha (eds.), *Kala Azar in South Asia*,
DOI 10.1007/978-3-319-47101-3_4

4.1 Introduction

The global incidence of kala-azar, or visceral leishmaniasis (VL), is approximately 500,000 new cases per year. Ninety percent of these cases are from India and the neighboring countries of Bangladesh and Nepal, as well as Sudan and Brazil. VL practically disappeared from the Indian subcontinent after dichlorodiphenyltrichloroethane (DDT) spraying was begun to eradicate malaria between the 1950s and the late 1960s. The resurgence of this disease began in the mid-1970s in India and in the late 1970s in Bangladesh, where in the 1980s an outbreak occurred in the Pabna district. The first outbreak of VL was reported in 1980 from the plains of Nepal bordering northern districts of the Indian state of Bihar. Fifty percent of the cases reported from India are from Bihar. Thirty-three of 38 districts of Bihar are affected and 200,000 deaths from VL have been reported from Bihar since 1977 [1, 2]. At present, 111 districts of South East Asia are affected: 54 in India 26 in Bangladesh, 12 in Nepal, 7 in Bhutan, and 12 provinces of Thailand [3].

4.1.1 History of VL on the Indian Subcontinent

Kala-azar—the word "kala" means black, referring to the black pigmentation of the skin, and "azar" means prolonged fever or ailment—has been known to occur in the Indian subcontinent for more than 180 years. The first description of an epidemic of this disease was reported in 1824–25 in Jessore (now Bangladesh), and this outbreak involved approximately 7,500 fatalities. The disease spread to Bengal between 1830 and 1850, and it was carried via river travel to the Brahmaputra valley in Assam in 1875. In 1872, it traveled along the Ganges River from Bengal to the plains of Bihar, where the disease was given the name kala-dukh (black disease) [4]. An epidemic of the disease involving 9,400 cases was reported in Bihar in 1895 (Fig. 4.1) [4]. The resurgence of VL began in the mid-1970s. In a 1977 epidemic, 100,000 cases were reported. A major epidemic occurred in Bihar during the period 1991–1992, with 250,000 cases and a mortality rate of 35 % [5]. Of the total cases of visceral leishmaniasis reported in India, 70 % are from the Bihar state in which 33 of 38 districts are affected. Other affected Indian states are as follows: 4 districts of Jharkhand, 11 districts of West Bengal, and 6 districts of eastern Uttar Pradesh.

4.1.2 Incidence of VL in Bangladesh

The prevalence of VL was substantially reduced in India and Bangladesh after DDT spraying to kill malaria-infested mosquitoes was begun during the 1960s [6]. Sporadic cases were reported in the late 1970s, and an outbreak occurred in the

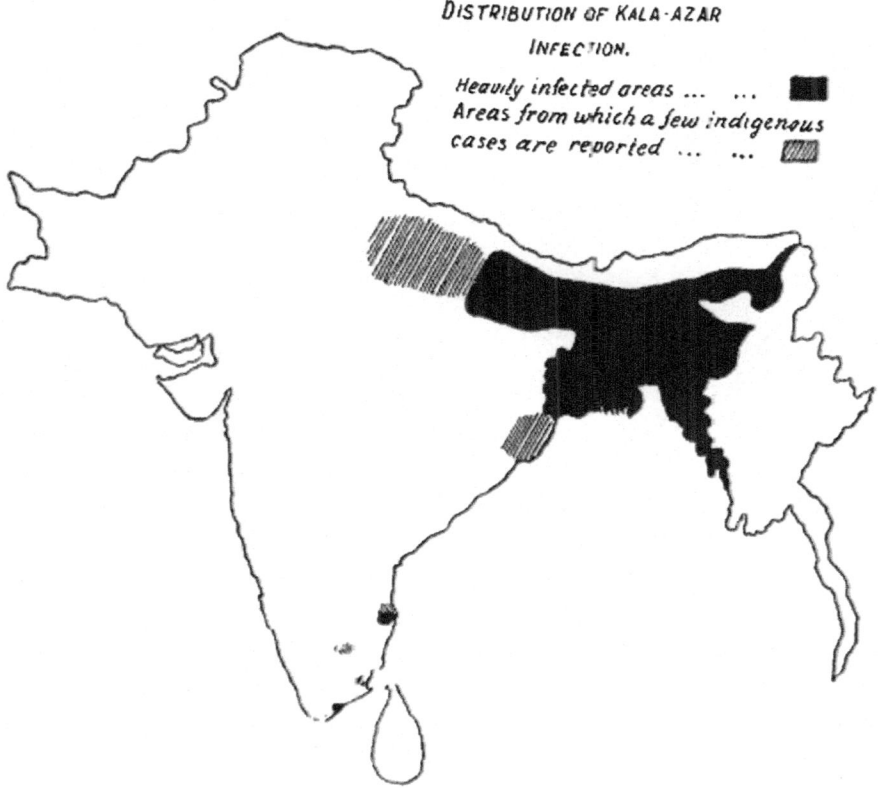

Fig. 4.1 Map of Indian subcontinent, 1903. *Source* Indian Med Gazette, May 1947, p. 281

Pabna district in 1980 [7]. Cases of VL have been reported every year since then. Thirty-four of 64 districts are affected, with 90 % of these cases found in a mere 10 districts (Fig. 4.2). In 1994, 3,965 cases were reported, and 5,920 cases in 2004. It is estimated that the actual number of cases is 5 times the number reported [8]. After launching of the kala azar elimination program in Bangladesh since 9th May, 2008, morbidity and mortality of kala-azar has been reduced considerably. Earlier reports estimated the VL burden in Bangladesh to be about 12,400–24,900 new cases annually with a mortality rate of 1.5 % [8]. In 2014, 1068 cases and 3 deaths were reported (DGHS 2014).

4.1.3 Incidence of VL in Nepal

Twelve districts of Nepal are endemic for VL (Fig. 4.3), with a population of 5.5 million at risk. Since 1980, 14,685 cases and 215 deaths from this disease have

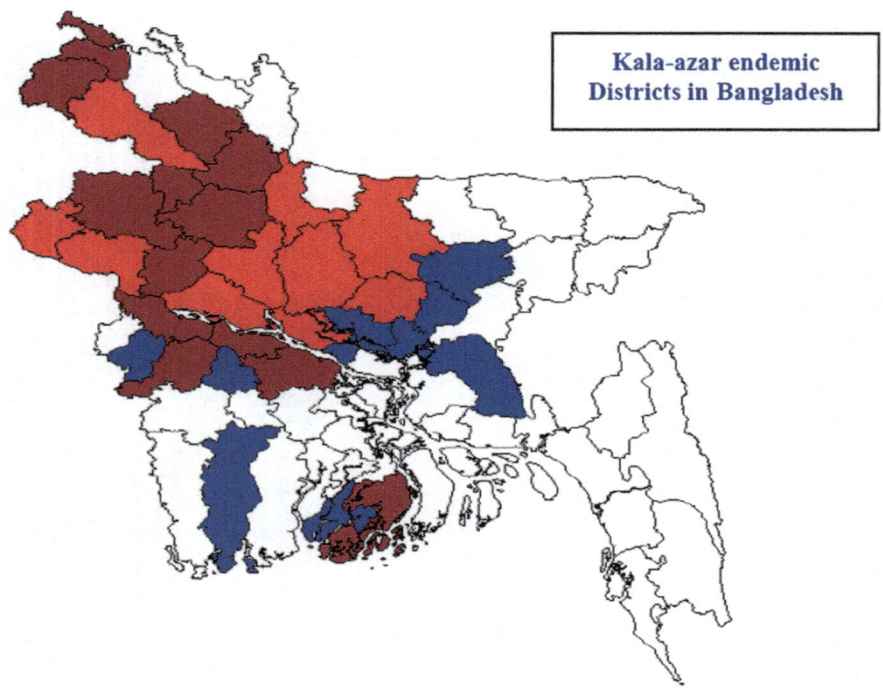

Fig. 4.2 Map of Bangladesh. *Source* http://www.searo.who.int/en/Section10/Section2163_13267.htm

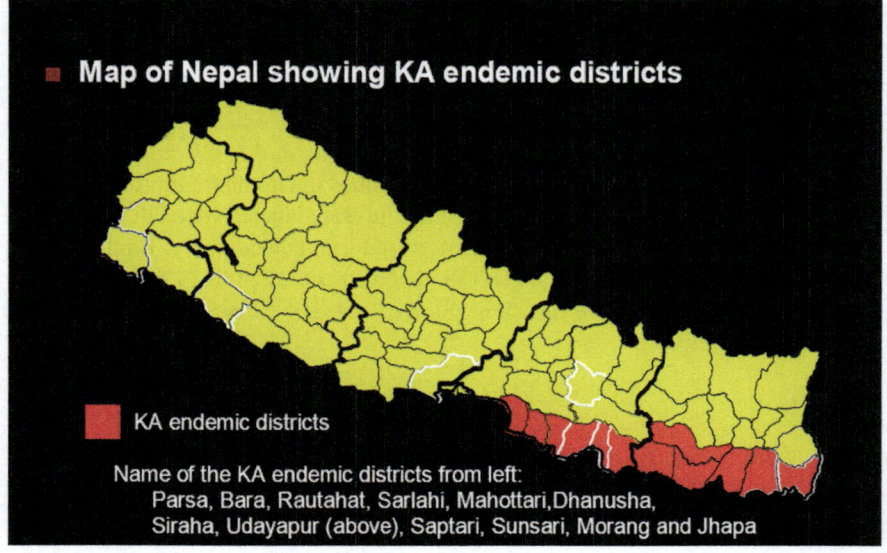

Fig. 4.3 Map of VL (KA) trends from 1987 to 2006. *Source* http://www.searo.who.int/en/Section10/Section2163_13267.htm

been reported, and this figure may be lower than the actual incidence. The case incidence is about 43 per 100,000 population at risk, and the fatality rate ranges from 0.84 to 1.75 % [9]. The incidence of morbidity and mortality has been reduced after launching of the kala-azar elimination program in 2008. The total number of cases reported in 2014 was 323 and 1 death. The incidence per 10,000 cases was 1.42 in 2009, which came down to 0.28 [3].

4.1.4 Incidence of VL in Bhutan

From 2013, the kala-azar elimination program has been extended to Bhutan and Thailand (after the Dhaka declaration of 2010). The total number of cases reported from 7 districts of Bhutan is 18, in between the years 2010 and 2013 (Fig. 4.4). However, no cases were reported in the year 2014 [3].

4.1.5 Incidence of VL in Thailand

Twenty cases of kala-azar and 3 cases of cutaneous leishmaniasis have been reported from Thailand up to December 2014 (Fig. 4.4). Of the 23 patients with leishmaniasis, 9 were co-infected with HIV [3].

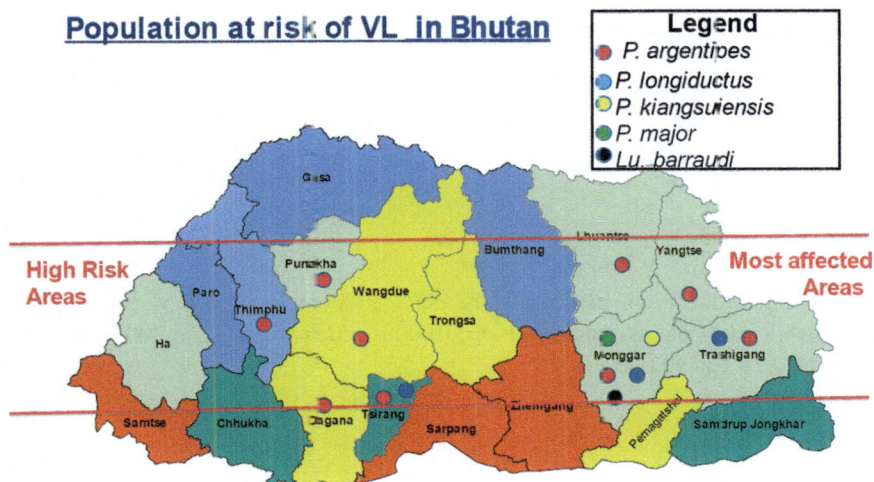

Fig. 4.4 Population at risk of visceral leishmaniasis (VL) in Bhutan. *Source* Informal expert consultation on VL WHO SEARO New Delhi, (25–27 March 2015)

4.1.6 The Current Scenario of VL in South East Asia

Historically, the VL epidemic cycle repeats regularly every 15–20 years [10]. In the Indian subcontinent, about 200 million people are at risk for VL. It is estimated that 25,000–40,000 cases and 200–300 deaths occur yearly in India, Bangladesh, and Nepal. However, these figures are considered to be grossly underestimated. One hundred eleven districts of South East Asia are affected: 54 in India, 26 in Bangladesh, 12 in Nepal, 7 in Bhutan, and 7 provinces of Thailand. The estimated numbers of VL cases in India, Bangladesh, Nepal, Bhutan, and Thailand are 2855 (April 2015), 1068, 323 (with 1 death), 18, and 20, respectively. The total burden of the disease in the subcontinent is 21 cases per 10,000 population [11]. The incidence of kala-azar in India after the launch of the elimination program is as shown in below Table: Source: Informal Expert Consultation on VL WHO SEARO, New Delhi, (25–27 March'2015) (Fig. 4.5).

Sl. No.	State	Total # of endemic districts	Total blocks	VL blocks < 1 case per 10,000 pop.	VL blocks > 1 case per 10,000 pop.
Visceral leishmaniasis cases per 10,000 population during the year 2013					
1	Bihar	33	426	266 (62 %)	160 (38 %)
2	Jharkhand	4	33	3 (9 %)	30 (91 %)
3	West Bengal	11	119	111 (93 %)	8 (6.7 %)
4	Uttar Pradesh	6	12	12 (100 %)	0
	Total	**54**	**590**	**392 (66 %)**	**198 (34 %)**
VL cases per 10,000 population during the year 2014					
1	Bihar	33	445	323 (73 %)	122 (27 %)
2	Jharkhand	4	33	9 (27 %)	24 (73 %)
3	West Bengal	11	121	110 (91 %)	11 (9 %)
4	Uttar Pradesh	6	12	12 (100 %)	0
	Total	**54**	**611**	**454 (74 %)**	**157 (26 %)**

In Bangladesh, the incidence of 1.57 per 10,000 declined to 0.34 per 10,000 in 2014, while in Nepal, the incidence per 10,000, which was 1.42 in 2007, has come down to 0.28 per 10,000 [3].

Fig. 4.5 Leishmaniasis cases by province in Thailand. *Source* Informal expert consultation on VL WHO SEARO, New Delhi, (25–27 March 2015)

4.2 History of the Efficacy of Antileishmanial Drugs in South East Asia

4.2.1 Sodium Stibogluconate (SSG) Resistance in Nepal

A prospective study was conducted in a tertiary-level hospital located in south-eastern Nepal from July 1999 to January 2001 on new parasitologically positive cases of VL. One hundred twenty cases were enrolled in the study, of which 110 patients completed SSG therapy. SSG was used in a dosage of 20 mg/kg daily for 30 days, which was extended to 40 days for patients who were still parasitologically positive. Of the 110 patients completing therapy, the definite cure rate at 6 month follow up was 90 % (99 patients), and 10 % had treatment failure. However, a definite cure rate of only 76 % was achieved in those patients coming from the plains of Nepal bordering North Bihar. Four patients (3.3 %) died, and overall cardiac toxicity was observed in 3.3 % [12].

In another study conducted in a tertiary hospital in Nepal between 1999 and 2004, 659 VL patients were enrolled; 526 cases came from districts outside the SSG-resistant zone of North Bihar, and 133 cases came from the plains of Nepal bordering known SSG-resistant districts such as North Bihar. Of these, 480 and 133 cases, respectively, were assessed. Final cure rates of 86.1 and 59.9 %, respectively, were achieved [13]. Miltefosine has been used in Nepal since the kala-azar elimination program was launched.

4.2.2 Efficacy of Miltefosine in Nepal

A phase 4 trial of miltefosine in Nepal was conducted during September 2003–March 2004 with 125 patients, of which 122 completed treatment. A final cure rate of 88 % (105 of 119 cases) was achieved (unpublished data). Miltefosine was used in the kala-azar elimination program till 2014.

4.2.3 Efficacy of SSG in Bangladesh

No data from any controlled trial testing the effectiveness of SSG have been reported so far from Bangladesh. There is a shortage of the standard preparation of SSG in this country. However, as per official recommendations, SSG is used at a dose of 20 mg/kg daily for 28 days. In 1 study (personal communication), 148 patients were enrolled, of which 144 had an initial clinical cure at the end of treatment. Four patients died during the treatment. The final cure rate at 6-month follow up could not be ascertained. A phase 4 study of miltefosine was conducted in Bangladesh by the Special Programme for Research and Training in Tropical Diseases of the World Health Organization (WHO) in cooperation with Zentaris (Germany). Miltefosine has been used in the kala-azar elimination program launched in Bangladesh from 9th May 2008.

4.2.4 Changing Response to Different Antileishmanial Agents in India

Pentavalent antimonials have been the main first-line treatment for all forms of leishmaniasis, including VL. After its discovery by Schmidt, Kikuth, and others, antimony gluconate (Solustibosan R) was synthesized in 1937. Several pentavalent antimonials have been synthesized since then, and these are available in the world market. Meglumine antimoniate is marketed as Glucantine and Prostib. SSG or sodium antimony gluconate (SAG) is available commercially as Pentostam,

Solustibosan, and Stibanate. Generic sodium antimony gluconate (Albert David, Calcutta, India) is 20 times cheaper than other products, costing US$ 13 versus US$ 200 per patient. There is some variation in the effectiveness of SSG in different states in India. Patients from eastern Uttar Pradesh are still responsive to SSG [14], and in the Dumka district of Jharkhand, where VL is common, SSG remains effective, as well.

SSG has been in use in India for treating VL since 1940. Used at a dose of 10 mg/kg daily for 10 days and administered parenterally (intramuscularly or intravenously), it has proved effective in curing 95 % of cases. Thus, a disease with over 90 % mortality in untreated patients could be cured in 95 % of them [4]. Until 1970, SSG used in the conventional dosage was effective in curing more than 90 % of cases of Indian VL. In an earlier resurgence of VL, which assumed epidemic proportions by 1977, an estimated 250,000 patients were affected in Bihar. Of those, 30 % of cases were nonresponsive to SSG [15].

During the late 1970s and early 1980s, healthcare workers used different regimens of SSG to counter the problem of unresponsiveness before pharmacokinetic studies of SSG were conducted (Table 4.1) [16–18]. A pharmacokinetic study of SSG was done in 1980. It was observed that the use of SSG can be safely prolonged up to 60 days when administered in a daily dose of 10 mg/kg. Tolerance in children was found to be better than that seen in adults [19]. In 1982, WHO recommended SSG in a daily dose of 20 mg/kg (maximum dose of 850 mg) for 20–30 days in new cases and for 40–60 days in relapse cases. Subsequently, WHO in 1990 recommended SSG in a daily dose of 20 mg/kg for 30 days [20]. Changes in the response to the WHO regimen were observed in Bihar by different healthcare workers (Table 4.2) [21–26].

However, nonresponders to the higher and prolonged dosage schedule recommended by WHO caused treatment failure in up to 60 % of cases of kala-azar from the hyperendemic districts of North Bihar (Fig. 4.4).

Table 4.1 Efficacy of sodium stibogluconate (SSG) at different dosages and durations between 1975 and 1981

Healthcare worker	Dose (mg/kg per day)	Duration in days	Number of courses	Number of cases	Number of nonresponders (%)
Jha (1975–76) [15]	10	10	1	200	17
Jha (1977–80)[a]	10	10	3	520	30
Thakur (1980–84) [16]	10	20	1	125	9
Jha (1980–81) [17]	10 adult 20 child	30-Fresh 42-Slow responder 60-Relapse	1	90	1.1

[a]Unpublished data

Table 4.2 Changing response to sodium stibogluconate (SSG) used in WHO regimen

Healthcare worker	Dose (mg/kg per day)	Duration in days	Number of courses	Number of cases	Number of nonresponders (%)
Thakur (1984–87) [20]	20	40	1	371	8
Jha (1989) [21]	20	30	1	252	26.1
Jha (1991) [22]	20	30	1	32	25
Jha et al. (1996) BMJ [23]	20	30	1	30	36.7
Thakur et al. (1998) [24]	20	30	1	80	54
Sundar et al. [25]	20	30	1	184	60

Drug-induced toxicities such as myocarditis, pancreatitis, renal failure, and deaths following WHO's new SSG regimen were reported. Even before this, there was controversy over whether the treatment failure was due to immunological failure or drug resistance. However, a 100 % treatment failure after SSG was administered in the WHO-recommended dosage was observed in 2 rural villages located in North Bihar. A total of 241 cases were found in the 2 villages [26]. Pharmacological failure was suggested to be the likely cause of treatment failure [27].

In 1991, Jha defined the therapeutic response zones of North Bihar (Fig. 4.6) [28].

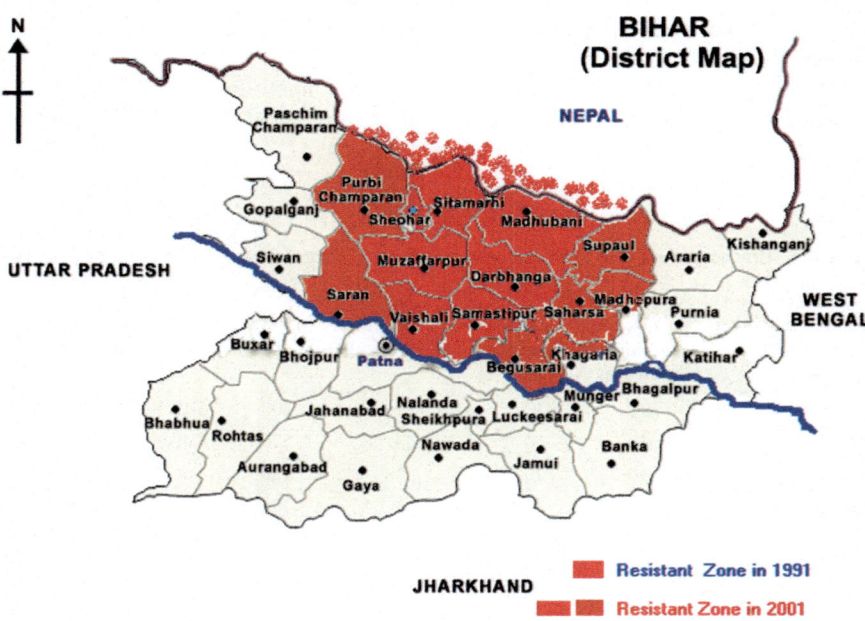

Fig. 4.6 Map of Bihar showing therapeutic status of sodium stibogluconate SSG in Bihar

4.2.5 Changing Response to Diamidine Compounds

Among diamidine compounds, pentamidine isethionate (May and Baker) and pentamidine demethane sulphonate (Lomidine-Specia, France) have been used in SSG-nonresponder cases of VL from the late 1970s to 2003 as a second-line treatment. In one study, 82 patients who were unresponsive to SSG were treated with pentamidine isethionate in a dose of 4 mg/kg for 10–12 intravenous injections on alternate days, with a cure rate of 98.2 % at 6 month follow up. Two of the 82 patients died of sudden cardiac arrest and 4 (4.81 %) developed a condition resembling diabetes mellitus [29].

In another study, 233 SSG-resistant cases were treated with pentamidine in a dose of 4 mg/kg administered intramuscularly or intravenously on alternate days. One hundred seventy-four cases received pentamidine isetheonate, while 59 were treated with pentamidine methane sulphonate (Lomidine). Clinical and parasitological evaluation was done every 10th day until 2 successive bone-marrow or splenic aspirates were negative for parasites. Eighteen to twenty injections were required for an initial cure. A full cure was achieved in 75.2 % of the cases treated with pentamidine isethionate, while a cure rate of 100 % was observed in the Lomidine group. Drug-induced diabetes mellitus was observed in 1.14 and 18.4 %, respectively. An overall cure rate of 81.5 % and relapse rate of 2.08 % were observed. Mortality rates of 0.57 and 9.23 % were observed in the 2 groups [30] (Fig. 4.7).

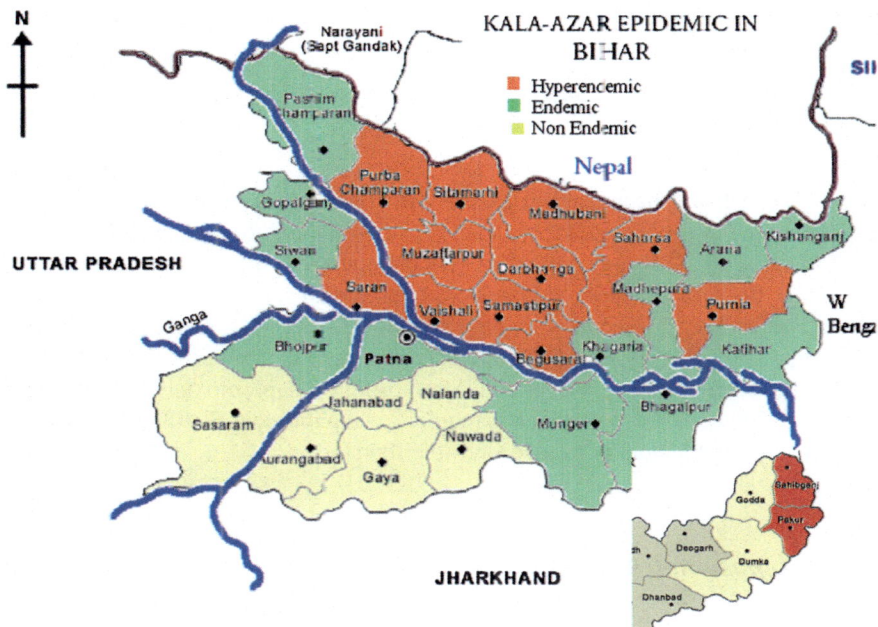

Fig. 4.7 Map of Bihar showing endemic zones

Table 4.3 Changing response to sodium stibogluconate (SSG) used in WHO regimen

Healthcare worker	Dose (mg/kg per day)	Duration in days	Number of courses	Number of cases	Number of nonresponders (%)
Thakur (1984–87) [20]	20	40	1	371	8
Jha (1989) [21]	20	30	1	252	26.1
Jha (1991) [22]	20	30	1	32	25
Jha et al. (1996) BMJ [23]	20	30	1	30	36.7
Thakur et al. (1998) [24]	20	30	1	80	54
Sundar et al. [25]	20	30	1	184	60

In another randomized study, pentamidine isethionate was used as a monotherapy in a dose of 4 mg/kg administered parenterally in 3 injections per week until a parasitological cure was achieved. Thirty-three or more injections were required to achieve a cure rate of 98 %. However, toxicities such as hyperglycemia were observed in 10 % of cases (6 % reversible, 4 % irreversible), and cardiac toxicity was also observed [31].

Both the diamidine compounds are imported, expensive, and highly toxic. Drug-induced irreversible insulin-dependent diabetes is reported in 5–12 % of cases. Cardiac toxicities leading to fatal arrhythmias have been observed. This drug is no longer used to treat Indian VL due to the declining response, serious toxicities, and high cost [32].

4.2.6 Amphotericin B

Amphotericin B has been used extensively in Bihar for all SSG-nonresponder VL cases. In some of the hyperendemic districts of North Bihar where between 30 and 60 % of cases are SSG-nonresponders, it is being used as the first-line therapy. It is used in a dose of 1 mg/kg administered intravenously with 5 % dextrose on alternate days for 15 injections. A cure rate of 98–100 % is achieved at 6-month follow up. This is a very effective but costly treatment that requires hospitalization, and febrile reactions are very common. Monitoring for nephrotoxicity and checking serum electrolytes for hypokalemia is required. Amphotericin B still remains a very effective drug for kala-azar as a second-line therapy.

4.2.7 Liposomal Amphotericin B (AmBisome)

There are three liposomal preparations: liposomal amphotericin B sold as AmBisome (Gilead Sciences; Dimas, California, USA); amphotericin B lipid complex sold as

Abelcet (Enzon Pharma; Fairfield, New Jersey, USA); and amphotericin B colloidal dispersion, sold as Amphotec (Intermune; Brishane, California, USA).

AmBisome used to treat resistant VL in a cumulative dose of 10–15 mg/kg cured 90–100 % of cases. The duration of therapy could be reduced to 2–5 days [33]. A total dose of 20 mg/kg, divided into 4 or 5 doses given intravenously on day 1, day 3, day 5, and day 7 resulted in a 100 % cure rate when used as a rescue drug in resistant cases of VL in different drug trials (unpublished data). At present, this is the most effective drug for VL that can be used in all resistant cases. However, it is too costly (one 50 mg vial of AmBisome costs about US$ 135). However, the cost of AmBisome has been reduced substantially through the negotiation of WHO with GLEAD, reducing the price to US$ 20 per 50 mg. AmBisome given in a single dose of 10 mg per kilogram of body weight daily (MKD) has been found to have up to a 96 % cure rate [34]. This single dose of AmBisome has been recommended as a first-line therapy in the kala-azar elimination program effective from 2015. A dose of 5 MKD is also being used in combination therapy with miltefosine and/or IM paromomycin.

4.2.8 Paromomycin (Aminosidine)

Paromomycin is an aminoglycoside found to have a potent antileishmanial action. Previously, it was tried in combination with SSG to improve cure rates [35]. In a phase 2 randomized controlled trial, paromomycin was used in a daily dose of 16 or 20 mg/kg given intramuscularly for 21 days. A cure rate of 93 and 97 %, respectively, was achieved at the 6-month follow up [36]. In a phase 3 multicenter, double-blind paromomycin study involving 500 cases of Indian VL, a cure rate of 94.6 % was achieved at 6-month follow up [37]. In phase 4 trial conducted at 17 sites, a cure rate of 94.2 % was achieved [38]. Paromomycin is a potential first line treatment for VL in the Indian subcontinent when used as a monotherapy. Problems of relapse and treatment failure can be anticipated, as with other aminoglycosides, when used as a monotherapy. A combination of paromomycin 11 MKD intramuscularly has been used in combination with liposomal amphotericin B in a dose of 5 MKD, and paromomycin 11 MKD intramuscularly with miltefosine 50 mg for 10 days.

4.2.9 Status of Oral Treatments for VL

Allopurinol and ketoconazole have been tried in cases of Indian VL both singularly and in combination with SSG without much success [39, 40].

4.2.10 Miltefosine as a Treatment for VL

Miltefosine is an active alkylphospholipid derivative found to have potent antileishmanial activity. It has undergone several clinical trials, including phase 1/2 [41], phase 2 [42], and phase 3 [43] in adults, phase 3 in children [44], and phase 4 in both children and adults [45]. Altogether more than 1,700 new and resistant cases of VL in adults and children were enrolled in different multicenter trials in Bihar [45]. A final cure rate of 94 % at 6-month follow up was achieved. Miltefosine is used in a dose of 50 mg twice daily (100 mg/day) for 28 days in patients weighing more than 25 kg; 50 mg daily for 28 days for children weighing 25 kg; and 2.5 mg/kg daily for 28 days for children aged 2–11. However, the cure rate dropped to about 82–87 % in phase 4 trials when used in outpatients, due to noncompliance related to gastrointestinal disturbance.

Miltefosine has a long half-life of about 150 h, which may create problems of relapse and treatment failure in the future. It is not recommended for use in females of child-bearing age unless they use contraception during and for 2 months after the end of treatment. Of 18,462 VL cases seen between 1975 and 2009 at the Kalazar Research Centre (Brahmpura, Muzaffarpur, Bihar, India), 18.56 % of cases were females of child-bearing age (unpublished data). This female population is likely to be excluded from treatment with Miltefosine. However, Miltefosine has been used as first-line treatment in the kala-azar elimination program since 2008 in India, Nepal, and Bangladesh.

4.2.11 Sitamaquin (WR 6026)

Sitamaquine is a primaquine analogue with effective anti-leishmanial activity. In a multicenter phase 2 dose-finding study on 120 Indian patients with VL, sitamaquine was used at a dose of 1.5, 1.75, 2.0, and 2.5 mg/kg daily for 28 days. A final cure rate of 81, 89, 100, and 80 %, respectively, was achieved. However, 5 % of patients had nephrotoxicity [46]. Sitamaquine can potentially be used as an oral drug in combination with another first-line anti-leishmanial agent.

4.3 Combination Therapy for Indian VL

Combining 1 potent antileishmanial drug with a short half-life and another drug with a long half-life may improve the efficacy and cure rate in VL as with tuberculosis or leprosy. The second drug may have a synergistic or additive effect without having an adverse drug interaction. SSG has been used with Th1-activating cytokine (interferon-gamma [IFN-γ]) as a form of immunochemotherapy to treat Indian VL. IFN-γ in a daily dose of 100 µg/m^2 for 30 days combined with SSG in a daily dose of 20 mg/kg for 30 days was used, but a full cure was achieved in only 49 % of cases [47]. In India and the plains of Nepal, where nonresponders to SSG are very high, this combination therapy is ineffective. In 1995, when SSG was used in combination with allopurinol or ketoconazole to treat Indian VL, no significant increase in efficacy was observed [40].

In one study, a combination of SSG 20 MKD and paromomycin 11 MKD was used, for 20 days with a cure rate of 81.8 % [34].

A study on the combination therapy of AmBisome single dose 5 MKD + miltefosine 50 mg daily for 7 days ($n = 160$), combination of AmBisome 5 MKD single dose + paromomycin 11 MKD for 10 days ($n = 158$), and paromomycin 11 MKD for 10 days with miltefosine 50 mg for 10 days ($n = 159$), was compared with amphotericin B 1 MKD intravenous on alternate days for 30 days (total dose 15 mg/kg). A definitive cure rate at 6-months follow up was found in 97.5, 97.5, and 98.7 %, respectively, while in the amphotericin B group 93.0 % was achieved [48].

4.3.1 Future Option of Combination Therapy (Immunotherapy)

Future options for combination therapy (immunotherapy) are as follows: (i) use of Th1 cytokines like IL-12, and others IFNγ and TNF + Sb; (ii) granuloma remodeling exogenous cytokines IL-2 or GM-CSF (granulocyte-macrophage colony stimulating factor) + a potent anti-leishmanial drug like paromomycin; (iii) IL-10 receptor blocker + chemotherapy + Sb/paromomycin, and (iv) amphotericin B + miltefosine—are directly microbicidal towards intracellular L. donovani amastigotes in vitro [49]. Both act independently of the immune response.

Recombinant protein anti-leishmanial vaccine plus effective chemotherapy is a future option.

4.4 Combination Therapy in VL Associated with HIV/AIDS

There is increasing incidence of coinfection of HIV/AIDS with VL in India, Bangladesh, Nepal, and Thailand. In 1994, not a single case of HIV/AIDS associated with VL was reported [50]. In 1997, 3 cases of coinfection were reported [50–52].

A study on VL associated with HIV/AIDS is being conducted at Kalazar Research Centre. There were 61 cases of coinfection reported in 573 cases of HIV/AIDS during the period of January 1995–January 2009 [53], and 45 cases of coinfection have been reported in 385 HIV/AIDS cases at Kalazar Research Centre between February 2009 and May 2015. The total number of cases with coinfection between the period of 1995 and May 2015 was 106 of 7,434 VL cases (1.43 %) whereas the incidence of coinfection in HIV/AIDS was 11.06 % (unpublished).

Different regimens of antimonial therapy have been tried to treat cases of coinfection, but resistance and relapses are common. In one study on 159 patients with VL/HIV, the coinfection was treated with AmBisome 20–25 mg/kg (bw) in 4 or 5 doses of 5 MKD over 5–10 days and antiretroviral treatment (ART) was initiated in 23 (14.5 %), 39 (24.5 %), and 61 (38.4 %) before, during, and after admission, respectively. Initial cure was achieved in all patients, but 36 patients died during the follow up and 6 died shortly after admission. The mortality rate was found to be 14.3 % at 6 months, 22.4 % at 2 years, and 29.7 % at 4 years of follow up. Of 153 patients receiving treatment, 26 relapsed during the follow up within 10 months. The estimated relapse was found to be 16.1 % at 1 year, 20.4 % at 2 years, and 25.9 % at 4 years. The incidence of relapse and death was found to be greater with CD4+ counts lower than 200 with low hemoglobin or when associated with concurrent tuberculosis [54]. Currently the WHO guideline for AmBisome in HIV/VL co-infection recommends AmBisome at a dose of 40 MKD in divided doses together with ART in all cases.

4.5 Conclusion

The number of nonresponders to SSG has been increasing in India and the plains of Nepal. In Bangladesh, the status of SSG efficacy is not known, but SSG is being used in Bhutan and Thailand. SSG as a first-line therapy is not recommended for the kala-azar elimination program in India and the neighboring countries like Nepal and Bangladesh. In May 2005, a memorandum of understanding was signed by 3 countries: India, Nepal, and Bangladesh. The burden of VL in these countries was 21 cases per 10,000 population. The target was to reduce this incidence to 1 case per 10,000 population by 2015. The London declaration in 2012 and the World Health Assembly resolution in 2013 extended this period of elimination to 2017 and also included Bhutan and Thailand in the elimination program. The memorandum of understanding was signed to this effect. Oral miltefosine was recommended as a first-line therapy for the elimination program. It would be ideal to use as a direct-observed treatment (DOT) to counter the problem of noncompliance. An injectable Depo-Provera contraceptive preparation should be given a few days before starting miltefosine in females of child-bearing age on the Indian subcontinent to avoid any teratogenic complications. Alternatively oral contraceptive pills should be given 1 week before starting the treatment, during the treatment, and 2 months after the therapy. During the active phase of the elimination program between 2008 and 2014, it was observed this was not possible in the clinic and the compliance rate was poor. At present, AmBisome in a single intravenous dose at 10 MKD has been recommended to be used in the elimination program from the latter part of 2014. This has been followed in all the endemic districts of kala-azar in South East Asia.

Acknowledgments Gratitude is expressed to Mr. Sarvajeet Kumar, Clinical Research Coordinator, Kalazar Research Centre, Brahmpura, Muzaffarpur, for secretarial assistance, and also to Dr. C.P.N. Thakur, Director, Pathological Research Lab, for reviewing the manuscript.

References

1. Bora D. Epidemiology of visceral leishmaniasis in India. Natl Med J India. 1999;12:62–8 (Review).
2. The control of leishmaniasis. Report of an expert committee world health organization, WHO. Tech Rep Ser. 1990;12:62–8.
3. Informal Expert Consultation on Kala-azar WHO SEARO, New Delhi, (25–27 March'2015).
4. Sen Gupta PC. History of kala-azar in India: Sir U N Brahmchari memorial lecture, royal asciatic society of Bengal May 1947. Indian Med Gaz. 1947;281–6.
5. Thakur CP. Leishmaniasis research—the challenge ahead. Indian J Med Res. 2006;123:193–4.
6. Desjeux P. Human leishmaniases: epidemiology and public health aspects. World Health Stat Q. 1992;45:267–75.
7. Elias M, Rahman AJ, Khan NI. Visceral leishmaniasis and its control in Bangladesh. Bull World Health Organ. 1989;67:43–9.
8. Alvar J, Vélez ID, Bern C, Herrero M, Desjeux P, et al. Leishmaniasis worldwide and global estimates of its incidence. PLoS ONE. 2012;7(5):e35671. doi:10.1371/journal.pone.0035671.
9. Gurubacharya RL, Gurubacharya SM, Gurubacharya DL, et al. Prevalence of visceral leishmania and HIV co-infection in Nepal. Indian J Med Res. 2006;123:473–5.
10. Alvar J, Yactayo S, Bern C. Leishmaniasis and poverty. Trends Parasitol. 2006;22:552–7.
11. Joshi A, Narain JP, Prasittisuk C, et al. Can visceral leishmaniasis be eliminated from Asia? J Vector Borne Dis. 2008;45:105–11.
12. Rijal S, Chappuis F, Singh R, et al. Treatment of visceral leishmaniasis in south-eastern Nepal: decreasing efficacy of sodium stibogluconate and need for a policy to limit further decline. Trans R Soc Trop Med Hyg. 2003;97:350–4.
13. Rijal S, Van der Stuyft P, Chappuis F, et al. Visceral leishmaniasis in Nepal: spreading endemicity and emerging antimony resistance Chapter 3: clinical and epidemiological pattern and burden of disease due to kala-azar in Nepal: kala-azar in Nepal: from clinical evidence to control Editor: Suman RIJAL Proefschrift voorgelegd aan de Faculteit Geneeskunde en Gezonheidswetenschappen tot het verkrijgen van de graad van Doctor in de Medische Wetenschappen Gent, 15 februari. Printed by Academia Press Scientific Publishers; 2006. p. 52–6. ISBN 9078344016 D/2006/4531/2.
14. Sundar S, More DK, Singh MK, et al. Failure of pentavalent antimony in visceral leishmaniasis in India: report from the center of the Indian epidemic. Clin Infect Dis. 2000;31:1104–7.
15. Peter W. The treatment of kala-azar—new approach to an old problem. Indian J Med Res. 1981;73:1–18.
16. Jha TK. Study on early diagnostic futures of kala-azar occurring in North Bihar. In: Abstracts of 10th international congress on tropical medicine and Malaria, Manila Philipines, November 9-15-1980:206 (Abst. No.197).

17. Thakur CP, Kumar M, Singh SK. Comparison of regimens of treatment with stibogluconate in kala-azar. BMJ. 1984;288:295–7.
18. Jha TK, Sharma VK. Prolonged sodium stibogluconate therapy in Indian kala-azar. J Assoc Physician India. 1986;34:469–71.
19. Rees PN, Keating MI, Kager PA, Hockmeyer WT. Renal clearance of pentavalent antimony (sodium stibogluconate). Lancet. 1980;ii:226–9.
20. Report of the informal meeting on the chemotherapy of visceral leishmaniasis. WHO/TDR/Chem/Leish/VL/1982-83/Narobi.
21. Thakur CP, Kumar M, Kumar P, et al. Rationalization of regimens of treatment of kala-azar with sodium stibogluconate in India: a randomized study. BMJ. 1988;4:463–7.
22. Jha TK, Singh NKP, Jha S. Therapeutic use of sodium stibogluconate in kala-azar from some hyperendemic districts of North Bihar, India. J Assoc Physicians India. 1992;40:868.
23. Jha TK, Singh NKP, Singh IJ, Jha S. Combination therapy in kala-azar. J Assoc Physicians India. 1995;43:319–20.
24. Jha TK, Olliaro P, Thakur CPN, et al. Randomised controlled trial of aminosidine (Paromomycin) v sodium stibogluconate for treating visceral leishmaniasis in North Bihar India. BMJ. 1998;316:1200–5.
25. Thakur CP, Sinha GP, Pandey AK, et al. Do the diminishing efficacy and increasing toxicity of sodium stibogluconate in the treatment of visceral leishmaniasis in Bihar, India, justify its continued use as a first line drug? An observation study of 80 cases. Ann Trop Med Parasitol. 1998;92:561–9.
26. Sundar S. Drug resistance in Indian visceral leishmaniasis. Trop Med Int Health. 2001;6:849–54.
27. Jha TK. Refractory kala-azar diagnosis and management. In: Das AK, editor. The association of physicians of India: medicine update; vol. 8; 1998. p. 137–46.
28. Jha TK. Management of kala-azar: API-post graduate medicine volume 6, 1992. In: Chugh KS editor. Proceeding of CME program XLVII Joint Annual conference of API. p. 11–8.
29. Jha TK, Sharma V. Evaluation of diamidine compound in the treatment of resistant cases of kala-azar occurring in North Bihar India. Trans R Soc Med Hyg. 1983;77:66–70.
30. Jha SN, Singh NKP, Jha TK. Changing response to diamidine compound in cases of kala-azar unresponsive to antimonials. J Assoc Physicians India. 1991;39:314–6.
31. Thakur CP, Kumar M, Pandey AK. Comparison of regimens of treatment of antimony resistant kala-azar patients: a randomized study. Am J Trop Med Hyg. 1991;45:435–41.
32. Sundar S: Treatment of kala-azar trends and research in leishmaniasis with particular reference to kala-azar. In: Sir Dorabji Tata centre for tropical disease—Published Jan 2005, Tata McGraw–Hill publishing company Ltd, New Delhi p. 157–65.
33. Sundar S, Agrawal NK, Sinha PR, et al. Short-course, low-dose amphotericin B lipid complex therapy for visceral leishmaniasis unresponsive to antimony. Ann Intern Med. 1997;127:133–7.
34. Thakur CP, Bhowmick S, Dolfi L, Olliaro P. Aminosidine plus sodium stibogluconate for the treatment of Indian kala-azar: a randomized dose-finding clinical trial. Trans R Soc Trop Med Hyg.1995;89(2):219–23.
35. Thakur CP, Bhowmick S, Dolfi L, Olliaro P. Aminosidine plus sodium stibogluconate for the treatment of Indian kala-azar: a randomized dose-finding clinical trial. Trans R Soc Trop Med Hyg. 1995;89:219–23.
36. Jha TK, Olliaro P, Thakur CPN, et al. Randomised controlled trial of aminosidine (Paromomycin) v sodium stibogluconate for treating visceral leishmaniasis in North Bihar India. BMJ. 1998;316:1200–5.
37. Sundar S, Jha TK, Thakur CP, et al. Injectable paromomycin for visceral leishmaniasis in India. N Eng J of Med. 2007;356:2571–81.
38. Sinha PK, Jha TK, Thakur CP, Nath D, Mukherjee S, Aditya AK, Sundar S. Phase 4 pharmacovigilance trial of paromomycin injection for the treatment of visceral leishmaniasis in India. J Trop Med. 2011; Article ID 645203:7.

39. Jha TK. Evaluation of allopurinol in the treatment of kala-azar occurring in North Bihar, India. Trans R Soc Trop Med Hyg. 1983;77:204–7.
40. Jha TK, Singh NKP, Singh IJ, Jha S. Combination therapy in kala-azar. J Assoc Physicians India. 1995;43:319–20.
41. Sundar S, Resenkaimer F, Makharia MK, et al. Trial of oral miltefosine for visceral leishmaniasis. Lancet. 1998;352:1821–3.
42. Jha TK, Sundar S, Thakur CP, et al. Miltefosine, an oral agent, for the treatment of Indian visceral leishmaniasis. N Engl J Med. 1999;341:1795–800.
43. Sundar S, Jha TK, Thakur CP, et al. Oral miltefosine for Indian visceral leishmaniasis. N Engl J Med. 2002;347:1739–46.
44. Bhattacharya SK, Jha TK, Sundar S, et al. Efficacy and tolerability of miltefosine for childhood visceral leishmaniasis in India. Clin Infect Dis. 2004;38:217–21.
45. Bhattacharya SK, Sinha PK, Sundar S, et al. Phase 4 trial of miltefosine for the treatment of Indian visceral leishmaniasis. J Infect Dis. 2007;196:591–8.
46. Jha TK, Sundar S, Thakur CP, et al. A phase II dose-ranging study of sitamaquine for the treatment of visceral Leishmaniasis in India. Am J Trop Med Hyg. 2005;73:1005–11.
47. Sundar S, Singh VP, Sharma S, et al. Response to interferon gamma plus pentavalent antimony in Indian visceral leishmaniasis. J Inf Dis. 1997;176:1117–9.
48. Sundar S, Sinha PK, Rai M, Verma DK, Nawin K, Alam S, Chakravarty J, Vaillant M, Verma N, Pandey K, Kumar P, Lal CS, Arora R, Sharma B, Ellis S, Strub-Wourgaft N, Balasegaram M, Olliaro P, Das P, Modabber F. Comparison of short-course multidrug treatment with standard therapy for visceral leishmaniasis in India: an open-label, non-inferiority, randomised controlled trial. Lancet. 2011 Feb 5;377(9764):477–86.
49. Murray HW. Clinical and experimental advances in treatment of visceral leishmaniasis. Antimicrob Agents Chemother. 2001;45(8):2185–97.
50. Jha TK, Singh NK, Thakur CPN, Singh TK. Incidence of HIV infection in resistant cases of Indian kala-azar. J Assoc Physicians India. 1994;42:263.
51. Dey AB, Shakti C, Kalpana N, et al. Visceral Leishmaniasis in AIDS. J Assoc Physicians India. 1997;45:63–4.
52. Vishwash LA, Singh S, Wali JP. Visceral leishmaniasis in AIDS. J Assoc Phy sicians Inida. 1997;45:582.
53. Thakur CP, Sinha PK, Singh RK, et al. Miltefosine in a case of visceral leishmaniasis with HIV co-infection; and rising incidence of this disease in India. Trans R Soc Trop Med Hyg. 2000;94:696–7.
54. Burza S, Mahajan R, Sinha PK, van Griensven J, Pandey K, Lima MA, Sanz MG, Sunyoto T, Kumar S, Mitra G, Kumar R, Verma N, Das P.: Visceral leishmaniasis and HIV co-infection in Bihar, India: long-term effectiveness and treatment outcomes with liposomal amphotericin B (AmBisome).: PLoS Negl Trop Dis. 2014 Aug 7;8(8):e3053.
55. Bern C, Chowdhury R. The epidemiology of visceral leishmaniasis in Bangladesh: prospects for improved control. Indian J Med Res. 2006;123:275–88.
56. Jha TK. Kala-azar with HIV/AIDS infected population of North Bihar India, Abstract No.557: 4th World Congress on Leishmnaisis (Worldleish4); 2009. p. 197.

Chapter 5
A Therapeutic Strategy for Treating Visceral Leishmaniasis in Regions with Drug Resistance

Shyam Sundar and Dipti Agarwal

Abstract Visceral leishmaniasis (VL) affects 500,000 people annually worldwide, and Bihar alone accounts for approximately 45 % of that burden. For the last 2 decades there has been a steady decline in response to pentavalent antimonial (Sb^v), the drug that has been used for treating VL for 7 decades. Oral miltefosine has been chosen as an alternative drug for use in the kala-azar elimination program in India, Nepal, and Bangladesh. There are only 4 approved antileishmanial drugs: Sb^v, miltefosine, paromomycin, and amphotericin B and its lipid formulations. Except for liposomal amphotericin B, all the other drugs have to be administered for 21–30 days and have frequent side effects, leading to noncompliance and early discontinuation of treatment. These factors, along with the intrinsic characteristics of the drugs (e.g., a long half-life in the case of miltefosine), are conducive to the development of drug resistance. Thus, alternative strategies must be developed to prolong the effective life span of these drugs, such as the use of single-dose liposomal amphotericin B, or directly observed therapy when longer-duration treatments are used. Finally, combination therapy with multiple drugs should be implemented as early as possible, because these are likely to shorten the duration of therapy, improve compliance, and decrease both toxicity and cost of treatment.

Keywords Drug resistance · Compliance · Liposomal amphotericin B · Combination therapy · Directly observed therapy

---

S. Sundar (✉)
Department of Medicine, Institute of Medical Sciences,
Banaras Hindu University, Varanasi, India
e-mail: drshyamsundar@hotmail.com

5.1 Introduction

Visceral leishmaniasis (VL) is the systemic and disseminated form of a disease also known as kala-azar. This syndrome is typically characterized by fever (often with chills and rigor), splenomegaly, hypergammaglobulinemia, and pancytopenia. Once clinical VL sets in, patients become emaciated and prone to develop secondary infections such as malaria, pneumonia, tuberculosis, and amoebic or bacillary dysentery. VL is a lethal form of leishmaniasis that is uniformly fatal unless treated. The incidence of VL is 500,000 cases per year. Ninety percent of the annual global burden of VL cases occurs in India, Nepal, Bangladesh, and Brazil. In these countries, VL is endemic, and epidemics are quite frequent, which leads to considerable mortality. It is thus a significant public health problem. In India, about 100,000 cases of VL are estimated to occur annually, with the state of Bihar accounting for >90 % of these cases.

Unfortunately, in Northern Bihar in India, >50 % of patients are refractory to pentavalent antimonial (Sb^V) therapy, and thus alternative drugs have to be used [1, 2]. The failure of Sb^V in this area is attributed to its widespread misuse at an anthroponotic focus with intense transmission [3]. A similar potential for resistance exists in East Africa, especially in Sudan, another anthroponotic focus of VL with intense transmission, where poverty, illiteracy, and poor healthcare facilities portend the misuse of the drug and consequent emergence of resistance [3]. Resistance develops as the epidemic turns endemic in foci where Sb^V has been used as a monotherapy for long periods, often with poor supervision and compliance [4]. In other parts of the world, Sb^V continues to be effective [5]. The antileishmanial drugs discussed in the following sections are available for use to plan a strategy for treatment in regions with drug resistance.

5.2 Antileishmanial Drugs

The treatment of VL has evolved significantly over time as a consequence of emerging resistance patterns and newer drug delivery systems.

5.2.1 Sodium Stibogluconate

Urea stibamine was the first antimonial drug, introduced over 70 years ago. It was replaced in the 1950s by sodium stibogluconate, which became the first-line treatment for VL. Initially, the drug was used at very low doses (e.g., 10 mg/kg per day for 6–10 days). For more than 7 decades, Sb^V compounds have been used for the treatment of all forms of leishmaniasis. The drug was inexpensive, and, at that time, effective and well tolerated. Then treatment failures started to occur, and a routine of gradually increasing the dose and duration of therapy began, in what has proven to be an unsuccessful attempt at catching up with the resistance. As a result of the increased doses, adverse events became common, including arthralgia, myalgia, and raised levels of hepatic transaminases. In a significant proportion of patients, severe toxicity such as chemical pancreatitis, especially in patients co-infected with human immunodeficiency virus (HIV) [6] and cardiotoxicity, manifested by concave ST-segment elevation, prolongation of the QT interval to >0.5 ms, ventricular ectopic beats, runs of ventricular tachycardia, torsades de pointes, ventricular fibrillation, and sudden death. With substandard Sb^V compounds, cardiotoxicity resulting in death occurs very frequently.

5.2.2 Amphotericin B

Amphotericin B is a polyene antibiotic isolated from *Streptomyces nodosus* that is used as a first-line drug in regions with high levels of unresponsiveness to Sb^V [7, 8]. While it is conventionally an antifungal drug, its antileishmanial activity was first shown in the early 1960s. In VL, amphotericin B should be used in doses of 0.75–1 mg/kg body weight on alternate days for at least 15 injections or until a cure is achieved. It is a polyene antibiotic that has high affinity to ergosterol-like sterols in the cell membrane, inhibiting their incorporation into the membrane-forming micropores and leading to increased membrane permeability and ultimately the death of the *Leishmania*. Amphotericin B infusions can cause renal dysfunction, hypokalemia, hepatic dysfunction, bone marrow suppression, and myocarditis, any of which can be fatal. Fever with chills, shock, aches and pains all over the body, nausea, and vomiting is common and can occur acutely during each infusion. Thrombophlebitis of the injected vein is also common.

5.2.3 Liposomal Drug Delivery System

The therapeutic options described earlier are far from ideal because of prolonged courses of treatment, large-scale treatment failures (except with amphotericin B), and serious toxicities. Fatal toxicities due to Sb^V, pentamidine, and amphotericin B may be as high as 7–10 %. The search for a better therapeutic option has led to the development of a targeted drug delivery system in the form of liposomal preparations: the "magic bullet" approach.

The "magic bullet" concept is based on the premise that the selective delivery of drugs to the tissues where they are to exert their pharmacological effects will not only enhance the desired therapeutic result but also minimize the occurrence of unrelated responses or toxic side effects. To achieve this, researchers have employed a variety of strategies, including drug design (e.g., prodrugs), carrier molecules (e.g., antibodies), and the incorporation of drugs into macromolecules (e.g., liposomes). Vesicular, cellular, and particulate carriers are removed from the blood rapidly and almost exclusively by the reticuloendothelial system, in particular macrophages of the liver and spleen. This makes VL an ideal disease for passive drug targeting. Liposomes are microscopic vesicles 20–100 nm in diameter that have the ability to carry the encapsulated drugs to specific sites. The intraphagocytic nature of the parasite renders it highly susceptible to targeted drug delivery using liposomes, as the drug-laden liposomes are selectively taken up by the reticuloendothelial cells, especially in the liver and spleen, where they end up in the lysosomal apparatus. Here they are disrupted, and the drug is released, acting either locally or after diffusion outside the organelle in other cell compartments. As the drug is targeted to those very cells that host the parasite, it becomes several times more effective. Moreover, there is a striking decrement in systemic toxicity, as only minimal quantities of the drug are released in the free form.

Liposomal amphotericin B (AmBisome; Gilead Sciences, Foster City, CA, USA) is the most commonly used drug worldwide. In India, several studies have shown that small doses of AmBisome can cure a large proportion of patients. In a pilot study, 6 mg/kg of AmBisome cured 100 % of 10 patients [9]. In another study in patients with refractory VL, AmBisome given in doses of 0.75, 1.5, and 3 mg/kg on 5 consecutive days cured 89, 93, and 97 % of patients, respectively [10]. No significant adverse events were observed. In another study, a single dose of 5 mg cured 91 % of patients [11]. Similarly, in a multicenter trial, 90 % of 203 patients were cured with a single 7.5 mg/kg dose of AmBisome [12].

Unfortunately, due to the high cost of AmBisome (25–30 times the cost of conventional amphotericin B), even a highly attractive single-dose, low-dose regimen remains beyond the reach of most affected patients. Fortunately, the World Health Organization (WHO) has now negotiated a preferential price of US$20—offering a 90 % discount of the usual US$200—for a 50 mg vial of AmBisome for use in developing countries. This cost reduction makes the use of AmBisome now feasible throughout most endemic regions of South Asia.

5.2.4 Paromomycin

Paromomycin (aminosidine), obtained from cultures of *Streptomyces rimosus*, belongs to the class of aminocyclitol-aminoglycosides and possesses not only antibacterial but additional antiprotozoal activity effective against *Leishmania*, *Entamoeba*, and *Cryptosporidium*. Aminoglycosides as a rule have poor intestinal absorption, and paromomycin is no exception. Therefore, parenteral preparations have been developed and used in VL. The general consensus is that a 21-day course of paromomycin (11 mg/kg per day) is well tolerated. It has been strongly considered as a first-line treatment in Bihar. In the phase 3 trial in India in VL, the most common adverse event with paromomycin was injection site pain (55 %). However, most of these cases were common toxicity criteria grade 1, and only 1 patient discontinued the drug due to injection site pain. None of the patients developed nephrotoxicity. Seven patients (2 %) in the paromomycin group had ototoxicity manifested as transient threshold shifts at high frequency, which returned to near baseline values during follow up. An unexpected adverse event was a rise in levels of hepatic transaminases in 31 patients (6 %) [13]. The drug is now being produced in India, and an adult treatment course is expected to cost US $15.00. Its extremely affordable cost, coupled with a 3-week treatment schedule, makes it an attractive treatment option.

5.2.5 Miltefosine

A phase 1/2 dose escalation trial of miltefosine in India established that, in adults, a daily dose between 100 and 150 mg for 28 days is well tolerated and will cure most patients [14]. This was followed by a series of phase 2 studies confirming the results of the pilot study [15–17]. This led to a multicenter pivotal phase 3 study in which a high cure rate (94 %) unquestionably established miltefosine as the first orally effective antileishmanial agent, thus revolutionizing antileishmanial therapy [18]. Its efficacy has also been reported in Sb^V-resistant cases. Depending on the individual weight, the recommended therapeutic regimen for patients weighing less than 25 kg is a single oral dose of 50 mg for 28 days, whereas individuals weighing more than 25 kg require a twice-daily dose of 50 mg for 28 days. However, miltefosine has its limitations in that it induces renal toxicity and mild to moderate gastrointestinal disturbances, such as vomiting and diarrhea in 40 and 15–20 % of patients, respectively. Nonetheless, these symptoms are reversible and are not a major cause for concern. As miltefosine is potentially teratogenic, it is contraindicated in pregnancy and in women of child-bearing age who do not use contraception.

5.3 Regulations and Policy

Irresponsible drug use is a potential threat to the useful lifespan of any drug, and this has probably contributed to the high level of treatment failure with Sb^V in Bihar, India [19]. The unrestricted availability of Sb^V in India has resulted in widespread misuse by unqualified practitioners, leading to incomplete treatment courses. According to a survey of drug resistance in India, only 26 % of patients were treated according to WHO guidelines, and patients often stopped treatment on their own initiative [4]. The high reliance on the private sector and local pharmacists on the Indian subcontinent even today highlights the need for tighter regulation of the modalities of VL treatment, and for treatment to be made available at no cost [20, 21]. When policymakers opt for combination therapy, they should take measures to limit the use of monotherapy, particularly where incomplete courses of treatment may result. The fact that miltefosine is available in India without prescription or regulation is worrisome, because this could facilitate the development of drug resistance [21]. At present, a single dose of 10 mg/kg of liposomal amphotericin B (L-AmB) or combination therapy consisting of either a single injection of 5 mg/kg L-AmB and 7-day 50 mg oral miltefosine or single injection of 5 mg/kg L-AmB and 10-day 11 mg/kg intramuscular paromomycin (PM) or 10 days each of miltefosine and PM are the preferred treatment options in the Indian subcontinent [22, 23].

5.4 Prevention of Drug Resistance

The problem of drug resistance in VL has been extensively reviewed elsewhere [24]. Treatment failure can manifest as initial treatment failure (failure to clear parasites at the end of the treatment course) or relapse (reappearance of parasites after initial cure, usually within 6 months of follow-up). Although Sb^v compounds have been successfully used throughout the world for decades, poor treatment response (mainly due to initial treatment failure) has increasingly been reported since the 1980s from Bihar, India, with geographical and temporal clustering in several hyperendemic districts [1, 25]. Although treatment outcomes could initially be improved with higher total doses, the improvement was only temporary [26–28]. In subsequent reports, therapy failed in up to 60 % of patients that were newly diagnosed [2, 29]. At the same time, misuse of the drugs was reported [4]. Increased treatment failure has also been reported in Nepal, in districts that border Bihar [30]. Although treatment failure can be due to several causes—including factors related to the drug, host, and parasite—substantial evidence exists that acquired drug resistance is a key issue. Reduced drug sensitivity has been reported with *Leishmania donovani* strains from unresponsive cases in vitro [31–33]. Reduced susceptibility to Sb^v has also been reported with *Leishmania infantum* in both human beings and animals [33–35]. In these studies, post-treatment isolates had reduced sensitivity compared with that seen in pretreatment isolates, supporting the notion of acquired drug resistance. However, more recent studies have reported less clear associations of in vitro susceptibility and clinical outcomes, underscoring the need for improved and standardized methods [36]. Limited understanding of the mechanism of resistance toward Sb^v and the shortcomings of drug sensitivity assays make it difficult to predict the risk of acquired resistance in other regions or with other drugs, and to assess the need for combination therapy to help prevent resistance. Given the risk of the development of resistance to established and new medicines, the Regional Strategic Framework for Elimination of Kalaazar from the South-East Asia Region recommends that monotherapy other than L-AmB should be avoided in this region [37]. A single dose of L-AmB and short-course combination therapy showing excellent efficacy have been major breakthroughs in the treatment of VL in this region.

However, on the basis of the available evidence, acquired drug resistance should be considered a potentially serious threat to VL control, and comprehensive strategies should be developed, including the use of combination therapy [1, 3, 20, 24].

5.5 Rationale for Combination Chemotherapy

For individual drugs, the ease with which resistance develops will mainly depend on the parasite burden, the probability of spontaneous development of resistance mutations, and the fitness cost associated with those mutations [38]. The level and pattern of drug use in a population constitutes the selection force for the development of resistance, and intact host immunity is generally thought to be protective. The potency of the drug, therapeutic index, and pharmacokinetic properties of the drug also play a part [38]. Combination therapy can delay resistance if 2 drugs with different modes of action and mechanisms of resistance are used. The combination of synergistic drugs is preferred, because if more effective replication can be inhibited then resistance is less likely during treatment. For resistance prevention, both drugs should ideally have similar pharmacokinetics. If parasites always confront both drugs, the probability of the emergence of double-resistant parasites would be expected to be extremely rare (i.e., the product of their individual per-parasite probabilities). A rapid elimination phase minimizes the duration of subtherapeutic drug concentrations that can provide an opportunity for amplification or selection of resistant parasites [38, 39]. In studies of malaria, the combination of 1 very active drug with a short half-life with a slower-acting drug with a longer half-life to clear the remaining parasites has been explored as a way to shorten treatment duration and improve treatment compliance [40]. However, recent studies have focused on the terminal elimination phase of the second drug, which can act as a selective filter for resistant malaria parasites [41, 42]. Finally, drugs can be combined to target different biological stages of the infectious agent. This has been done for tuberculosis and malaria, although the drugs are essentially targeted at preventing relapse and only indirectly prevent or delay the development of resistance.

5.6 Pharmacological Considerations for Combination Therapy

Although the mechanisms of action and resistance remain poorly understood for all antileishmanial drugs currently in use (except amphotericin B), they are all thought to act on different targets [43]. Recent findings from India suggest that field isolates from areas with high-level resistance to Sb^V show reduced sensitivity toward other antileishmanial drugs such as amphotericin B and miltefosine [44]. However, true cross-resistance between the various drugs has not been reported so far. Several combinations have shown activity enhancement in animal experiments [45]. Clear differences in pharmacokinetics exist. Miltefosine might be particularly vulnerable to the emergence of resistance, because of its narrow therapeutic index and long half-life, which has been estimated at around 7 days [46, 47]. Recent data from patients with cutaneous leishmaniasis suggest a terminal half-life of 31 days, with miltefosine still detectable 5–6 months after the end of treatment [48]. Resistant strains might be selected and amplified during this period because of subtherapeutic drug concentrations, either from relapsing patients, or from newly acquired infections [46]. If confirmed, this situation might have important repercussions for the risk of emerging resistance and the duration of contraceptive measures. Paromomycin has a short half-life (2–3 h in patients without VL), but has a low therapeutic index. Resistance can easily be induced in vitro [49], and clinical resistance has been noted with its antibacterial use [19, 50]. Most of a Sb^V compound (about 99 %) is eliminated within a few hours, followed by a slower elimination phase with a half-life of 76 h [51]. At least in East Africa, these drugs remain highly effective. Amphotericin B could be thought less likely to induce resistance, given its high efficacy and mechanism of action [24, 52–54]. Although resistance can been induced in vitro, clinical cases of amphotericin-B resistance are rare. Liposomal amphotericin B has a bioavailability in tissues for several weeks despite a relatively shorter plasma half-life [55–57]. Given this long tissue half-life, a single dose of liposomal amphotericin B followed by daily administration of a second drug (e.g., sodium stibogluconate, paromomycin, or miltefosine) would result in simultaneous exposure of the parasite to both drugs. The use of a single dose of liposomal amphotericin B (10 mg/kg) in monotherapy has proven to be extremely effective in India [57]. In a randomized, non-comparative, group-sequential, triangular multidrug therapy study in India, 181 subjects were assigned to treatment with 5 mg/kg of L-AmB alone, 5 mg/kg of L-AmB followed by miltefosine for 10 days or 14 days, or 3.75 mg/kg of L-AmB followed by miltefosine for 14 days. When it became apparent that all regimens were effective, 45 additional, nonrandomized patients were assigned to receive 5 mg/kg of L-AmB followed by miltefosine for 7 days. Final CRs were high (>95 %) and similar in all the groups. These results suggest that a single infusion of L-AmB (in most instances, administered in an outpatient setting) followed by a brief self-administered course of miltefosine could be an excellent option against Indian kala-azar [58].

5.7 Compliance with Treatment

Besides the intrinsic characteristics of the combination regimen, compliance with the regimen also affects the risk of drug resistance. All conventional monotherapies (apart from liposomal amphotericin B) require a long treatment duration

(21–30 days), making compliance more challenging. This is of particular concern for treatment with miltefosine, the only oral drug, for which the risk of premature treatment interruption is high. Even in a phase 4 trial, 4.5 % of patients were lost to follow up before the end of treatment, and 14.5 % were not available for assessment 6 months after treatment [59]. Thus, except in the case of high-dose single-injection AmBisome, monotherapy should be completely eliminated, including in the private sector. The shorter treatment duration might help to increase compliance, as has been the case for patients receiving combination regimens for malaria [3]. The lower costs to patients associated with shortened combination therapy could also improve access to and acceptance of VL treatment. A directly observed treatment strategy, which has been successfully used for tuberculosis, must be employed to ensure good compliance with treatment, although this will increase the indirect and direct costs marginally [60]. The elimination program for VL in South Asia has opted for miltefosine as a first-line drug, but the long treatment regimen, domiciliary treatment, and adverse events are strong deterrents to good compliance. It will be necessary to engage in the monitoring of clinical treatment outcomes and pharmacovigilance to ensure effective management of the VL elimination program [60].

5.8 Response to Treatment in AIDS-Related Kala-Azar

Treatment results in T-cell- and cytokine-deficient animals suggest that CD41 T-cell-depleted patients who have VL associated with acquired immune deficiency syndrome (AIDS) would respond poorly to Sb^v but satisfactorily to amphotericin B, and that they would also be likely to relapse if the initial treatment successfully induced an apparent clinical response and the drug was then discontinued. Taken together (but with variability in treatment regimens and definitions of efficacy), most reports from southern Europe, where coinfection has been best demonstrated [61], appear to confirm the following: (a) overall, approximately 50 % of patients fail to respond initially to Sb^v in a region where 0–5 % of otherwise healthy individuals are Sb^v unresponsive [62]; (b) of a total of approximately 50 co-infected patients treated with some form of amphotericin B, 90 % showed initial responses; and (c) relapse rates in HIV-related kala-azar after any treatment is discontinued are 50 % and up to 90–100 % [63]. Results from Spain, however, in the only randomized controlled study in HIV-associated kala-azar [6], provided a different finding in that the initial efficacies of both Sb^v (66 % response) and amphotericin B (62 % response) were reduced. Because this study did not include secondary prophylaxis, the majority of initial responders to either treatment relapsed. However, while once-monthly injections of amphotericin B lipid complex or AmBisome may prevent symptomatic recurrences [64, 65], no consensus has been reached about what constitutes optimal maintenance treatment in such patients.

Lipid formulations infused at a dose of 3–5 mg/kg per day or intermittently for 10 doses (on days 1–5, 10, 17, 24, 31, and 38) up to a total dose of 40 mg/kg are recommended. Antiretroviral therapy should be initiated and secondary prophylaxis should be given till the CD4 counts are >200/μl [66].

5.9 Investigational Drugs

Among all the investigational drugs, fexinidazole, a nitroimidazole, has reached the stage of phase II clinical trial for VL [67]. Fexinidazole has excellent in vitro and in vivo antileishmanial activities. The oral advantage, comparable leishmanicidal activity to miltefosine, and safety reiterates the potential of fexinidazole as a much-needed additional oral therapy for VL. Sitamaquine is the second orally active antileishmanial drug after miltefosine that has reached phase II trials [68]. Unfortunately, due to its low efficacy, development of this drug has been stopped for VL. Naphthoquinone buparvaquone, except for the liposomal formulation, is more active for CL than VL, but its other derivatives could be explored for activity against VL [69]. Doxorubicin, an anticancer drug, has shown strong antileishmanial activity at low doses. Active targeting of this drug to infected macrophages is being explored; however, at present its toxicity is its major drawback [70].

5.10 Conclusion

In addition to intrinsic pharmacologic features, there are a number of human parameters that may favor the emergence and spread of leishmanial resistance. These include poor compliance, expensive treatment, availability of antileishmanial drugs over the counter, and limited access to healthcare facilities for early diagnosis and treatment. Given the current situation of widespread emergence of antimonial resistance in India, there is growing concern about preserving the efficacy of novel antileishmanials. Such a strategy should focus on the following approaches:

(1) Treatment of VL should be based on guidelines for prompt diagnosis, selection of first-line drugs, and management of cases unresponsive to antimonials and HIV-co-infected cases;
(2) Directly observed therapy for antileishmanials should be implemented, as in tuberculosis control programs, to enhance compliance;
(3) VL cases should be treated early to avoid further transmission of resistant parasites to the community;
(4) Distribution of and clinical response to antileishmanials should be monitored;
(5) Antileishmanial treatment should be provided free of charge through the healthcare system;

(6) The emergence and spread of antileishmanial resistance should be monitored; and
(7) The efficacy and safety of combination regimens should be evaluated in large trials, and if successful, such regimens should be immediately implemented.

Acknowledgments This work was supported by NIAID, NIH TMRC, Grant No. I-P50AI074321.

References

1. Sundar S. Drug resistance in Indian visceral leishmaniasis. Trop Med Int Health. 2001;6:849–54.
2. Sundar S, More DK, Singh MK, et al. Failure of pentavalent antimony in visceral leishmaniasis in India: report from the center of the Indian epidemic. Clin Infect Dis. 2000;31:1104–7.
3. Bryceson A. A policy for leishmaniasis with respect to the prevention and control of drug resistance. Trop Med Int Health. 2001;6:928–34.
4. Sundar S, Thakur BB, Tandon AK, et al. Clinicoepidemiological study of drug resistance in Indian kala-azar. BMJ. 1994;308:307.
5. Berman JD. Human leishmaniasis: clinical, diagnostic, and chemotherapeutic developments in the last 10 years. Clin Infect Dis. 1997;24:684–703.
6. Laguna F, Lopez-Velez R, Pulido F, et al. Treatment of visceral leishmaniasis in HIV-infected patients: a randomized trial comparing meglumine antimoniate with amphotericin B. Spanish HIV-Leishmania Study Group. Aids. 1999;13:1063–9.
7. Mishra M, Biswas UK, Jha AM, Khan AB. Amphotericin versus sodium stibogluconate in first-line treatment of Indian kala-azar. Lancet. 1994;344:1599–600.
8. Mishra M, Biswas UK, Jha DN, Khan AB. Amphotericin versus pentamidine in antimony-unresponsive kala-azar. Lancet. 1992;340:1256–7.
9. Thakur CP, Pandey AK, Sinha GP, Roy S, Behbehani K, Olliaro P. Comparison of three treatment regimens with liposomal amphotericin B (AmBisome) for visceral leishmaniasis in India: a randomized dose-finding study. Trans R Soc Trop Med Hyg. 1996;90:319–22.
10. Sundar S, Jha TK, Thakur CP, Mishra M, Singh VR, Buffels R. Low-dose liposomal amphotericin B in refractory Indian visceral leishmaniasis: a multicenter study. Am J Trop Med Hyg. 2002;66:143–6.
11. Sundar S, Agrawal G, Rai M, Makharia MK, Murray HW. Treatment of Indian visceral leishmaniasis with single or daily infusions of low dose liposomal amphotericin B: randomised trial. BMJ. 2001;323:419–22.
12. Sundar S, Jha TK, Thakur CP, Mishra M, Singh VP, Buffels R. Single-dose liposomal amphotericin B in the treatment of visceral leishmaniasis in India: a multicenter study. Clin Infect Dis. 2003;37:800–4.
13. Sundar S, Jha TK, Thakur CP, Sinha PK, Bhattacharya SK. Injectable paromomycin for visceral leishmaniasis in India. N Engl J Med. 2007;356:2571–81.
14. Sundar S, Rosenkaimer F, Makharia MK, et al. Trial of oral miltefosine for visceral leishmaniasis. Lancet. 1998;352:1821–3.
15. Jha TK, Sundar S, Thakur CP, et al. Miltefosine, an oral agent, for the treatment of Indian visceral leishmaniasis. N Engl J Med. 1999;341:1795–800.
16. Sundar S, Gupta LB, Makharia MK, et al. Oral treatment of visceral leishmaniasis with miltefosine. Ann Trop Med Parasitol. 1999;93:589–97.
17. Sundar S, Makharia A, More DK, et al. Short-course of oral miltefosine for treatment of visceral leishmaniasis. Clin Infect Dis. 2000;31:1110–3.

18. Sundar S, Jha TK, Thakur CP, et al. Oral miltefosine for Indian visceral leishmaniasis. N Engl J Med. 2002;347:1739–46.
19. Teklemariam S, Hiwot AG, Frommel D, Miko TL, Ganlov G, Bryceson A. Aminosidine and its combination with sodium stibogluconate in the treatment of diffuse cutaneous leishmaniasis caused by leishmania aethiopica. Trans R Soc Trop Med Hyg. 1994;88:334–9.
20. den Boer ML, Alvar J, Davidson RN, Ritmeijer K, Balasegaram M. Developments in the treatment of visceral leishmaniasis. Expert Opin Emerg Drugs. 2009;14:395–410.
21. Sundar S, Murray HW. Availability of miltefosine for the treatment of kala-azar in India. Bull World Health Organ. 2005;83:394–5.
22. Sundar S, Chakravarty J, Agarwal D, et al. Single-dose liposomal amphotericin B for visceral leishmaniasis in India. N Engl J Med. 2010;362(6):504–12.
23. Sundar S, Sinha PK, Rai M, et al. Comparison of short-course multidrugtreatment with standard therapy for visceral leishmaniasis in India: an openlabel, non-inferiority, randomised controlled trial. Lancet. 2011;377(9764):477–86.
24. Croft SL, Sundar S, Fairlamb AH. Drug resistance in leishmaniasis. Clin Microbiol Rev. 2006;19:111–26.
25. Peters W. The treatment of kala-azar—new approaches to an old problem. Indian J Med Res. 1981;73(Suppl):1–18.
26. Thakur CP, Kumar M, Kumar P, Mishra BN, Pandey AK. Rationalisation of regimens of treatment of kala-azar with sodium stibogluconate in India: a randomised study. Br Med J (Clin Res Ed). 1988;296:1557–61.
27. Thakur CP, Kumar M, Pandey AK. Evaluation of efficacy of longer durations of therapy of fresh cases of kala-azar with sodium stibogluconate. Indian J Med Res. 1991;93:103–10.
28. Thakur CP, Kumar M, Singh SK, et al. Comparison of regimens of treatment with sodium stibogluconate in kala-azar. Br Med J (Clin Res Ed). 1984;288:895–7.
29. Sundar S, Singh VP, Sharma S, Makharia MK, Murray HW. Response to interferon-gamma plus pentavalent antimony in Indian visceral leishmaniasis. J Infect Dis. 1997;176:1117–9.
30. Rijal S, Chappuis F, Singh R, et al. Treatment of visceral leishmaniasis in south-eastern Nepal: decreasing efficacy of sodium stibogluconate and need for a policy to limit further decline. Trans R Soc Trop Med Hyg. 2003;97:350–4.
31. Dube A, Singh N, Sundar S. Refractoriness to the treatment of sodium stibogluconate in Indian kala-azar field isolates persist in in vitro and in vivo experimental models. Parasitol Res. 2005;96:216–23.
32. Laurent T, Rijal S, Yardley V, et al. Epidemiological dynamics of antimonial resistance in Leishmania donovani: genotyping reveals a polyclonal population structure among naturally-resistant clinical isolates from Nepal. Infect Genet Evol. 2007;7:206–12.
33. Carrio J, Riera C, Gallego M, Ribera E, Portus M. In vitro susceptibility of Leishmania infantum to meglumine antimoniate in isolates from repeated leishmaniasis episodes in HIV-coinfected patients. J Antimicrob Chemother. 2001;47:120–1.
34. Carrio J, Portus M. In vitro susceptibility to pentavalent antimony in Leishmania infantum strains is not modified during in vitro or in vivo passages but is modified after host treatment with meglumine antimoniate. BMC Pharmacol. 2002;2:11.
35. Faraut-Gambarelli F, Piarroux R, Deniau M, et al. In vitro and in vivo resistance of leishmania infantum to meglumine antimoniate: a study of 37 strains collected from patients with visceral leishmaniasis. Antimicrob Agents Chemother. 1997;41:827–30.
36. Rijal S, Bhandari S, Koirala S, et al. Clinical risk factors for therapeutic failure in kala-azar patients treated with pentavalent antimonials in Nepal. Trans R Soc Trop Med Hyg. 104:225–9.
37. World health Organization. Regional office for South East Asia. The regional strategic framework for elimination of kala-azar from the South-East Asia region 2011–15, New Delhi; 2012.
38. Hastings IM. Modelling parasite drug resistance: lessons for management and control strategies. Trop Med Int Health. 2001;6:883–90.

39. Stepniewska K, White NJ. Pharmacokinetic determinants of the window of selection for antimalarial drug resistance. Antimicrob Agents Chemother 2008;52:1589–96.
40. Nosten F, White NJ. Artemisinin-based combination treatment of falciparum malaria. Am J Trop Med Hyg. 2007;77:181–92.
41. Pongtavornpinyo W, Yeung S, Hastings IM, Dondorp AM, Day NP, White NJ. Spread of anti-malarial drug resistance: mathematical model with implications for ACT drug policies. Malar J. 2008;7:229.
42. White NJ. Antimalarial drug resistance. J Clin Invest. 2004;113:1084–92.
43. Alvar J, Croft S, Olliaro P. Chemotherapy in the treatment and control of leishmaniasis. Adv Parasitol. 2006;61:223–74.
44. Kumar D, Kulshrestha A, Singh R, Salotra P. In vitro susceptibility of field isolates of Leishmania donovani to miltefosine and amphotericin B: correlation with sodium antimony gluconate susceptibility and implications for treatment in areas of endemicity. Antimicrob Agents Chemother. 2009;53 835–8.
45. Seifert K, Croft SL. In vitro and in vivo interactions between miltefosine and other antileishmanial drugs. Antimicrob Agents Chemother. 2006;50:73–9.
46. Perez-Victoria FJ, Sanchez-Canete MP, Seifert K, et al. Mechanisms of experimental resistance of leishmania to miltefosine: Implications for clinical use. Drug Resist Updat. 2006;9:26–39.
47. Berman J, Bryceson AD, Croft S, et al. Miltefosine: issues to be addressed in the future. Trans R Soc Trop Med Hyg. 2006;100(Suppl 1):S41–4.
48. Dorlo TP, van Thiel PP, Huitema AD, et al. Pharmacokinetics of miltefosine in old world cutaneous leishmaniasis patients. Antimicrob Agents Chemother. 2008;52:2855–60.
49. Maarouf M, Adeline MT, Solignac M, Vautrin D, Robert-Gero M. Development and characterization of paromomycin-resistant leishmania donovani promastigotes. Parasite. 1998;5:167–73.
50. Davidson RN, den Boer M, Ritmeijer K. Paromomycin. Trans R Soc Trop Med Hyg. 2009;103:653–60.
51. Chulay JD, Fleckenstein L, Smith DH. Pharmacokinetics of antimony during treatment of visceral leishmaniasis with sodium stibogluconate or meglumine antimoniate. Trans R Soc Trop Med Hyg. 1988;82:69–72.
52. Di Giorgio C, Faraut-Gambarelli F, Imbert A, Minodier P, Gasquet M, Dumon H. Flow cytometric assessment of amphotericin B susceptibility in leishmania infantum isolates from patients with visceral leishmaniasis. J Antimicrob Chemother. 1999;44:71–6.
53. Durand R, Paul M, Pratlong F, et al. Leishmania infantum: lack of parasite resistance to amphotericin B in a clinically resistant visceral leishmaniasis. Antimicrob Agents Chemother. 1998;42:2141–3.
54. Lachaud L, Bourgeois N, Plourde M, Leprohon P, Bastien P, Ouellette M. Parasite susceptibility to amphotericin B in failures of treatment for visceral leishmaniasis in patients coinfected with HIV type 1 and leishmania infantum. Clin Infect Dis. 2009;48:e16–22.
55. Lee JW, Amantea MA, Francis FA, et al. Pharmacokinetics and safety of a unilamellar liposomal formulation of amphotericin B (AmBisome) in rabbits. Antimicrob Agents Chemother. 1994;38:713–8.
56. Bekersky I, Fielding RM, Dressler DE, Lee JW, Buell DN, Walsh TJ. Pharmacokinetics, excretion, and mass balance of liposomal amphotericin B (AmBisome) and amphotericin B deoxycholate in humans. Antimicrob Agents Chemother. 2002;46:828–33.
57. Sundar S, Chakravarty J, Agarwal D, Rai M, Murray HW. Single-dose liposomal amphotericin B for visceral leishmaniasis in India. N Engl J Med. 362:504–12.
58. Sundar S, Rai M, Chakravarty J, et al. New treatment approach in Indian visceral leishmaniasis: single-dose liposomal amphotericin B followed by short-course oral miltefosine. Clin Infect Dis. 2008;47(8):1000–6.
59. Bhattacharya SK, Sinha PK, Sundar S, et al. Phase 4 trial of miltefosine for the treatment of Indian visceral leishmaniasis. J Infect Dis. 2007;196(4):591–8

60. Sundar S, Olliaro PL. Miltefosine in the treatment of leishmaniasis: clinical evidence for informed clinical risk management. Ther Clin Risk Manag. 2007;3:733–40.
61. Alvar J, Canavate C, Gutierrez-Solar B, et al. Leishmania and human immunodeficiency virus coinfection: the first 10 years. Clin Microbiol Rev. 1997;10:298–319.
62. Gradoni L, Bryceson A, Desjeux P. Treatment of mediterranean visceral leishmaniasis. Bull World Health Organ. 1995;73:191–7.
63. Murray HW. Treatment of visceral leishmaniasis (kala-azar): a decade of progress and future approaches. Int J Infect Dis. 2000;4:158–77.
64. Lopez-Velez R, Videla S, Marquez M, et al. Amphotericin B lipid complex versus no treatment in the secondary prophylaxis of visceral leishmaniasis in HIV-infected patients. J Antimicrob Chemother. 2004;53:540–3.
65. Montana M, Chochoi N, Monges P, et al. Liposomal amphotericin B in secondary prophylaxis of visceral leishmaniasis in HIV-infected patients: report of five clinical cases. Pathol Biol (Paris). 2004;52:66–75.
66. Report of a meeting of the WHO Expert Committee on the Control of Leishmaniases. 22-March, 2010. http://whqlibdoc.who.int/trs/WHO_TRS_949_eng.pdf.
67. Fexinidazole (1200 mg) bioavailability under different food intake conditions. http://clinicaltrials.gov/ct2/show/NCT01340157. Accessed 21 Feb 2014.
68. Sherwood JA, Gachihi GS, Muigai RK,et al. Phase 2 efficacy trial of an oral 8-aminoquinoline (WR6026) for treatmentof visceral leishmaniasis. Clin Infect Dis.19(6):1034–9.
69. Garnier T, Mantyla A, Jarvinen T, et al. In vivo studies on the antileishmanial activity of buparvaquone and its prodrugs. J Antimicrob Chemother. 2007;60(4):802–10.
70. Das SL, Kole L, et al. Targeting of parasite-specific immunoliposome-encapsulated doxorubicin in the treatment of experimental visceral leishmaniasis. J Infect Dis. 04;189(6):1024–34.

Chapter 6
Treatment of Post-kala-azar Dermal Leishmaniasis

V. Ramesh and Prashant Verma

Abstract Post-kala-azar dermal leishmaniasis (PKDL), a parasitological disease, is a cutaneous affliction which follows the systemic counterpart namely visceral leishmaniasis (VL)/kala-azar. It is a scourge of human community as it constitutes the reservoir of leishmania parasites, which upon transmission to another individual can result in visceral leishmaniasis. So far there hasn't been much development as far as treatment of PKDL is concerned. This chapter is an endeavour to give a brief of the various drugs used for treating PKDL and a management guide for both Indian and Sudanese forms of PKDL.

Keywords Post-kala-azar dermal leishmaniasis · Treatment · Miltefosine · Amphotericin · Sodium antimony gluconate

6.1 Introduction

Aptly termed, post-kala-azar dermal leishmaniasis (PKDL) is a well known sequel to visceral leishmaniasis (VL) also known as kala-azar. Clinically, it presents as macular, maculopapular, or nodular rash in a patient who has recovered from VL. PKDL, with differences in clinical presentation and natural course, is in particular clustered in India and Sudan where *Leishmania donovani* is the main species causing VL [1].

Notoriously, PKDL acts as a reservoir of *Leishmania* parasites during the inter-epidemic periods of VL. Alterations in the cytokine milieu, IFN gamma, IL-10, and FOXP3 in particular, are the root mechanisms underlying PKDL [2].

Diagnosis often relies on the clinical presentation; demonstration of parasites in skin smears and cultures is confirmatory, though not sensitive. Molecular techniques, namely polymerase chain reaction (PCR) and monoclonal antibodies, are

---

V. Ramesh (✉)
Department of Dermatology & STD, Vardhman Mahavir Medical College & Safdarjang Hospital, New Delhi 110 023, India
e-mail: weramesh@gmail.com

more sensitive and are recommended when available [3]. Because PKDL remains a reservoir for the spread of VL, early diagnosis and prompt treatment are paramount. Though treatment and research into drugs for PKDL still remain unsatisfactory, some advances have made it possible to devise effective regimens for this condition.

6.2 Antimonials

Pentavalent antimony complexes were the first successfully used chemotherapy recommended in almost all forms of leishmaniasis. Sodium stibogluconate (SSG) and meglumine antimoniate have been popular choices. The mechanism of action of this group of drugs is largely unknown. T-cell based mechanisms, cytokine alterations, intracellular signaling, and reactive oxygen species are proposed antileishmanial actions.

Effects on trypanothione and glutathione metabolism have also been speculated [4]. It is important to keep in mind that the functional immune requirements for the drug to act and the clinical evidence accumulated so far suggest antimonials might be less effective as well as toxic in immunocompromised subjects [5].

SSG is administered parenterally, intramuscularly, or intravenously, at a dose of 20 mg/kg per day for at least 120 days to effect a cure in Indian PKDL [6]. In contrast, 2 months of therapy at a dose of 20 mg/kg per day is recommended in Sudan [7].

Although inexpensive, antimonials have a plethora of adverse effects including nausea, vomiting, abdominal pain, headache, phlebitis, transaminitis, nephrotoxicity, fever, rash, pancytopenia, and neurotoxicity. In addition, pancreatitis and cardiac conduction defects are potentially dangerous side effects. Accordingly, proper monitoring with hematological, biochemical, and cardiac assessments is warranted. Unfortunately, in India resistant parasites have been encountered and the cure rate of antimonial compounds in VL has plummeted over the past 30 years to as low as 42 %. Indiscriminate and inappropriate use of antimonials is considered the cause of the increasing treatment failure observed with this drug [8].

Pentamidine, ketoconazole, allopurinol, rifampicin, and paromomycin have all been anecdotally used for PKDL treatment without any consistent results. Pentamidine, an aromatic diamidine, competitively inhibits arginine transport and non-competitively inhibits putrescine and spermidine. The inherent potential of causing irreversible insulin-dependent diabetes mellitus and death, and low cure rates, has not favored further use.

Ketoconazole belongs to the azole group of drugs. Its leishmanicidal action is attributed to interference with ergosterol synthesis in the cell membrane. It has been used at a dose of 800 mg orally per day for 9 months; however, it has a poor cure rate, in addition to causing hepatotoxicity [9]. Allopurinol, a pyrazolopyrimidine, converts to ribonucleotide triphosphate and incorporates into the RNA of *Leishmania*, hence disrupting macromolecule synthesis. Allopurinol when used orally in a dose of 800 mg/day in divided doses, took an unacceptably long period of nearly 2 years to bring about a cure, indicating poor leishmanicidal activity [10]. It has shown synergism when given along with SSG, and the combination has been used in relapse cases of PKDL in Sudan following SSG therapy [11].

Allopurinol can cause serious hypersensitivity reactions, drug reaction with eosinophilia and systemic symptoms, and Stevens-Johnson syndrome. Rifampicin in combination with SSG was reported to be effective in Indian PKDL [12]. In a larger retrospective study on Indian PKDL, both allopurinol and rifampicin showed less encouraging results when given in combination with SSG [13]. Paromomycin, an aminoglycoside, is a protein synthesis inhibitor. It was used in PKDL at a dose of 11 mg/kg given intramuscularly for 45 days, but the cure rates were low [14].

Thus till recently, antimonials have remained in a way the "Hobson's choice" [15]. Of late, the nonavailability of SSG across India is a serious issue because it still remains the drug of choice in a related condition, cutaneous leishmaniasis (CL).

6.3 Amphotericin B

Originally extracted from *Streptomyces nodosus*, the name of this drug is derived from its amphoteric properties. It is available as a plain formulation, cholesteryl sulfated complex, lipid complex, and a liposomal form. Amphotericin makes the cell membrane of the parasite more permeable by interfering with sterol synthesis in the cell membranes of the organism. Conventional plain amphotericin is fraught with undesirable effects, nephrotoxicity and hypokalemia in particular. Infusion reactions may also develop, manifesting as fever, chills, headache, rash, hypotension, nausea, vomiting, and drowsiness.

However, liposomal amphotericin B (LAMB) containing amphotericin packaged in lipid vesicles has been shown to be less nephrotoxic because of its affinity for macrophages. Further, LAMB has a better safety profile compared with lipid formulations. Response to amphotericin B appears to be far superior to that to SSG in terms of speed and cure. Amphotericin B deoxycholate is administered as a slow intravenous infusion over 2 h in 5 % dextrose at a dose of 1 mg/kg per day given as 320-day courses separated by 20 days of drug-free intervals [16].

In a large cohort of 1303 patients in Bangladesh, LAMB was used successfully at a dosage of 5 mg/kg twice a week for 3 weeks [17]. Later, 223 patients, 110 in Bangladesh and 113 in India were treated with the same dose of 30 mg/kg over 3 weeks in divided doses. In India, of the 50 patients who completed 12 months of follow up, 42 (84 %) cases showed substantial or complete cure, with excellent tolerance and safety. In Bangladesh, of 88 patients who completed 12 months of follow up, 59 (67 %) showed substantial or complete cure; however, 6.5 % developed severe hypokalaemia. No patient in either group developed rhabdomyolysis or further sequelae. In view of hypokalemia, a lower dose of 15 mg/kg over 3 weeks is being tried in Bangladesh, which appears encouraging, but the final report is awaited. *(unpublished, BurzaS, den Boer M, Mahajan R, Das AK, Mitra G,*

Almeida P, Lima MA, Be-Nazir A, Sunyoto T, Ritmeijer K. Post Kala-Azar Dermal Leishmaniasis treated with Liposomal Amphotericin B (AmBisome)).

Because Sudanese PKDL, unlike Indian PKDL, is often self-limiting and/or mild, LAMB is given at a dose of 2.5 mg/kg per day intravenously for 20 days. Complete clearance of the skin rash was observed in 83 % of the Sudanese PKDL patients with no detectable toxic effects [18].

6.4 Miltefosine

Miltefosine, a derivative of alkylphosphocholine compounds, exerts its leishmanicidal action though multiple mechanisms including membrane phospholipid alteration, apoptosis-like cell death, mitochondrial dysfunction, and immunomodulation of the host response facilitating the elimination of the parasite. The rate of parasite clearance has been demonstrated to be 1 log/week for *Leishmania major* [19] *and Leishmania infantum* [20] infections. In an adult patient receiving 100 mg/day, miltefosine reaches a maximal concentration of 70 μg/mL on the 23rd day of treatment; in children receiving 2.5 mg/kg a maximal concentration of 24 μg/mL is reached between day 23 and 28 of treatment [21]. This may have resulted in a higher relapse rate in children leading a study to recommend allometric dosing instead of linear dosing (2.5 mg/kg) of miltefosine, which would optimize the drug exposure of children with leishmaniasis [22]. In India it is given at a dose of 50 mg once daily for those weighing <25 kg, and 100 mg daily for those weighing 25 kg or more, for a period of 12 weeks (Figs. 6.1 and 6.2).

The clinical pharmacodynamic properties of miltefosine are ill defined, and the lower limit of the therapeutic range of miltefosine is unknown [23]. From a pharmacokinetic viewpoint, a dose higher than the presently recommended 100 mg/day in adults does not appear to be beneficial because of the extremely slow elimination and significant accumulation of miltefosine.

Being a relatively innocuous orally administered drug, miltefosine offers advantages over both SSG and amphotericin B. This is evidenced by the cure rates seen with miltefosine. It has also been used successfully in HIV/AIDS coinfected patients [24] and antimony unresponsive patients. Unfortunately, its availability is far from reliable and this situation definitely needs to be improved. Gastrointestinal intolerance, nausea, vomiting, and diarrhea are the principal side effects, which may result in noncompliance with the treatment. In our experience anorexia is an important side effect that can be severe in some patients.

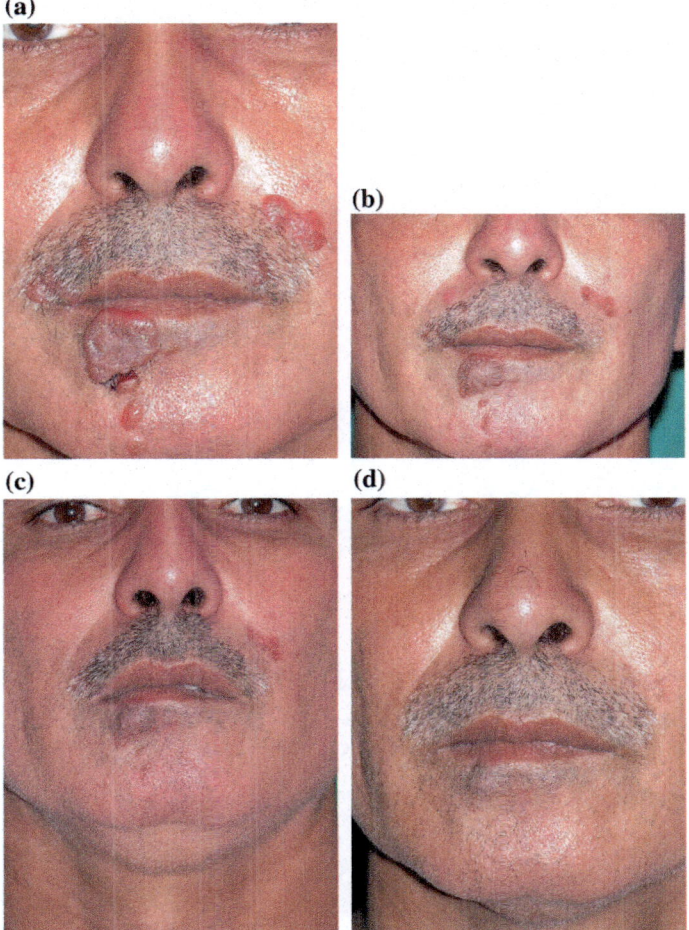

Fig. 6.1 Facial plaques of post-kala-azar dermal leishmaniasis, **a** before miltefosine therapy, **b** mid-miltefosine therapy, **c** at end of therapy, and **d** normal face at 1 year follow up

6.5 Immunotherapy

In an endeavor to reduce the duration and side effects of chemotherapy, immunotherapy has been tried in PKDL. The *Mycobacterium welchii* vaccine, which enhances cell-mediated immunity, did not produce any noticeable change in the regression of lesions when used along with SSG in Indian PKDL as compared to results for those taking only SSG [13].

In Sudan, a combination of alum-precipitated *Leishmania major* vaccine plus Bacille Calmette-Guerin and SSG was found to be effective [25]

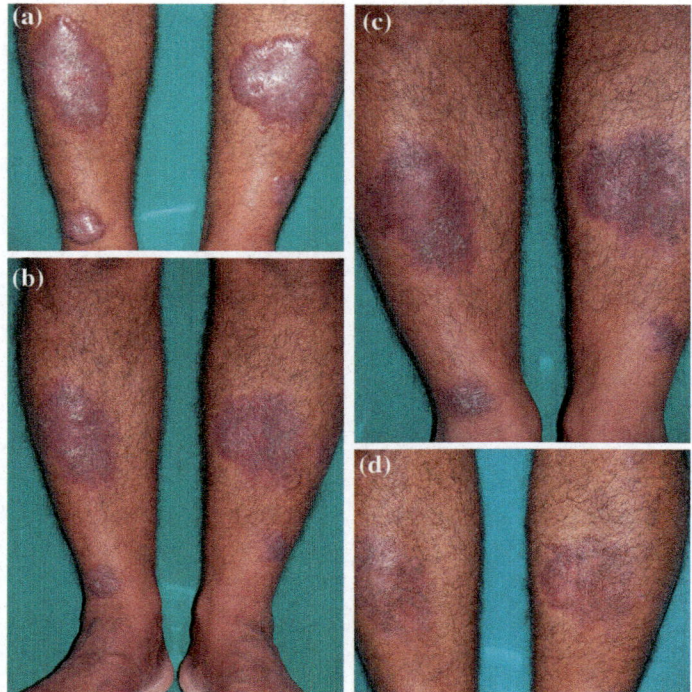

Fig. 6.2 Large plaques on the legs of a patient with post-kala-azar dermal leishmaniasis, **a** before miltefosine therapy, **b** mid-miltefosine therapy, **c** at end of therapy, and **d** at 1 year follow up

6.6 Resistance

A rising concern regarding the treatment of leishmaniasis is increasing resistance to the drugs currently in use. Antimonials are facing this problem particularly in Bihar, India. Inadequate dosing, short courses of therapy, and poor compliance have been blamed. Decline in SSG responsiveness of PKDL patients from high endemicity areas of Bihar was noted and highly resistant parasites of PKDL origin were proposed to contribute to increasing antimony resistance in VL via anthroponotic transmission [26]. In addition, it has been postulated that chronic exposure to environmental arsenic in Bihar results in cross-resistance to the related metalloid antimony [27]. It has been speculated that impaired thiol metabolism results in inhibition of conversion of pentavalent SSG to the active trivalent form by amastigotes [28]. Increased levels of trypanothione have been observed in some lines selected for resistance to trivalent SSG [29]. However, resistance to amphotericin B in VL has been a rarity, although some unresponsiveness has been seen in HIV-coinfected patients [30].

Membrane-bound ATP-binding cassette (ABC) transporters on the cell membrane, and a thiol metabolic pathway have been shown to be altered in amphotericin-resistant parasites [31].

A matter of serious concern is that within a short time after its introduction, miltefosine, raising everyone's hopes as the much sought after oral drug, is also showing signs of emerging resistance. Lately, relapse of VL following treatment with miltefosine has been reported [32]. In our experience of treating PKDL with miltefosine, high relapse rates were found in those with high pretreatment parasite loads [33] (Fig. 6.3).

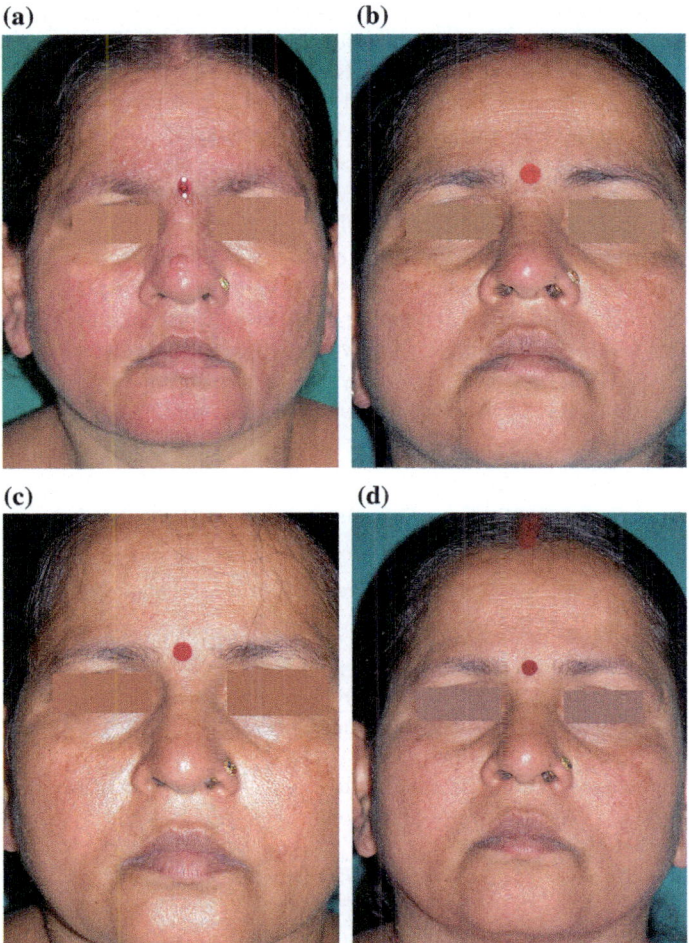

Fig. 6.3 Facial papulonodules on a patient with post-kala-azar dermal leishmaniasis, **a** at start of miltefosine therapy, **b** relapse after 4 months seen as nasal and facial erythema with chin induration, **c** clearance following liposomal amphotericin B, and **d** normal face at end of 1 year follow up

A decrease in intracellular drug accumulation is the most important mechanism of resistance to miltefosine. The inactivation of miltefosine transporter protein (LdMT) in *L. donovani* leads to impaired translocation of miltefosine [34].

6.7 Combination Regimes

A combination of leishmanicidal drugs has been found effective in VL and CL. Co-administration of glycyrrhizic acid with the antileishmanial drug has been found to be curative in SSG-resistant VL [35]. Likewise, a miltefosine-pentoxifylline combination has been found to be more effective than miltefosine alone in CL [36]. A similar attempt has been made in PKDL. A combination of miltefosine and amphotericin B has been effectively used [37] and is particularly recommended in patients with extensive lesions or severe involvement.

As a concluding note, it is relevant to reiterate that the management of Indian and Sudanese PKDL differs (Table 6.1). What is effective in one may not show the same efficacy in the other. Sudanese PKDL is self limiting in 85 % of cases over the course of 1 year while Indian PKDL often follows a prolonged, retracted course. Therefore, the former needs treatment only in severe deforming cases, children with oral lesions posing difficulty in feeding, and those with an unremitting course. WHO recommends SSG 20 mg/kg intravenously or intramuscularly daily for 30–60 days or LAMB 2.5 mg/kg a day intravenously for 20 days [38]. Another expert panel has recommended LAMB 2.5 mg/kg per day for 20 days as the first-line

Table 6.1 Treatment of post-kala-azar dermal leishmaniasis (PKDL) in the Indian Subcontinent

Drug	Dose	Duration
Miltefosine	50 mg/day orally for those weighing less than 25 kg 100 mg/day (50 mg twice in a day) orally for those weighing more than 25 kg	Minimum 12 weeks
Amphotericin B (liposomal)	30 mg/kg (5 mg/kg twice a week) intravenously	3 weeks
Amphotericin B deoxycholate (non-liposomal)	1 mg/kg per day intravenously	60–80 doses over 4 months
Treatment of East African PKDL		
Sodium stibogluconate	20 mg/kg intravenously or intramuscularly daily	30-60 days
Amphotericin B (liposomal)	2.5 mg/kg per day intravenously	20 days
Sodium stibogluconate + *Leishmania major* based vaccine	Sodium stibogluconate 20 mg/kg intravenously + *Leishmania major* based vaccine intradermally	Sodium stibogluconate for 40 days and weekly alum-treated *L. major* vaccine for 4 weeks

treatment and SSG 20 mg/kg for 40 days along with a weekly intradermal alum-treated *Leishmania major*-based vaccine for 4 weeks as the 2nd line treatment for Sudanese PKDL [7]. In addition, WHO also recommends miltefosine as a treatment option for Sudanese PKDL patients coinfected with HIV [39].

The Indian form of PKDL needs treatment in the context of the elimination of VL, and miltefosine has demonstrated to be the best evidence-based treatment. In adults weighing more than 25 kg a dose of 100 mg/day, and for those weighing less than 25 kg a dose of 50 mg/day, is recommended. In children, a dose of 2.5 mg/kg per day is recommended [38]. Miltefosine is given for a duration of 12 weeks. Occasionally, therapy may have to be extended to ensure complete cure [40].

Amphotericin B deoxycholate intravenously in a dose of 1 mg/kg per day for 60–80 doses over 4 months is the second line of treatment [38].

Trials using a combination of liposomal amphotericin and miltefosine are underway to define the optimum dose of each drug and this may well be the choice in future.

Acknowledgment Dr. Ruchi Singh Paliwal, Scientist D, National Institute of Pathology (ICMR), reviewed the chapter.

References

1. Zijlstra EE1, Musa AM, Khalil EA, el-Hassan IM, el-Hassan AM. Post-kala-azar dermal leishmaniasis. Lancet Infect Dis. 2003;3(2):87–98.
2. Katara GK, Ansari NA, Verma S, Ramesh V, Salotra P. Foxp3 and IL-10 expression correlates with parasite burden in lesional tissues of post kala azardermal leishmaniasis (PKDL) patients. PLoS Negl Trop Dis. 2011;5(5):e1171.
3. Salotra P, Singh R. Challenges in the diagnosis of post kala-azar dermal. leishmaniasis. Indian J Med Res. 2006;123(3):295–310.
4. Frézard F, Demicheli C, Ribeiro RR. Pentavalent antimonials: new perspectives for old drugs. Molecules. 2009;14(7):2317–36.
5. Diro E, Lynen L, Mohammed R, Boelaert M, Hailu A, van Griensven J. High parasitological failure rate of visceral leishmaniasis to sodium stibogluconate among HIV co-infected adults in Ethiopia. PLoS Negl Trop Dis. 2014;8(5):e2875.
6. World Health Organisation. Control of the leishmaniases. World Health Org Tech Rep Ser. 1990;793:54.
7. Musa AM, Khalil EA, Younis BM, Elfaki ME, Elamin MY, Adam AO, Mohamed HA, Dafalla MM, Abuzaid AA, El-Hassan AM. Treatment-based strategy for the management of post-kala-azar dermal leishmaniasis patients in the Sudan. J Trop Med. 2013;2013:708391. doi:10.1155/2013/708391. Epub 2013 Apr 15.
8. Das VN, Ranjan A, Bimal S. Siddique NA, Pandey K, Kumar N, Verma N, Singh VP, Sinha PK, Bhattacharya SK. Magnitude of unresponsiveness to sodium stibogluconate in the treatment of visceral leishmaniasis in Bihar. Natl Med J India. 2005;18(3):131–3.
9. Ramesh V, Saxena U, Misra RS. Efficacy of ketoconazole in post-kala-azar dermal leishmaniasis. Arch Dermatol. 1992;128:411–2.
10. Ramesh V. Allopurinol therapy in post-kala-azar dermal leishmaniasis. Acta Derm Venereol. 1996;76:328–9.

11. Muigai R, Gachihi GS, Oster CN, Were JB, Nyakundi PM, Chunge CN, et al. Post-kala-azar dermal leishmaniasis: the Kenyan experience. East Afr Med J. 1991;68:801–6.
12. Sharma VK, Prasad HR, Sethuraman G, Khaitan BK. Combination of sodium stibogluconate and rifampicin in post-kala-azar dermal leishmaniasis. Indian J Dermatol Venereol Leprol. 2007;73:53–4.
13. Ramesh V, Kumar J, Kumar D, Salotra P. A retrospective study of intravenous sodium stibogluconate alone and in combinations with allopurinol, rifampicin, and an immunomodulator in the treatment of Indian post-kala-azar dermal leishmaniasis. Indian J Dermatol Venereol Leprol. 2010;76(2):138–44.
14. Sundar S, Singh A, Tiwari A, Shukla S, Chakravarty J, Rai M. Efficacy and safety of paromomycin in treatment of post-kala-azar dermal leishmaniasis. ISRN Parasitol. 2014;2014 (548010):4. doi:10.1155/2014/548010.
15. Ramesh V, Kumar D, Salotra P. Antimonial Therapy in Post-kala-azar dermal Leishmaniasis-A Hobson's choice. In: Tandon S, Mehrotra S, editors. Drugs and pharmaceuticals-current R & D highlights. Lucknow: Publisher Central Drug Research Institute; 2008. p. 7–10.
16. Thakur CP, Narain S, Kumar N, Hassan SM, Jha DK, Kumar A. Amphotericin B is superior to sodium antimony gluconate in the treatment of Indian post-kala-azar dermal leishmaniasis. Ann Trop Med Parasitol. 1997;91(6):611–6.
17. Balasegaram M, Ritmeijer K, Lima MA, Burza S, Ortiz Genovese G, Milani B, Gaspani S, Potet J, Chappuis F. Liposomal amphotericin B as a treatment for human leishmaniasis. Expert Opin Emerg Drugs. 2012;17(4):493–510.
18. Musa AM, Khalil EA, Mahgoub FA, Hamad S, Elkadaru AM, El Hassan AM. Efficacy of liposomal amphotericin B (AmBisome) in the treatment of persistent post-kala-azar dermal leishmaniasis (PKDL). Ann Trop Med Parasitol. 2005;99(6):563–9.
19. Dorlo TPC, van Thiel PPAM, Schoone GJ, et al. Dynamics of parasite clearance in cutaneous leishmaniasis patients treated with miltefosine. PLoS Negl Trop Dis. 2011;5:e1436.
20. de Vries PJ, van der Meide WF, Godfried MH, et al. Quantification of the response to miltefosine treatment for visceral leishmaniasis by QT-NASBA. Trans R Soc Trop Med Hyg. 2006;100:1183–6.
21. Dorlo TP, Balasegaram M, Beijnen JH. de Vries PJ Miltefosine: a review of its pharmacology and therapeutic efficacy in the treatment of leishmaniasis. J Antimicrob Chemother. 2012;67 (11):2576–97.
22. Dorlo TP, Huitema AD, Beijnen JH, de Vries PJ. Optimal dosing of miltefosine in children and adults with visceral leishmaniasis. Antimicrob Agents Chemother. 2012;56(7):3864–72.
23. Dorlo TP, van Thiel PP, Huitema AD, Keizer RJ, de Vries HJ, Beijnen JH, de Vries PJ. Pharmacokinetics of miltefosine in Old World cutaneous leishmaniasis patients. Antimicrob Agents Chemother. 2008;52(8):2855–60.
24. Belay AD, Asafa Y, Mesure J, Davidson RN. Successful miltefosine treatment of post-kala-azar dermal leishmaniasis occurring during antiretroviral therapy. Ann Trop Med Parasitol. 2006;100(3):223–7.
25. Musa AM, Khalil EA, Mahgoub FA, Elgawi SH, Modabber F, Elkadaru AE, et al. Immunochemotherapy of persistent post-kala-azar dermal leishmaniasis: a novel approach to treatment. Trans R Soc Trop Med Hyg. 2008:58–63.
26. Singh R, Kumar D, Ramesh V, Negi NS, Singh S, Salotra P. Visceral. leishmaniasis, or kala azar (KA): high incidence of refractoriness to antimony is contributed by anthroponotic transmission via post-KA dermal leishmaniasis. J Infect Dis. 2006;194(3):302–6.
27. Perry MR, Prajapati VK, Menten J, Raab A, Feldmann J, Chakraborti D, Sundar S, Fairlamb AH, Boelaert M, Picado A. Arsenic exposure and outcomes of antimonial treatment in visceral leishmaniasis patients in Bihar, India: a retrospective cohort study. PLoS Negl Trop Dis. 2015;9(3):e0003518.
28. Grondin K, Haimeur A, Mukhopadhyay R, Rosen BP, Ouellette M. Co-amplification of the gamma-glutamylcysteine synthetase gene gsh1 and of the ABC transporter gene pgpA in arsenite-resistant Leishmania tarentolae. EMBO J. 1997;16:3057–65.

29. Mukhopadhyay R, Dey S, Xu N, Gage D, Lightbody J, Ouellette M, Rosen BP. Trypanothione overproduction and resistance to antimonials and arsenicals in Leishmania. Proc Natl Acad Sci USA. 1996;93:10383–7.
30. Lachaud L, Bourgeois N, Plourde M, Leprohon P, Bastien P, Ouellette M. Parasite susceptibility to amphotericin B in failures of treatment for visceral leishmaniasis in patients coinfected with HIV type 1 and Leishmania infantum. Clin Infect Dis. 2009;48(2):e16–22.
31. Purkait B1, Kumar A, Nandi N, Sardar AH, Das S, Kumar S, Pandey K, Ravidas V, Kumar M, De T, Singh D, Das P. Mechanism of amphotericin B resistance in clinical isolates of Leishmania donovani. Antimicrob Agents Chemother. 2012;56:1031–41.
32. Pandey BD1, Pandey K, Kaneko O, Yanagi T, Hirayama K. Relapse of visceral leishmaniasis after miltefosine treatment in a Nepalese patient. Am J Trop Med Hyg. 2009;80:580–2.
33. Ramesh V, Singh R, Avishek K, Verma A, Deep DK, Verma S, Salotra P. Decline in clinical efficacy of oral miltefosine in treatment of post Kala-azar dermal leishmaniasis (PKDL) in India. PLOS Neg Trop Dis. 2015;9(10):e0004093.
34. Seifert K, Pérez-Victoria FJ, Stettler M, Sánchez-Cañete MP, Castanys S, Gamarro F, et al. Inactivation of the miltefosine transporter, LdMT, causes miltefosine resistance that is conferred to the amastigote stage of Leishmania donovani and persists in vivo. Int J Antimicrob Agents. 2007;30:229–35.
35. Bhattacharjee A, Majumder S, Majumdar SB, Choudhuri SK, Roy S, Majumdar S. Co-administration of glycyrrhizic acid with the antileishmanial drug sodium antimony gluconate (sodium stibogluconate) cures sodium stibogluconate-resistant visceral leishmaniasis. Int J Antimicrob Agents. 2015;45(3):268–77.
36. Santarem AA, Greggianin GF, Debastiani RG, Ribeiro JB, Polli DA, Sampaio RN. Effectiveness of miltefosine-pentoxifylline compared to miltefosine in the treatment of cutaneous leishmaniasis in C57BL/6 mice. Rev Soc Bras Med Trop. 2014;47:517–20.
37. Ramesh V, Avishek K, Sharma V, Salotra P. Combination therapy with amphotericin-B and miltefosine for post-kala-azar dermal leishmaniasis: a preliminary report. Acta Derm Venereol. 2014;94(2):242–3.
38. World Health Organization. Control of the leishmaniases: report of a meeting of the WHO Expert Committee on the Control of Leishmaniasis, Geneva, 22–26 March 2010. Geneva, World Health Organization, 2010 (WHO Technical Report Series, No. 949).
39. World Health Organisation. Post Kala Azar Dermal leishmaniasis: a manual for case management and control: report of a WHO consultative meeting, Kolkata, India, 2–3 July 2012.
40. Ramesh V, Katara GK, Verma S, Salotra P. Miltefosine as an effective choice in the treatment of post-kala-azar dermal leishmaniasis. Br J Dermatol. 2011;165(2):411–4.

Chapter 7
Combination Therapy for Leishmaniases

Farrokh Modabber

Abstract After more than 50 years of antimonial treatment of visceral leishmaniasis (VL), new drugs recently available include miltefosine, paromomycin, and amphotericin-B deoxycholate (Ampho) and its liposomal formulations (LamB). Miltefosine requires 28 days of treatment, is potentially teratogenic, and causes gastrointestinal disturbances. Because of its high cost, side effects, and because patients feel relieved from VL symptoms after a few days of treatment, compliance is low and increased resistance is highly anticipated. Paromomycin is inexpensive, but requires 3 weeks of daily injections with mild or moderate pain that can result in low compliance, and thereby increase the probability of resistance. Strains resistant to miltefosine or paromomycin have already been isolated from patients refractory to monotherapy with these 2 drugs. Ampho is effective if patients can tolerate it, but requires 15 intravenous infusions during 30 days of hospitalization. LamB is highly effective with minimal side effects and can cure more than 95 % of VL patients with a single infusion, but is very expensive. Hence, the Drugs for Neglected Diseases *initiative* in collaboration with the Indian Medical Research Council and the Kala-azar Medical Research Center initiated studies to assess the safety and efficacy of short-course, 2-drug combination therapies for the treatment of VL in India. The results are better than expected, with high tolerability and higher efficacy than that seen with any monotherapy. In addition to better safety, the duration of treatment is shorter (7–11 days) with the combination of low-dose LamB plus paromomycin or miltefosine than with either drug alone.

Keywords Combination therapy · Immunochemotherapy · Immunotherapy · Leishmaniasis vaccine · Leishmaniasis · Prophylaxis · Leishmaniasis treatment

---

F. Modabber (✉)
Honorary Faculty, Center for Research and Training on Skin Diseases and Leprosy (CRTSDL), Tehran University of Medical Sciences,
79 Taleghani Ave, 14155-6383 Tehran, Iran
e-mail: fmodabber@gmail.com

7.1 Introduction

Because of rising resistance to antimonials—the first-line drug for visceral leishmaniasis (VL) in Bihar, India [1]—it became necessary to change treatment policies. Amphotericin B (Ampho) replaced antimonials in Bihar; however, due to its potential toxicity, the required 1 month hospitalization, and the 15 intravenous infusions, it is not practical for the VL Elimination Program (VLEP) of the Indian subcontinent. Subsequently, when miltefosine, the first oral drug for VL was developed, VLEP adopted miltefosine [2]. However, because of its potential teratogenicity for humans, miltefosine cannot be used in pregnant women and in women of child-bearing age without a pregnancy test and contraception for at least 3 months. In addition, miltefosine requires 28 days of treatment. Compliance was shown to be low due to the long duration of treatment, cost of the drug, and side effects, especially gastrointestinal disturbances. With the ease of developing resistant parasites in vitro [3], it is expected that miltefosine monotherapy, particularly without a directly observed treatment (DOT) program, would shortly lead to a rise of resistant parasites.

Paromomycin, an old drug in the class of aminosidines, was developed in an injectable form [4] and was registered in India in 2006. Although inexpensive and relatively safe, paromomycin requires 21 daily injections that can be painful. Also, based on laboratory evidence, paromomycin could become ineffective for the treatment of VL if used as monotherapy due to rise of resistant parasites [5].

Although the efficacy of AmBisome against VL was demonstrated in the 1990s [6], and its safety established through its wide use in fungal infections, this drug has been inaccessible to patients because of its extremely high price. However, a reduced price negotiated by WHO has made it possible for AmBisome to be considered for use in VL-endemic countries. Although the price is still high, its efficacy and lack of toxicity (compared to Ampho) make it a very attractive drug for the treatment of VL, at least in India, where a single dose has been shown to cure about 96 % of patients [7]. So far, no *Leishmania* resistant to Ampho has been shown either in a clinic or laboratory setting [8]; however, its use in leishmaniases has been very limited. Therefore, predicting how long it will take for resistant parasites to emerge is impossible. To prevent the rise of drug-resistant *Leishmania* and to reduce the duration of treatment, and thereby the toxicity and side effects of these drugs (used as monotherapy), the Drugs for Neglected Diseases initiative (DND*i*) assessed the safety and efficacy of available drugs given in combination for shorter periods.

7.2 Drugs in Use

7.2.1 Pentavalent Antimonials

Two pentavalent antimonials are currently available: meglumine antimoniate and sodium stibogluconate. They are chemically similar and their toxicity and efficacy in VL are related to their active ingredient, pentavalent antimony (Sb^{5+}). Meglumine antimoniate solution contains about 8.1 % Sb^{5+} (81 mg/mL), whereas the sodium stibogluconate solution contains about 10 % Sb^{5+} (100 mg/mL). In general, children are more tolerant than adults to antimonials. Common side effects are anorexia, vomiting, nausea, malaise, myalgia, headache, and lethargy. Electrocardiographic changes depend on dosage and duration of treatment (cardiotoxicity is dose dependent and usually occurs after 2 weeks of treatment), but the most common are T-wave inversion, prolonged Q-T interval, and, rarely, arrhythmia. A rarely reported side effect is nephrotoxicity.

Initial treatment of a parasitologically proven case of VL should be based on a daily injection of 20 mg of Sb^{5+} per kg of body weight. Injections are normally given for 4 weeks. The duration of treatment varies from one endemic area to another, but should be individually determined for each country and each endemic focus. Treatment should be given under the supervision of medical personnel. Treatment with antimonials is usually well tolerated, but if serious side effects arise (in most cases related to hepatotoxicity and cardiotoxicity) it should be interrupted temporarily. Rescue treatment is advised if other drugs are available. If relapses occur, patients should be treated with other available anti-leishmaniasis drugs if possible.

7.2.2 Amphotericin B

Amphotericin B was used as a second-line treatment for patients not responding to antimonials; however, in recent years it is recommended as first-line treatment in Bihar, India, due to parasite resistance to antimonials. The recommended dose is 15 mg/kg given at 1 mg/kg as a slow (4–6 h) infusion every other day for 30 days. Adverse events, primarily hepatotoxicity, may occur in 5–7 % of patients. Relapse or lack of response occurs in 1–2 % of cases. Treatment should always be given in hospital because of the nephrotoxicity and cardiotoxicity of the drug.

7.2.3 Liposomal Amphotericin B (LamB)

The toxicity of amphotericin B is greatly reduced when given in association with lipids. Several different formulations exist, but liposomal amphotericin B (LamB) is the most frequently used (Fig. 7.1). Doses of up to 21 mg/kg body weight over 10 days have been given with over 98 % efficacy. However, a single dose of 10 mg/kg body weight has been reported to cure about 95 % of patients in a trial in India [7]. Patients in Africa and Latin America respond differently than those in India and may require higher doses of LamB. The guideline for treatment of VL in Sudan is 3 mg/kg on alternating days for 14 days (total 21 mg/kg) and for Kenya is 14 mg/kg total given in 7 daily infusions of 2 mg/kg. For VL in Latin America

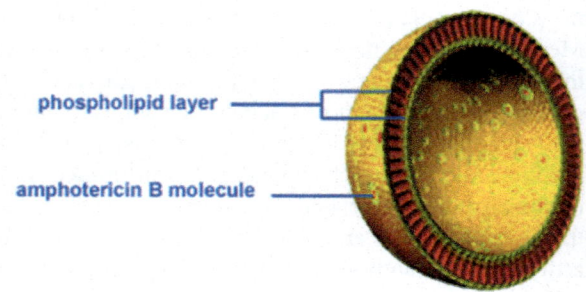

Fig. 7.1 Schematic presentation of liposomal amphotericin B. From the package insert: http://www.fujisawa.com/medinfo/pi/pi_page_amb.htm

caused by *L. chagasi* (*L. infantum*), the total suggested dose is 20 mg/kg administered over 5–7 days. However, further studies are required to find the most cost-effective dose.

7.2.4 Paromomycin

Produced from culture filtrates of *Streptomyces chrestomyceticus* and active against various micro-organisms, the aminosidine paromomycin has been registered in India for the treatment of VL but is not yet widely available. Intramuscular injections of 11 mg/kg paromomycin, equivalent to 15 mg paromomycin sulphate, given for 3 weeks have been efficacious in India and have shown a similar safety profile to other drugs of its class. The most commonly reported adverse drug reactions are injection-site pain, aspartate aminotransferase (AST) and alanine aminotransferase (ALT) value elevations (above 3 times the upper limit of the normal range), pyrexia, and an abnormal audiogram. These effects are usually mild to moderate and transient or reversible after the end of treatment [9].

In Sudan, the efficacy of paromomycin at the same dose used in India was less than 50 %. In a dose–response study in Sudan to define the required dose, an increased dose of 15 mg paromomycin (equivalent to 20 mg paromomycin sulphate) gave an efficacy of 80 % [10]. The advantages of paromomycin are its low cost and high stability. However, continued use of paromomycin as monotherapy will eventually become useless due to development of drug resistant parasites, as seen with other aminoglycosides as mentioned earlier [5]. No data exists on paromomycin efficacy in Latin America.

7.2.5 Miltefosine

Miltefosine (hexadecylphosphocholine) is the first oral drug active against VL and is available as 10 mg or 50 mg capsules. The recommended dose for adults and children over 2 years of age is 1.5–2.5 mg/kg (maximum total dosage of 100 mg a day). Miltefosine efficacy against VL is high in the Indian sub-continent [11], but being a potential teratogenic drug, treatment cannot be given to pregnant women and contraception must be used by women of child-bearing age. Compliance is a concern because the drug has some gastrointestinal toxicity and patients usually feel better after a few days of treatment. Other common side effects include vomiting, diarrhea, elevation of liver enzymes, and serum creatinine. These effects are usually mild to moderate and transient or reversible at the end of treatment. Less common side effects include thrombocytopenia, anorexia, and abdominal pain. Data are limited from other disease foci in Africa and Latin America. One phase 3 study in Ethiopia suggested a similar safety and efficacy profile to that of sodium stibogluconate in non-HIV-positive patients [12].

7.3 Combination Therapies

Monotherapy options with available drugs are mostly long-duration therapies: 21 days for paromomycin, 28 days for miltefosine or antimonials, and 30 days for amphotericin B; liposomal amphotericin B treatment is short (1–10 days depending on the country). Toxicity and side effects usually increase with duration of treatment. Because of side effects, the cost of treatment, and the fact that patients feel better after a few days of taking the medications (particularly miltefosine), compliance can be low without a DOT program. Low compliance can increase the chance of resistance emerging. The cost of LamB and miltefosine are still high despite WHO price negotiations with manufacturers; therefore, a combination of lower doses of these drugs with paromomycin (the most affordable anti-leishmanial drug) would reduce the cost of treatment. Combination treatments have the following advantages:

- Shortening the duration of treatment for available monotherapies, thereby reducing side effects and cost (except for paromomycin) for patients and increasing compliance;
- Reducing the overall dose of drugs, thereby reducing toxic effects;
- Reducing the chance of resistance developing, thereby prolonging the effective life of the available drugs.

Several combination treatment trials have had favourable results (for a recent review see [13]). The combination of paromomycin and sodium stibogluconate increased the cure rate in VL patients in Bihar, India, compared with sodium stibogluconate alone (unresponsiveness of 40–60 % [14]). Combination of a single dose of 5 mg/kg of LamB and various doses of miltefosine had over 95 % efficacy in a phase 2 trial with 40 patients per arm [15]. A definitive, randomised, open-label, hospital-based, non-inferiority ($\Delta > 7$) trial was completed for comparing the standard treatment with amphotericin B (1 mg/kg every other day for 30 days) with combination treatments using 2 drugs: a single dose of LamB

(5 mg/kg) with 7 days of miltefosine (instead of 28 days); or a single dose of LamB (5 mg/kg) and paromomycin for 10 days (instead of 21 days); or a combination of miltefosine and paromomycin for 10 days instead of 21–28 days (Fig. 7.2). A total of 634 parasitologically confirmed VL patients from 2 centres in Bihar (Patna and Muzaffarpur) were enrolled, treated and followed up for 6 months. All 3 combinations showed a safety profile better than that of the standard [16]. In the amphotericin group, adverse reactions included elevated levels of liver enzymes or serum creatinine. There was a fatal cardiac infarct (a described rare reaction to amphotericin B) in a high-risk 59-year-old male patient. In the combination groups, there was only 1 patient who showed an immediate hypersensitivity reaction to the test injection of LamB. Laboratory parameters measured during and after treatment (haematology and biochemistry) showed no clinically significant variations in the combination groups. Laboratory parameters (haematology and biochemistry) during the treatment period and up to day 45 of follow up showed no significant differences between the 3 combination groups, but they all showed significantly lower increases of creatinine, aspartate aminotransferase, alanine transaminase, blood urea nitrogen, and significantly better recovery of anaemia than that seen with the standard amphotericin B treatment. If these results can be confirmed in field conditions, no monitoring of patients undergoing combination therapy might be required based on the safety profile. If this is confirmed, with a very short hospitalization (1–2 days) followed by outpatient treatment, the cost effectiveness of the

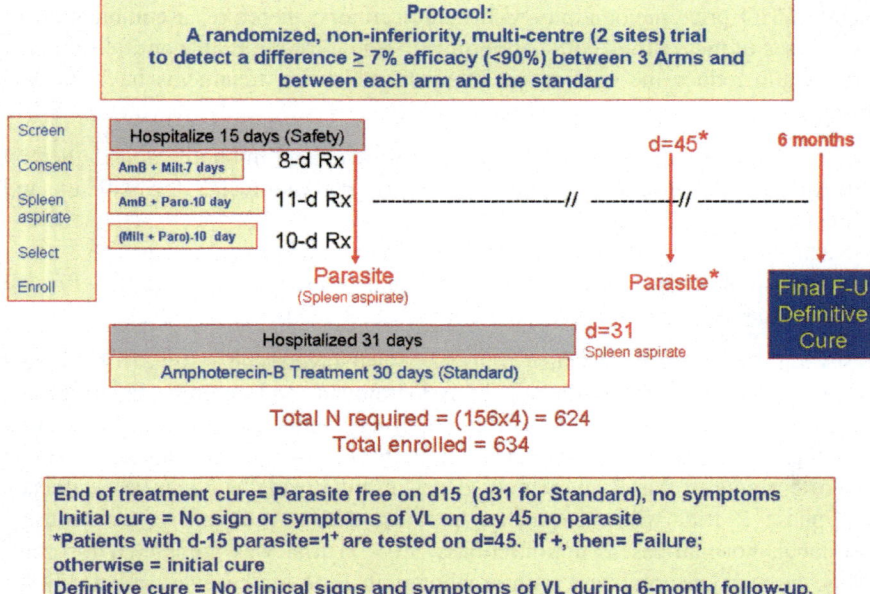

Fig. 7.2 Description of the trial protocol

combination would be improved compared with that of other modalities of treatment presently available. However, further studies with larger patient populations under field conditions are ongoing to support this finding.

7.4 Future Trends for the Treatment of Leishmaniases

A strong immunity is usually developed after recovery from most forms of leishmaniases in immunocompetent individuals. In immunocompromised hosts, when chemotherapy is stopped, the disease returns, hence maintenance treatment is required (Fig. 7.3a). In immunocompetent hosts, a protective immune response is developed after chemotherapy that will maintain the parasite under control (Fig. 7.3b). Indeed, it is generally believed that sterile immunity does not exist in leishmaniasis and that the parasites are controlled by the host's effector immune responses. This is supported by development of fulminating leishmaniasis when, after recovery from an infection, the person becomes infected with HIV or is otherwise immunocompromised by immunosuppressive drugs used for cancer therapy. Therefore, the protective immune response is an important part of recovery from leishmaniases. The concept of immunochemotherapy is to use a vaccine or immunomodulators to induce the protective response rapidly to cure the disease (Fig. 7.3c). The concept of immunostimulation during chemotherapy is not new

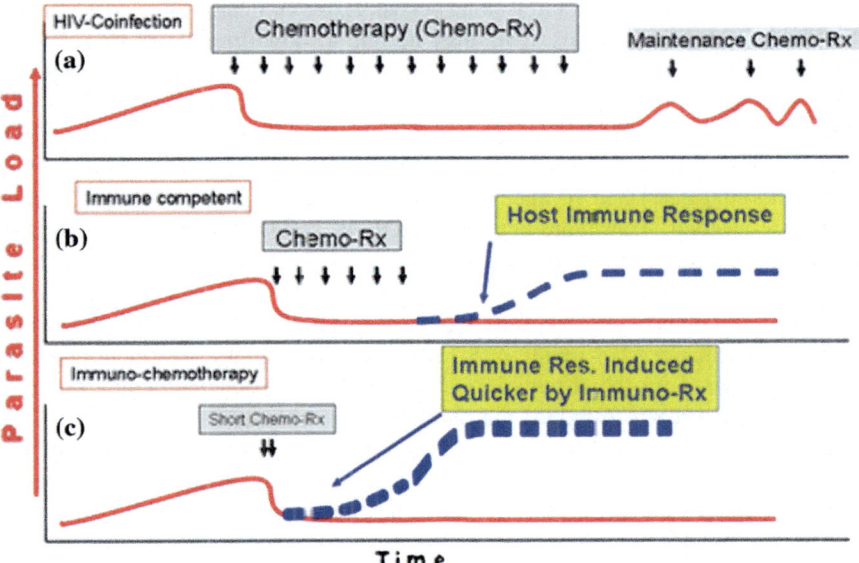

Fig. 7.3 Conceptual representation of chemotherapy plus immunotherapy in different forms of leishmaniasis (from [22])

and not specific for leishmaniasis. Immunomodulators have been used for cancer therapy and new ones are being developed.

Immunotherapy with or without chemotherapy has been used for treatment of cutaneous leishmaniasis. In Venezuela, killed *L. mexicana* plus live Bacille Calmette-Guerin (BCG) was used instead of antimonials to reduce cost and side effects [17]. In Brazil, to reduce the dose of antimonials to minimize toxicity and cost, daily doses of killed *L. amazonensis* vaccine (Mayrink's vaccine) together with a low dose of antimonial (8 mg/kg body weight per day) cured 100 % of patients compared to 4 % with low-dose antimonial alone after 4 cycles of treatment [18]. A cycle is 10 daily injections followed by 10 days of rest. The vaccine was registered in Brazil as an adjunct for low-dose chemotherapy. Several case studies of immunochemotherapy for mucosal leishmaniasis exist, which use first-generation vaccines or recombinant antigens of *Leishmania* plus granulocyte-macrophage-colony stimulating factor and antimonials [19]. However, more studies are needed for this approach.

An exploratory hospital-based, randomized, controlled trial in Sudan [20] on patients with persistent post-kala-azar dermal leishmaniasis (PKDL) (lesions >6 months—difficult to cure with drugs alone), showed that the cure rate by immunochemotherapy was significantly better than that by chemotherapy alone (final cure rates at day 90 were 100 vs. 40 %; $p < 0.004$). The vaccine was a mixture of killed *L. major* adsorbed to alum plus BCG, given 4 times at weekly intervals. The exact mechanism involved in enhanced the cure rate is not fully understood. However, in PKDL lesions, there is an increase of $\alpha\beta$ T-cells but a reduction of CD1 antigen-presenting cells. Even in the presence of effector Th-1 cells, parasites are not killed and lesions persist. External interferon-γ does not cure all patients with leishmaniasis. This may be due to down-regulation of B7-1 and up-regulation of B7-2 by IL-10, thereby leading to a predominant Th-2 response, which can be diverted by immunomodulation using a vaccine or possibly other immunomodulators [20]. It should be noted that a minority of PKDL patients attained a cure without showing a pure Th-1 response. Therefore, the classical Th-1/Th-2 dichotomy described in mice does not necessarily apply to humans; further studies are needed. However, because PKDL is believed to be a reservoir of infection and without treatment the lesions may last for years, treatment should be given. Patients with PKDL, which may develop months or years after VL but in some cases simultaneously, generally feel healthy and do not seek treatment, which is long and expensive (2–4 months). An affordable short-course treatment of PKDL is urgently needed for the elimination of VL, and immunochemotherapy seems to be a possible approach to reduce the duration of treatment.

A few new immunomodulators are being developed for cancer therapy; with adequate knowledge of their mechanisms of action, selected agents should be tested for the treatment of VL and other forms of leishmaniases.

7.5 Recommendation

To protect the available drugs against the development of parasite resistance and to reduce treatment time and cost (hence increase compliance), we recommend monotherapy with antimonials, paromomycin, and miltefosine should be replaced by combination treatment in the Indian Subcontinent. Monotherapy with LamB, however, has been recommended by the WHO, mainly because based on the mode of action resistance is unlikely to occur in the immediate future. The choice of combination treatment depends on the ease of pregnancy testing and contraceptive use in women of child-bearing age for miltefosine, and the availability of the infrastructure for maintenance and administration of LamB. At present, the combination of miltefosine and paromomycin is the most cost-effective treatment [21].

References

1. Alvar J, Croft S, Olliaro P. Chemotherapy in the treatment and control of leishmaniasis. Adv Parasitol. 2006;61:224–61.
2. Sundar S, Jha TK, Thakur CP. Oral miltefosine for Indian visceral leishmaniasis. N Engl J Med. 2002;347:1739–46.
3. Seifert K, Pérez-Victoria FJ, Stettler M, et al. Inactivation of the miltefosine transporter, LdMT, causes miltefosine resistance that is conferred to the amastigote stage of *L. donovani* and persists in vivo. Int J Antimicrob Agents. 2007;30:229–35.
4. Jha TK, Olliaro P, Thakur CP, et al. Randomised controlled trial of aminosidine (paromomycin) v sodium stibogluconate for treating visceral leishmaniasis in North Bihar, India. BMJ. 1998;316:1200–5.
5. Maarouf M, Adeline MT, Solignac M, et al. Development and characterization of paromomycin-resistant *Leishmania donovani* promastigotes. Parasite. 1998;5:167–73.
6. Berman JD, Badaro R, Thakur CP, et al. Efficacy and safety of liposomal amphotericin B (AmBisome) for visceral leishmaniasis in endemic developing countries. Bull World Health Organ. 1998;76:25–32.
7. Sundar S, Chakravarty J, Agarwal D, et al. Single-dose liposomal amphotericin B for visceral leishmaniasis in India. N Engl J Med. 2010;362:504–12.
8. Lachaud L, Bourgeois N, Plourde M. Parasite susceptibility to amphotericin B in failures of treatment for visceral leishmaniasis in patients coinfected with HIV type 1 and *Leishmania infantum*. Clin Infect Dis. 2009;48:e16–22.
9. Sundar S, Jha TK, Thakur CP, et al. Injectable paromomycin for visceral leishmaniasis in India. N Engl J Med. 2007;356:2571–81.
10. Davidson RN, den Boer M, Ritmeijer K. Paromomycin. Trans R Soc Trop Med Hyg. 2009;103:653–60.
11. Bhattacharya SK, Sinha PK, Sundar S, et al. Phase 4 trial of miltefosine for the treatment of Indian visceral leishmaniasis. J Infect Dis. 2007;196:591–8.
12. Ritmeijer K, Dejenie A, Assefa Y, et al. A comparison of miltefosine and sodium stibogluconate for treatment of visceral leishmaniasis in an Ethiopian population with high prevalence of HIV infection. Clin Infect Dis. 2006;43:357–64.
13. van Griensven J, Balasegaram M, Meheus F, et al. Combination therapy for visceral leishmaniasis. Lancet Infect Dis. 2010;10:184–94.

14. Thakur CP, Bhowmick S, Dolfi L, et al. Aminosidine plus sodium stibogluconate for the treatment of Indian kala-azar: a randomized dose-finding clinical trial. Trans R Soc Trop Med Hyg. 1995;89:219–23.
15. Sundar S, Rai M, Chakravarty J, et al. New treatment approach in Indian visceral leishmaniasis: single-dose liposomal amphotericin B followed by short-course oral miltefosine. Clin Infect Dis. 2008;47:1000–6.
16. Sundar S, Sinha PK, Rai M, et al. Comparison of short-course multidrug treatment with standard therapy for visceral leishmaniasis in India: an open-label, non-inferiority, randomised controlled trial. Lancet. 2011, 5;377 (9764):477–86. doi:10.1016/S0140-6736(10) 62050-8. Epub 2011 Jan 20.PMID:21255828.
17. Convit J, Ulrich M, Zerpa O, et al. Immunotherapy of American cutaneous leishmaniasis in Venezuela during the period 1990–99. Trans R Soc Trop Med Hyg. 2003;97:469–72.
18. Machado-Pinto J, Pinto J, da Costa CA, et al. Immunochemotherapy for cutaneous leishmaniasis: a controlled trial using killed Leishmania (Leishmania) amazonensis vaccine plus antimonial. Int J Dermatol. 2002;41:73–8.
19. Badaro R, Lobo I, Munos A, et al. Immunotherapy for drug-refractory mucosal leishmaniasis. J Infect Dis. 2006;194:1151–9.
20. Musa AM, Khalil EAG, Mahgoub FAE, et al. Immunochemotherapy of persistent post-kala-azar dermal leishmaniasis: a novel approach of treatment. Trans R Soc Trop Med Hyg. 2008;102:58–63.
21. Meheus F, Balasegaram, Olliaro P, et. al. Cost-effectiveness analysis of combination therapies for visceral leishmaniasis in the Indian subcontinent. PLoS Negl Trop Dis. 2010;4(9). pii: e818. doi:10.1371/journal.pntd.0000818. PMID: 20838649.
22. Musa AM, Noazin S, Khalil EAG. Modabber. F Trans Roy Soc Trop Med Hyg. 2010;104:1–2.

Chapter 8
Vaccine Development for Leishmaniasis

Yasuyuki Goto

Abstract Among control strategies for leishmaniasis, vaccination can be the most cost-effective through providing long term antileishmanial immunity to people with high risk of infection. The possibility of developing a vaccine against leishmaniasis has been historically indicated by the protective immunity acquired by people cured of the disease. This was the basis of the first generation of vaccines, known as 'leishmanization', but a safer and more efficient vaccine has been long desired. Recent advances in the immunological understanding of leishmaniasis, as well as the tools for vaccination, have opened a door to the clinical development of a vaccine against the disease. In addition, the vaccine target is no longer limited to amastigotes in mammalian hosts, enabling a multilateral prevention strategy. Furthermore, the use of a vaccine is now not only for prevention but also for treatment. Within this manuscript we review background information, the current status, and the future direction of vaccine development for leishmaniasis.

Keywords Leishmaniasis · Vaccine · Protective immunity

8.1 Protective Immunity Against *Leishmania* Infection

Leishmaniases range from self-healing cutaneous leishmaniasis (CL) to life-threatening visceral leishmaniasis (VL), depending on the parasite species causing the disease. Even in the case of VL, however, people acquire protective immunity against the disease after recovery from the initial incident by chemotherapy. Therefore, control of the disease by prophylactic vaccination must be a viable approach [1, 2]. Of course, gaining an understanding of the protective

---

Y. Goto (✉)
Laboratory of Molecular Immunology, Department of Animal Resource Sciences,
Graduate School of Agricultural and Life Sciences,
The University of Tokyo, Tokyo, Japan
e-mail: aygoto@mail.ecc.u-tokyo.ac.jp

immunity against leishmaniasis is the key component for advancing vaccine development.

Leishmaniasis is a vector-borne disease and *Leishmania* parasites are transmitted by phlebotomine sand flies (Fig. 8.1). The parasites proliferate as promastigotes within their insect host. Once transmitted to mammalian hosts through a bite by sand flies, they differentiate into and replicate as amastigotes. Because the parasites reside mainly within macrophages in their mammalian hosts, vaccines that stimulate cellular immunity are required for control of intracellular replication. Murray et al. reported that IFN-γ is a pivotal cytokine to activate human macrophages to kill intracellular *Leishmania donovani* [3]. The importance of IFN-γ was confirmed also in *Leishmania major* infection as recombinant IFN-γ up-regulates killing of parasites by murine macrophages [4].

IFN-γ needs to be produced by acquired immune cells, i.e., either T or B cells, for the cytokine to play a protective role in vaccination against leishmaniasis. Although there is now accumulating evidence that appropriate T cell responses correlate with protection against leishmaniasis in both humans and animal models, the discovery of the Th1/Th2 paradigm in animal models had a large impact on, and opened a door to, vaccine development [5]. Scott et al. has reported a distinct helper T cell population can either protect or exacerbate CL caused by *L. major* infection [6]. The former was

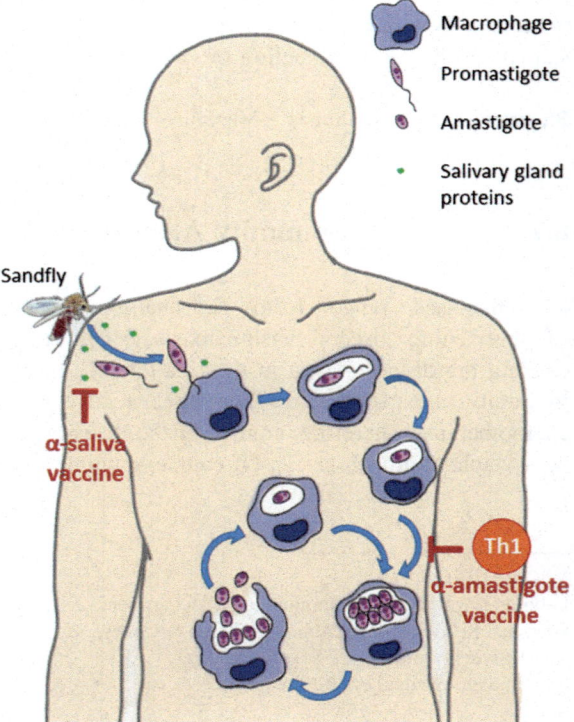

Fig. 8.1 Targets of vaccines under clinical development for leishmaniasis. Illustrated by Haruka Mizobuchi

characterized as Th1 according to production of IL-2 and IFN-γ upon antigen recall, and the latter as Th2 due to production of IL-4 and IL-5 [6]. IL-4 is one of the key cytokines for Th2 response and some reports demonstrated exacerbating roles of the cytokine in experimental *L. major* infection [7–10]. Reciprocal expression of IFN-γ and IL-4 in infected mice also supports protective and exacerbating roles of these cytokines, respectively [11]. However, there are some reports that IL-4 has no detrimental effect or has even a protective effect on *L. major* infection [12–14]. The Th1/Th2 paradigm becomes even more controversial in *L. donovani* infection [15]. Disruption of IL-4 does not contribute to protection against *L. donovani* infection [16, 17]; in fact, the cytokine is required for generation of protective $CD8^+$ T cell responses by vaccination [18]. Recently, IL-10 and TGF-β have also received attention as inhibitory factors during leishmaniasis [19–23].

Nonetheless, we can say there is consensus that induction of Th1 is important for vaccine-mediated protection against leishmaniasis. IL-12, a Th1 potentiator, demonstrates its protective effect in both therapeutic and prophylactic settings against leishmaniasis [24–27]. However, recombinant IL-12 may not be suitable for commercial use due to its chimeric nature and resulting production cost; the development of other potent Th1 inducers may play a central role in future leishmaniasis vaccine development.

8.2 History and Recent Advances in Vaccine Development

The history of leishmaniasis vaccine development overlaps with that for other diseases, as it started with live vaccines, followed by killed vaccines, and partly shifted to subunit vaccines. Live vaccination was historically practiced either by exposing one's skin to sand fly bites or by injecting the pus of an active lesion, and was later replaced by inoculation of live cultured promastigotes [28]. This process of 'leishmanization' is still being used in Uzbekistan [29], but it can be difficult to maintain the virulence of parasites and there are issues of standardization and safety [2]. Also, the practice has mostly been limited to CL, possibly because live vaccines for VL are very challenging due to the possibility of causing a fatal disease by the vaccine. First-generation vaccines using either parasite lysates or killed parasites were developed to overcome the safety issue, and have been shown to provide protection against CL and VL in experimental models. These vaccines have also been shown to be safe and immunogenic, and in some cases exhibited partial efficacy against leishmaniasis in humans [30–32]. However, protection has been inconsistent, and standardization still remains difficult.

Issues on these historical vaccines consequently generated 2 major paths of leishmaniasis vaccine development, i.e., safer live vaccines and recombinant vaccines. Safer live vaccines may be achievable by using naturally non- or low-virulent parasites. One example is *L. donovani* in Sri Lanka that causes mainly CL and not VL; there is an attempt to utilize these less virulent parasites as a live vaccine for life-threatening VL [33]. Another example is *Leishmania tarentolae* that is considered non-pathogenic to humans. Vaccination using the parasite confers protection against *L. donovani* infection in mice [34]. Genetic engineering approaches enabling *L. tarentolae* to express defined vaccine candidates for virulent species have been explored to enhance the efficacy [35–38]. In addition to naturally occurring non- or low-virulent parasites, genetically attenuated parasites can be used as one of the choices for safer live vaccines. There are molecular targets in *Leishmania* parasites including cysteine proteinase, DHFR-TS, biopterin transporter, ornithine decarboxylase, and centrin identified as indispensable for survival in mammalian hosts [39–43]. Mutants lacking such genes that do not cause disease and will not acquire the virulence back again are proposed to be used in live vaccines [39–41, 44–47]. Recent advances in CRISPR/Cas9-based genetic manipulations for *Leishmania* parasites may accelerate development of live attenuated vaccines for leishmaniasis [48, 49].

Although these naturally or genetically attenuated parasites may offer a reasonable level of safety to inoculated people, the safety and standardization issues will be still difficult to remove completely. This is why non-parasite-based vaccines are also being studied intensively. This approach includes subunit, DNA, viral vector, and prime-boost vaccines. Among these, subunit vaccines have been most advanced toward clinical application. Subunit vaccines comprise defined antigens and appropriate adjuvants. Single antigens including HASPB, FML/NH36, SMT, KMP11, A2, and CPB as well as fusion proteins such as Leish-111f and KSAC have been reported to confer protection against VL in animal models [50–57]. In addition to the effort on antigen discovery, recent advances in adjuvant research play a key role in the future success of the clinical development of leishmaniasis vaccines. Toll-like receptor agonists are especially attractive components for the adjuvant because they have strong potency for inducing Th1 responses. In fact, CpG oligonucleotides [58–60], monophosphoryl lipid A [55, 61], and glycopyranosyl lipid adjuvant [62] have shown promising results in experimental vaccine studies for CL and VL.

A groundbreaking approach to vaccine development for leishmaniasis that is completely different from the earlier mentioned vaccines (the so called anti-amastigote vaccines), has recently come from the fact that prior exposure of mice to bites of uninfected sand flies confers powerful protection against *Leishmania* infection [63]. It has been revealed that sand fly saliva contains immunomodulatory components that exacerbate *Leishmania* infection when inoculated together with the parasites [64–66]. These results have built a concept of anti-saliva vaccines for leishmaniasis. However, the immunomodulatory components of saliva contain antigens that induce opposite effects on disease outcome when used as vaccines [67]. Therefore, studies on the saliva components have been performed to define the protective antigens such as maxadilan, SP15, and LJM19 [68–70].

8.3 Current Status of Clinical Development

As of the end of 2015, there were 14 clinical studies found in ClinicalTrials.gov when the database was searched for 'leishmaniasis' and 'vaccine' (1 was excluded due to low relevance to vaccine development). Three are for killed parasite-based vaccines, 10 are for subunit vaccines from the Infectious Disease Research Institute (IDRI), and the remaining one is related to anti-saliva vaccines. The first subunit vaccine antigen that entered clinical evaluation was LEISH-F1, which is also known as Leish-111f for preclinical use [55, 61]. Multiple phase I/II trials have been conducted in the USA, Columbia, Brazil, Peru, and India, and have demonstrated that the LEISH-F1 + MPL-SE vaccine is safe, immunogenic, and well tolerated in both naïve and pre-immune people [71–74]. LEISH-F2, which is an improved version of LEISH-F1 due to the manufacturing process and is also known as Leish-110f [62], has been also evaluated in phase I/II trails and is also safe, immunogenic, and well tolerated. IDRI, the developer of LEISH-F1 and LEISH-F2, now has developed the LEISH-F3 vaccine. LEISH-F1/Leish-111f was originally developed for CL and is a fusion of 3 antigens: TSA, LmSTI1, and LeIF, which were cloned from species responsible for CL [61]. It then became a candidate also for VL because it provided cross protection in experimental models of VL [55]. In contrast, the LEISH-F3 vaccine was designed for targeting VL from the beginning, and is comprised of 2 vaccine candidates cloned from *L. donovani* complex, i.e., NH and SMT [52, 57]. IDRI has also developed a new generation of Th1-inducing adjuvants, GLA-SE (also known as EM005), which is an oil-in-water emulsion of glucopyranosyl lipid A [75–78]. The LEISH-F3 + GLA-SE vaccine is capable of inducing Th1 responses and protection against *L. donovani* and *Leishmania infantum* in mice [79]. The vaccine has also shown its safety and immunogenicity in a phase I trial in the USA [79].

One trail is ongoing to evaluate the human immune response induced by uninfected laboratory-raised sand fly bites to select possible substances to use for a future vaccine to protect against the parasite leishmaniasis. A recent study has demonstrated that exposure to uninfected sand fly bites protects non-human primates against sand fly–transmitted CL and that PdSP15, one of the defined salivary gland antigens formulated with GLA-SE, also provides protection in the non-human primate model of CL [80]. Considering these data, we can say that clinical development of anti-saliva vaccines for leishmaniasis has become more realistic.

8.4 Conclusion/Future Direction

Although there is still no licensed vaccine for human leishmaniasis, tremendous efforts have been taken to make the vaccine a realistic possibility. Advances in target discovery as well as adjuvant development are serving as a strong driver to the success. Subunit vaccines might be an appropriate format for the first vaccine for leishmaniasis due to their relative ease for standardization. Of course, efficacy is one of the most important factors for vaccines together with safety and immunogenicity, and we cannot be too optimistic at this moment because efficacy trials have not yet been conducted. But, one of the clinical trials in patients with CL demonstrated that the LEISH-F1 + MPL-SE vaccine shortens the time to cure when used in combination with chemotherapy [72]. Therapeutic efficacy of the Leish-111f + MPL-SE vaccine has been also demonstrated in canine VL [81]. Therefore, vaccines can be relied on as one of the future control strategies for leishmaniasis. To maximize the efficacy, we believe that further studies will be necessary, especially on antigen optimization. Use of a single defined antigen as a vaccine component will be insufficient to cover populations with divergent MHC haplotypes. Also, leishmaniasis is a variety of diseases caused by a variety of different organisms. Therefore, use of multiple antigens may be a better strategy for broader coverage, and each antigen is favorable to be conserved between species and cross-protective like SMT [52, 82]. Antigen selection based not only on animal experiments but also on human immune responses may also be an important point when we aim at clinical development [83]. Recent success in anti-saliva vaccines will enable the development of multilateral vaccines, i.e., anti-amastigote/saliva vaccines controlling both the initial infection and proliferation stages simultaneously. Because GLA-SE serves as a useful adjuvant in both settings [62, 79, 80], it can be expected that vaccines like the LEISH-F1/F2/F3 + PdSP15 + GLA-SE vaccine will be the second generation of 'second-generation vaccines' for leishmaniasis.

References

1. Palatnik-de-Sousa CB. Vaccines for leishmaniasis in the fore coming 25 years. Vaccine. 2008;26:1709–24.
2. Coler RN, Reed SG. Second-generation vaccines against leishmaniasis. Trends Parasitol. 2005;21:244–9.
3. Murray HW, Rubin BY, Rothermel CD. Killing of intracellular *Leishmania donovani* by lymphokine-stimulated human mononuclear phagocytes. Evidence that interferon-gamma is the activating lymphokine. J Clin Investig. 1983;72:1506–10.
4. Nacy CA, Fortier AH, Meltzer MS, Buchmeier NA, Schreiber RD. Macrophage activation to kill *Leishmania major*: activation of macrophages for intracellular destruction of amastigotes can be induced by both recombinant interferon-gamma and non-interferon lymphokines. J Immunol. 1985;135:3505–11.
5. Scott P, Pearce E, Cheever AW, Coffman RL, Sher A. Role of cytokines and CD4+ T-cell subsets in the regulation of parasite immunity and disease. Immunol Rev. 1989;112:161–82.
6. Scott P, Natovitz P, Coffman RL, Pearce E, Sher A. Immunoregulation of cutaneous leishmaniasis. T cell lines that transfer protective immunity or exacerbation belong to different T helper subsets and respond to distinct parasite antigens. J Exp Med. 1988;168:1675–84.
7. Sadick MD, Heinzel FP, Holaday BJ, Pu RT, Dawkins RS, Locksley RM. Cure of murine leishmaniasis with anti-interleukin 4 monoclonal antibody. Evidence for a T cell-dependent, interferon gamma-independent mechanism. J Exp Med. 1990;171:115–27.
8. Chatelain R, Varkila K, Coffman RL. IL-4 induces a Th2 response in *Leishmania major*-infected mice. J Immunol. 1992;148:1182–7.

9. Leal LM, Moss DW, Kuhn R, Muller W, Liew FY. Interleukin-4 transgenic mice of resistant background are susceptible to *Leishmania major* infection. Eur J Immunol. 1993;23:566–9.
10. Kopf M, Brombacher F, Kohler G, Kienzle G, Widmann KH, Lefrang K, Humborg C, Ledermann B, Solbach W. IL-4-deficient Balb/c mice resist infection with *Leishmania major*. J Exp Med. 1996;184:1127–36.
11. Heinzel FP, Sadick MD, Holaday BJ, Coffman RL, Locksley RM. Reciprocal expression of interferon gamma or interleukin 4 during the resolution or progression of murine leishmaniasis. Evidence for expansion of distinct helper T cell subsets. J Exp Med. 1989;169:59–72.
12. Noben-Trauth, N, Kropf P. Muller I. Susceptibility to *Leishmania major* infection in interleukin-4-deficient mice. Science. 1996;271:987–990 (New York, NY).
13. Carter KC, Gallagher G, Baillie AJ, Alexander J. The induction of protective immunity to *Leishmania major* in the BALB/c mouse by interleukin 4 treatment. Eur J Immunol. 1989;19:779–82.
14. Biedermann T, Zimmermann S, Himmelrich H, Gumy A, Egeter O, Sakrauski AK, Seegmuller I, Voigt H, Launois P, Levine AD, Wagner H, Heeg K, Louis JA, Rocken M. IL-4 instructs TH1 responses and resistance to *Leishmania major* in susceptible BALB/c mice. Nat Immunol. 2001;2:1054–60.
15. Kaye PM, Curry AJ, Blackwell JM. Differential production of Th1- and Th2-derived cytokines does not determine the genetically controlled or vaccine-induced rate of cure in murine visceral leishmaniasis. J Immunol. 1991;146:2763–70.
16. Satoskar A, Bluethmann H, Alexander J. Disruption of the murine interleukin-4 gene inhibits disease progression during *Leishmania mexicana* infection but does not increase control of *Leishmania donovani* infection. Infect Immunol. 1995;63:4894–9.
17. Stager S, Alexander J, Carter KC, Brombacher F, Kaye PM. Both interleukin-4 (IL-4) and IL-4 receptor alpha signaling contribute to the development of hepatic granulomas with optimal antileishmanial activity. Infect Immunol. 2003;71:4804–7.
18. Stager S, Alexander J, Kirby AC, Botto M, Rooijen NV, Smith DF, Brombacher F, Kaye PM. Natural antibodies and complement are endogenous adjuvants for vaccine-induced CD8+ T-cell responses. Nat Med. 2003;9:1287–92.
19. Murphy ML, Wille U, Villegas EN, Hunter CA, Farrell JP. IL-10 mediates susceptibility to *Leishmania donovani* infection. Eur J Immunol. 2001;31:2848–56.
20. Wilson ME, Young BM, Davidson BL, Mente KA, McGowan SE. The importance of TGF-beta in murine visceral leishmaniasis. J Immunol. 1998;161:6148–55.
21. Barral-Netto, M, Barral A, Brownell CE, Skeiky YA, Ellingsworth LR, Twardzik DR, Reed SG. Transforming growth factor-beta in leishmanial infection: a parasite escape mechanism. Science. 1992;257:545–548 (New York, NY).
22. Ghalib HW, Piuvezam MR, Skeiky YA, Siddig M, Hashim FA, el-Hassan AM, Russo DM, Reed SG. Interleukin 10 production correlates with pathology in human *Leishmania donovani* infections. J Clin Investig. 1993;92:324–9.
23. Kane MM, Mosser DM. The role of IL-10 in promoting disease progression in leishmaniasis. J Immunol. 2001;166:1141–7.
24. Heinzel FP, Schoenhaut DS, Rerko RM, Rosser LE, Gately MK. Recombinant interleukin 12 cures mice infected with *Leishmania major*. J Exp Med. 1993;177:1505–9.
25. Sypek JP, Chung CL, Mayor SE, Subramanyam JM, Goldman SJ, Sieburth DS, Wolf SF, Schaub RG. Resolution of cutaneous leishmaniasis: interleukin 12 initiates a protective T helper type 1 immune response. J Exp Med. 1993;177:1797–802.
26. Afonso LC, Scharton TM, Vieira LQ, Wysocka M, Trinchieri G, Scott, P. The adjuvant effect of interleukin-12 in a vaccine against *Leishmania major*. Science. 1994;263:235–237 (New York, NY).
27. Murray HW, Hariprashad J. Interleukin 12 is effective treatment for an established systemic intracellular infection: experimental visceral leishmaniasis. J Exp Med. 1995;181:387–91.
28. Handman E. Leishmaniasis: current status of vaccine development. Clin Microbiol Rev. 2001;14:229–43.

29. Khamesipour A, Dowlati Y, Asilian A, Hashemi-Fesharki R, Javadi A, Noazin S, Modabber F. Leishmanization: use of an old method for evaluation of candidate vaccines against leishmaniasis. Vaccine. 2005;23:3642–8.
30. Sharifi I, FeKri AR, Aflatonian MR, Khamesipour A, Nadim A, Mousavi MR, Momeni AZ, Dowlati Y, Godal T, Zicker F, Smith PG, Modabber F. Randomised vaccine trial of single dose of killed *Leishmania major* plus BCG against anthroponotic cutaneous leishmaniasis in Bam Iran. Lancet. 1998;351:1540–3.
31. Armijos RX, Weigel MM, Aviles H, Maldonado R, Racines J. Field trial of a vaccine against New World cutaneous leishmaniasis in an at-risk child population: safety, immunogenicity, and efficacy during the first 12 months of follow-up. J Infect Dis. 1998;177:1352–7.
32. Khalil EA, El Hassan AM, Zijlstra EE, Mukhtar MM, Ghalib HW, Musa B, Ibrahim ME, Kamil AA, Elsheikh M, Babiker A, Modabber F. Autoclaved *Leishmania major* vaccine for prevention of visceral leishmaniasis: a randomised, double-blind, BCG-controlled trial in Sudan. Lancet. 2000;356:1565–9.
33. McCall LI, Zhang WW, Ranasinghe S, Matlashewski G. Leishmanization revisited: immunization with a naturally attenuated cutaneous *Leishmania donovani* isolate from Sri Lanka protects against visceral leishmaniasis. Vaccine. 2013;31:1420–5.
34. Breton M, Tremblay MJ, Ouellette M, Papadopoulou B. Live nonpathogenic parasitic vector as a candidate vaccine against visceral leishmaniasis. Infect Immunol. 2005;73:6372–82.
35. Mizbani A, Taheri T, Zahedifard F, Taslimi Y, Azizi H, Azadmanesh K, Papadopoulou B, Rafati S. Recombinant *Leishmania tarentolae* expressing the A2 virulence gene as a novel candidate vaccine against visceral leishmaniasis. Vaccine. 2009;28:53–62.
36. Shahbazi M, Zahedifard F, Taheri T, Taslimi Y, Jamshidi S, Shirian S, Mahdavi N, Hassankhani M, Daneshbod Y, Zarkesh-Esfahani SH, Papadopoulou B, Rafati S. Evaluation of live recombinant nonpathogenic *Leishmania tarentolae* expressing cysteine proteinase and A2 genes as a candidate vaccine against experimental canine visceral leishmaniasis. PLoS One. 2015;10:e0132794.
37. Zahedifard F, Gholami E, Taheri T, Taslimi Y, Doustdari F, Seyed N, Torkashvand F, Meneses C, Papadopoulou B, Kamhawi S, Valenzuela JG, Rafati S. Enhanced protective efficacy of nonpathogenic recombinant *leishmania tarentolae* expressing cysteine proteinases combined with a sand fly salivary antigen. PLoS Negl Trop Dis. 2014;8:e2751.
38. Katebi A, Gholami E, Taheri T, Zahedifard F, Habibzadeh S, Taslimi Y, Shokri F, Papadopoulou B, Kamhawi S, Valenzuela JG, Rafati S. *Leishmania tarentolae* secreting the sand fly salivary antigen PpSP15 confers protection against *Leishmania major* infection in a susceptible BALB/c mice model. Mol Immunol. 2015;67:501–11.
39. Titus RG, Gueiros-Filho FJ, de Freitas LA, Beverley SM. Development of a safe live *Leishmania* vaccine line by gene replacement. Proc Natl Acad Sci USA. 1995;92:10267–71.
40. Alexander J, Coombs GH, Mottram JC. *Leishmania mexicana* cysteine proteinase-deficient mutants have attenuated virulence for mice and potentiate a Th1 response. J Immunol. 1998;161:6794–801.
41. Papadopoulou B, Roy G, Breton M, Kundig C, Dumas C, Fillion I, Singh AK, Olivier M, Ouellette M. Reduced infectivity of a *Leishmania donovani* biopterin transporter genetic mutant and its use as an attenuated strain for vaccination. Infect Immunol. 2002;70:62–8.
42. Selvapandiyan A, Debrabant A, Duncan R, Muller J, Salotra P, Sreenivas G, Salisbury JL, Nakhasi HL. Centrin gene disruption impairs stage-specific basal body duplication and cell cycle progression in *Leishmania*. J Biol Chem. 2004;279:25703–10.
43. Boitz JM, Yates PA, Kline C, Gaur U, Wilson ME, Ullman B, Roberts SC. *Leishmania donovani* ornithine decarboxylase is indispensable for parasite survival in the mammalian host. Infect Immunol. 2009;77:756–63.
44. Selvapandiyan A, Dey R, Nylen S, Duncan R, Sacks D, Nakhasi HL. Intracellular replication-deficient *Leishmania donovani* induces long lasting protective immunity against visceral leishmaniasis. J immunol. 2009;183:1813–20.

45. Dey R, Dagur PK, Selvapandiyan A, McCoy JP, Salotra P, Duncan R, Nakhasi HL. Live attenuated *Leishmania donovani* p27 gene knockout parasites are nonpathogenic and elicit long-term protective immunity in BALB/c mice. J Immunol. 2013;190:2138–49.
46. Anand S, Madhubala R. Genetically Engineered Ascorbic acid-deficient live mutants of *Leishmania donovani* induce long lasting protective immunity against visceral leishmaniasis. Sci Rep. 2015;5:10706.
47. Fiuza JA, Gannavaram S, Santiago Hda C, Selvapandiyan A, Souza DM, Passos LS, de Mendonca LZ, Lemos-Giunchetti Dda S, Ricci ND, Bartholomeu DC, Giunchetti RC, Bueno LL, Correa-Oliveira R, Nakhasi HL, Fujiwara RT. Vaccination using live attenuated *Leishmania donovani* centrin deleted parasites induces protection in dogs against *Leishmania infantum*. Vaccine. 2015;33 280–8.
48. Sollelis L, Ghorbal M, MacPherson CR, Martins RM, Kuk N, Crobu L, Bastien P, Scherf A, Lopez-Rubio JJ, Sterkers Y. First efficient CRISPR-Cas9-mediated genome editing in *Leishmania* parasites. Cell Microbiol. 2015;17:1405–12.
49. Zhang WW, Matlashewski G. CRISPR-Cas9-Mediated Genome Editing in *Leishmania donovani*. MBio. 2015;6:e00861.
50. Stager S, Smith DF, Kaye PM. Immunization with a recombinant stage-regulated surface protein from *Leishmania donovani* induces protection against visceral leishmaniasis. J Immunol. 2000;165:7064–71.
51. Basu R, Bhaumik S, Basu JM, Naskar K, De T, Roy S. Kinetoplastid membrane protein-11 DNA vaccination induces complete protection against both pentavalent antimonial-sensitive and -resistant strains of *Leishmania donovani* that correlates with inducible nitric oxide synthase activity and IL-4 generation: evidence for mixed Th1- and Th2-like responses in visceral leishmaniasis. J Immunol. 2005;174:7160–71.
52. Goto Y, Bogatzki LY, Bertholet S, Coler RN, Reed SG. Protective immunization against visceral leishmaniasis using *Leishmania* sterol 24-c-methyltransferase formulated in adjuvant. Vaccine. 2007;25:7450–8.
53. Ghosh A, Zhang WW, Matlashewski G. Immunization with A2 protein results in a mixed Th1/Th2 and a humoral response which protects mice against *Leishmania donovani* infections. Vaccine. 2001;20:59–66.
54. Rafati S, Zahedifard F, Nazgouee F. Prime-boost vaccination using cysteine proteinases type I and II of *Leishmania infantum* confers protective immunity in murine visceral leishmaniasis. Vaccine. 2006;24:2169–75.
55. Coler RN, Goto Y, Bogatzki L, Raman V, Reed SG. Leish-111f, a recombinant polyprotein vaccine that protects against visceral Leishmaniasis by elicitation of CD(4+) T cells. Infect Immunol. 2007;75:4648–54.
56. Goto Y, Bhatia A, Raman VS, Liang H, Mohamath R, Piccone AF, Vidal SE, Vedvick TS, Howard RF, Reed SG. KSAC, the First Defined Polyprotein Vaccine Candidate for Visceral Leishmaniasis. Clin Vaccine Immunol: CVI. 2011;18:1118–24.
57. Aguilar-Be I, da Silva Zardo R, Paraguai de Souza E, Borja-Cabrera GP, Rosado-Vallado M, Mut-Martin M, Garcia-Miss Mdel R, Palatnik de Sousa CB, Dumonteil E. Cross-protective efficacy of a prophylactic *Leishmania donovani* DNA vaccine against visceral and cutaneous murine leishmaniasis. Infect Immunol. 2005;73:812–9.
58. Rhee EG, Mendez S, Shah JA, Wu CY, Kirman JR, Turon TN, Davey DF, Davis H, Klinman DM, Coler RN, Sacks DL, Seder RA. Vaccination with heat-killed leishmania antigen or recombinant leishmanial protein and CpG oligodeoxynucleotides induces long-term memory CD4+ and CD8+ T cell responses and protection against *Leishmania major* infection. J Exp Med. 2002;195:1565–73.
59. Stacey KJ, Blackwell JM. Immunostimulatory DNA as an adjuvant in vaccination against *Leishmania major*. Infect Immunol. 1999;67:3719–26.
60. Walker PS, Scharton-Kersten T, Krieg AM, Love-Homan L, Rowton ED, Udey MC, Vogel JC. Immunostimulatory oligodeoxynucleotides promote protective immunity and provide systemic therapy for leishmaniasis via IL-12- and IFN-gamma-dependent mechanisms. Proc Natl Acad Sci USA. 1999;96:6970–5.

61. Coler RN, Skeiky YA, Bernards K, Greeson K, Carter D, Cornellison CD, Modabber F, Campos-Neto A, Reed SG. Immunization with a polyprotein vaccine consisting of the T-Cell antigens thiol-specific antioxidant, *Leishmania major* stress-inducible protein 1, and *Leishmania* elongation initiation factor protects against leishmaniasis. Infect Immunol. 2002;70:4215–25.
62. Bertholet S, Goto Y, Carter L, Bhatia A, Howard RF, Carter D, Coler RN, Vedvick TS, Reed SG. Optimized subunit vaccine protects against experimental leishmaniasis. Vaccine. 2009;27:7036–45.
63. Kamhawi S, Belkaid Y, Modi G, Rowton E, Sacks D. Protection against cutaneous leishmaniasis resulting from bites of uninfected sand flies. Science. 2000;290:1351–1354 (New York, NY).
64. Bethony JM, Cole RN, Guo X, Kamhawi S, Lightowlers MW, Loukas A, Petri W, Reed S, Valenzuela JG, Hotez PJ. Vaccines to combat the neglected tropical diseases. Immunol Rev. 2011;239:237–70.
65. Titus RG, Ribeiro JM. Salivary gland lysates from the sand fly *Lutzomyia longipalpis* enhance *Leishmania* infectivity. Science. 1988;239:1306–1308 (New York, NY).
66. Belkaid Y, Kamhawi S, Modi G, Valenzuela J, Noben-Trauth N, Rowton E, Ribeiro J, Sacks DL. Development of a natural model of cutaneous leishmaniasis: powerful effects of vector saliva and saliva preexposure on the long-term outcome of *Leishmania major* infection in the mouse ear dermis. J Exp Med. 1998;188:1941–53.
67. Oliveira F, Lawyer PG, Kamhawi S, Valenzuela JG. Immunity to distinct sand fly salivary proteins primes the anti-*Leishmania* immune response towards protection or exacerbation of disease. PLoS Negl Trop Dis. 2008;2:e226.
68. Morris RV, Shoemaker CB, David JR, Lanzaro GC, Titus RG. Sandfly maxadilan exacerbates infection with *Leishmania major* and vaccinating against it protects against *L. major* infection. J Immunol. 2001;167:5226–30.
69. Valenzuela JG, Belkaid Y, Garfield MK, Mendez S, Kamhawi S, Rowton ED, Sacks DL, Ribeiro JM. Toward a defined anti-*Leishmania* vaccine targeting vector antigens: characterization of a protective salivary protein. J Exp Med. 2001;194:331–42.
70. Gomes R, Teixeira C, Teixeira MJ, Oliveira F, Menezes MJ, Silva C, de Oliveira CI, Miranda JC, Elnaiem DE, Kamhawi S, Valenzuela JG, Brodskyn CI. Immunity to a salivary protein of a sand fly vector protects against the fatal outcome of visceral leishmaniasis in a hamster model. Proc Natl Acad Sci USA. 2008;105:7845–50.
71. Velez ID, Gilchrist K, Martinez S, Ramirez-Pineda JR, Ashman JA, Alves FP, Coler RN, Bogatzki LY, Kahn SJ, Beckmann AM, Cowgill KD, Reed SG, Piazza FM. Safety and immunogenicity of a defined vaccine for the prevention of cutaneous leishmaniasis. Vaccine. 2009;28:329–37.
72. Nascimento E, Fernandes DF, Vieira EP, Campos-Neto A, Ashman JA, Alves FP, Coler RN, Bogatzki LY, Kahn SJ, Beckmann AM, Pine SO, Cowgill KD, Reed SG, Piazza FM. A clinical trial to evaluate the safety and immunogenicity of the LEISH-F1 + MPL-SE vaccine when used in combination with meglumine antimoniate for the treatment of cutaneous leishmaniasis. Vaccine. 2010;28:6581–7.
73. Llanos-Cuentas A, Calderon W, Cruz M, Ashman JA, Alves FP, Coler RN, Bogatzki LY, Bertholet S, Laughlin EM, Kahn SJ, Beckmann AM, Cowgill KD, Reed SG, Piazza FM. A clinical trial to evaluate the safety and immunogenicity of the LEISH-F1 + MPL-SE vaccine when used in combination with sodium stibogluconate for the treatment of mucosal leishmaniasis. Vaccine. 2010;28:7427–35.
74. Chakravarty J, Kumar S, Trivedi S, Rai VK, Singh A, Ashman JA, Laughlin EM, Coler RN, Kahn SJ, Beckmann AM, Cowgill KD, Reed SG, Sundar S, Piazza FM. A clinical trial to evaluate the safety and immunogenicity of the LEISH-F1 + MPL-SE vaccine for use in the prevention of visceral leishmaniasis. Vaccine. 2011;29:3531–7.
75. Bertholet S, Ireton GC, Ordway DJ, Windish HP, Pine SO, Kahn M, Phan T, Orme IM, Vedvick TS, Baldwin SL, Coler RN, Reed SG. A defined tuberculosis vaccine candidate

boosts BCG and protects against multidrug-resistant Mycobacterium tuberculosis. Sci Trans Med 2010;2:53ra74.
76. Anderson RC, Fox CB, Dutill TS, Shaverdian N, Evers TL, Poshusta GR, Chesko J, Coler RN, Friede M, Reed SG, Vedvick TS. Physicochemical characterization and biological activity of synthetic TLR4 agonist formulations. Colloids Surf B, Biointerfaces. 2010;75:123–32.
77. Baldwin SL, Shaverdian N, Goto Y, Duthie MS, Raman VS, Evers T, Mompoint F, Vedvick TS, Bertholet S, Coler RN, Reed SG. Enhanced humoral and Type 1 cellular immune responses with Fluzone adjuvanted with a synthetic TLR4 agonist formulated in an emulsion. Vaccine. 2009;27:5956–63.
78. Coler RN, Bertholet S, Moutaftsi M, Guderian JA, Windish HP, Baldwin SL, Laughlin EM, Duthie MS, Fox CB, Carter D, Friede M, Vedvick TS, Reed SG. Development and characterization of synthetic glucopyranosyl lipid adjuvant system as a vaccine adjuvant. PLoS One. 2011;6:e16333.
79. Coler RN, Duthie MS, Hofmeyer KA, Guderian J, Jayashankar L, Vergara J, Rolf T, Misquith A, Laurance JD, Raman VS, Bailor HR, Cauwelaert ND, Reed SJ, Vallur A, Favila M, Orr MT, Ashman J, Ghosh P, Mondal D, Reed SG. From mouse to man: safety, immunogenicity and efficacy of a candidate leishmaniasis vaccine LEISH-F3 + GLA-SE. Clin Transl Immunol. 2015;4:e35.
80. Oliveira F, Rowton E, Aslan H, Gomes R, Castrovinci PA, Alvarenga PH, Abdeladhim M, Teixeira C, Meneses C, Kleeman LT, Guimaraes-Costa AB, Rowland TE, Gilmore D, Doumbia S, Reed SG, Lawyer PG, Andersen JF, Kamhawi S, Valenzuela JG. A sand fly salivary protein vaccine shows efficacy against vector-transmitted cutaneous leishmaniasis in nonhuman primates. Sci Trans Med 2015;7:290ra290.
81. Trigo J, Abbehusen M, Netto EM, Nakatani M, Pedral-Sampaio G, de Jesus RS, Goto Y, Guderian J, Howard RF, Reed SG. Treatment of canine visceral leishmaniasis by the vaccine Leish-111f + MPL-SE. Vaccine. 2010;28:3333–40.
82. Goto Y, Bhatia A, Raman VS, Vidal SE, Bertholet S, Coler RN, Howard RF, Reed SG. *Leishmania infantum* sterol 24-c-methyltransferase formulated with MPL-SE induces cross-protection against *L. major* infection. Vaccine. 2009;27:2884–90.
83. Kumar R, Goto Y, Gidwani K, Cowgill KD, Sundar S, Reed SG. Evaluation of *ex vivo* human immune response against candidate antigens for a visceral leishmaniasis vaccine. Am J Trop Med Hyg. 2010;82:808–13.

Chapter 9
Siccanin Is a Novel Selective Inhibitor of Trypanosomatid Complex II (Succinate-Ubiquinone Reductase) and a Potent Broad-Spectrum Anti-trypanosomatid Drug Candidate

Nozomu Nihashi, Daniel Ken Inaoka, Chiaki Tsuge, Emmanuel Oluwadare Balogun, Yasutaka Osada, Yasuyuki Goto, Yoshitsugu Matsumoto, Takeshi Nara, Tatsushi Mogi, Shigeharu Harada and Kita Kiyoshi

Abstract Trypanosomiasis and leishmaniasis, neglected tropical diseases caused by protozoan trypanosomatid parasites, result in heavy socioeconomic impact. Currently available drugs are old with, narrow spectrum, and unreliable. We recently reported that the pathogen of American trypanosomiasis, *Trypanosoma cruzi*, has a noncanonical mitochondrial complex II (succinate-quinone oxidoreductase: SQR) that comprises 12-subunits and displays poor sensitivities to mammalian SQR inhibitors. Because their orthologs are only conserved in the genome of trypanosomatids, the SQR is a drug target candidate. Herein, we searched for selective inhibitors of trypanosomatid SQR by using the nonpathogenic trypanosomatid *Leishmania tarentolae* enzyme as model. Nano LC-MS/MS and biochemical studies revealed the SQRs of *L. tarentolae* and *T. cruzi* have similar subunits composition and sensitivity to established quinone-site SQR inhibitors. Further, we performed screening for selective LtSQR inhibitors and found that the old fungicidal compound siccanin, with an IC_{50} value of 190 nM, is presently the most potent trypanosomatid SQR inhibitor. Interestingly, it showed negligible inhibition against porcine SQR (selectivity index: 4,500-fold) but displayed mixed-type inhibition against quinone with K_{i1} and K_{i2} of 39 and 102 nM, respectively. Notably, siccanin specifically targets complex II and spares other quinone-utilizing enzymes of the respiratory chain. Siccanin is not species-specific; it inhibited SQR of *T. cruzi*, *T. brucei*, and *L. donovani* with IC_{50} of up to 0.368 µM. Remarkably, it caused in vitro growth inhibition across the clinically

K. Kiyoshi (✉)
Department of Biomedical Chemistry, Graduate School of Medicine,
The University of Tokyo, 7-3-1 Hongo, Bunkyo-ku, Tokyo 113-0033, Japan;
School of Tropical Medicine and Global Health, Nagasaki University,
Nagasaki 852-8523, Japan
e-mail: kitak@m.u-tokyo.ac.jp

relevant forms of the major trypanosomatid parasites at IC_{50} of 0.7–13 µM, and hence is a potential broad-spectrum drug candidate against trypanosomatids.

Keywords Respiratory chain · Inhibitor · Drug target · Trypanosomatids · Siccanin

9.1 Introduction

Trypanosomatids are monoflagelated protozoans that are of immense public health concern for being the causative agents of debilitating and potentially fatal human infectious diseases. These diseases include: African trypanosomiasis (known as sleeping sickness; caused by subspecies of *Trypanosoma brucei*), American trypanosomiasis (known as Chagas disease; caused by *Trypanosoma cruzi*), and leishmaniasis (caused by *Leishmania* spp.). The World Health Organization (WHO) estimates of annual infected patients stand at 30,000 people for sleeping sickness, 10 million people for Chagas disease, and 350 million people for leishmaniasis [1–3]. Owing to the lack of vaccines, chemotherapy is the major means for controlling these diseases. However, the currently used drugs such as pentamidine, benznidazole, pentavalent antimonials, and nifurtimox have severe side effects and limited efficacy. In addition, most of the drugs are species-specific and may not be helpful in the event of mixed infection, which is a common occurrence in endemic areas [3]. A recent study also indicated that drug-resistant strains to standard chemotherapies have been emerging in Africa [4–6]. Together, these situations prompt the need for the development of new and safe drugs that are effective across the pathogenic trypanosomatid organisms, and that possess a novel mechanism of action.

Mitochondria of trypanosomatids have energy metabolism systems that are distinct from those of the mammalian hosts, and hence are attractive targets of chemotherapy. As a whole, they lack the complete tricarboxylic acid (TCA) cycle and secrete acetate and succinate as endproducts [7–9]. In their respiratory chain, the canonical proton-pumping complex I is replaced by single-subunit NADH dehydrogenase (NDH2). Other respiratory chain complexes are present in trypanosomatids except for the blood stream form of *T. brucei*, which lacks the functional complex III and IV and produces ATP mainly via glycolysis, and re-oxidize NADH by using trypanosoma alternative oxidase (TAO). The leishmanicidal compounds miltefosine [10] and tafemoquine [11] target complex IV and III, respectively, validating the parasites respiratory chain enzyme complexes as potent drug targets. Interestingly, our recent characterization of respiratory chain complex II from *T. cruzi* shows that this enzyme has a subunit structure and inhibitor sensitivity that are quite different from the mammalian counterpart [12].

Mitochondrial respiratory chain complex II, also called succinate-ubiquinone oxidoreductase (SQR), is the enzyme that participates in both the respiratory chain and TCA cycle through electron transfer between succinate and ubiquinone. Usually this reaction is mediated by 4 catalytic subunits named SDH1 (Fp), SDH2 (Ip), SDH3 (CybL) and SDH4 (CybS) [13]. Fp is a hydrophilic subunit containing the flavin adenine dinucleotide (FAD) co-factor, which withdraws 2 electrons from succinate in the mitochondrial matrix. The electrons are transferred by the Ip subunit (another hydrophilic component of SQR) from FAD to ubiquinone through 3 iron sulfur clusters [2Fe-2S], [4Fe-4S], and [3Fe-4S]. The heterodimeric hydrophobic anchor (CybL and CybS) subunits with 6 transmembrane helices form the quinone-binding site in the vicinity of Ip [3Fe-4S].

As in other mitochondrial complexes, the subunit structure of SQR is divergent across organisms. For example, some Gram-positive bacteria have 3-subunit SQR with fused CybL and CybS as a membrane anchor [14], while SQR from plants exemplified by *Arabidopsis* is comprised of 8 subunits, 4 of which are plant-specific polypeptides with unknown functions [15–17]. An extreme case is the SQR from *T. cruzi* (TcSQR), the pathogen of Chagas disease, which comprises 6 hydrophilic and 6 hydrophobic subunits with heterodimeric Ip subunits [12]. TcSQR shows low sensitivity to known SQR inhibitors such as malonate, carboxin, and atpenin A5, suggesting its unusual inhibitor-binding mode. Notably, genomic analysis indicated that subunit compositions of SQR are conserved within pathogenic trypanosomatids such as *T. brucei*, *T. cruzi*, *Leishmania donovani*, and *L. major*. A recent study also showed that *Euglena gracilis*, phylogenetically akin to trypanosomatid, has unusual subunit-structure of SQR [18].

In this study, we searched for a highly selective inhibitor of trypanosomatid SQR and checked its trypanocidal effect on *T. cruzi*, *T. brucei*, *L. donovani*, and *L. major* in vitro. For the screening of SQR inhibitors, nonpathogenic trypanosomatid *L. tarentolae* was used as model organism, because large-scale preparation of SQR is difficult in pathogenic trypanosomatids. *L. tarentolae* is a parasite isolated from gecko lizards, and used as a model organism to study RNA editing systems among trypanosomatids [19]. Because the safety and large scale culture of this parasite has already been established [20], *L. tarentolae* is potentially suitable for enzyme preparation in a large amount. *L. tarentolae* has also been utilized in a screening system for leishmaniacidal compounds [21–24].

Our genomic analysis and MS/MS analysis of partially purified enzyme indicated that trypanosomatid SQR are also conserved in *L. tarentolae* at both the genomic and proteomic level. Two-dimensional native electrophoresis of partially purified LtSQR detected 12 bands, 8 of which were confirmed as LtSQR subunits by LC-MS/MS. Further, we compared sensitivities of trypanosomatid SQR against 5 classical SQR inhibitors including malonate, carboxin, TTFA, HQNO, and atpenin A5. The IC_{50} of these compounds were comparable between LtSQR and pathogenic trypanotomatids. Subsequently, we investigated the effects of 11 classical and nonclassical respiratory chain enzyme inhibitors on LtSQR, and found that a fungal SQR inhibitor siccanin inhibited LtSQR at a sub-micromolar order (190 nM). Siccanin was originally known for its fungicidal effect [25, 26] and clinically used as a drug for skin infection (US Patent 3974291). Interestingly, the IC_{50} of siccanin against porcine SQR was 861 μM [27], accounting for a difference in sensitivity that is approximately 4,500 times higher than that of LtSQR. Further, herein, it is shown that siccanin inhibited the activity of SQR of *T. brucei*, *T. cruzi*, and *L. donovani*, and killed *L. tarentolae*, *T. brucei*, *T. cruzi*, *L. donovani*, and *L. major* in vitro with IC_{50} values ranging between 0.7 and 13 μM.

9.2 Materials and Methods

Siccanin and its derivatives were kindly provided by Daiichi-Sankyo Co., Ltd. and Dr. Kazuro Shiomi, Kitasato University. Promastigotes of *L. major* Friedlin strain were provided by Dr. Steven Reed, Infectious Disease Research Institute, Seattle, USA. Promastigotes of *L. donovani* D10 were provided by the National BioResource Project, Institute of Tropical Medicine, Nagasaki University.

9.3 Preparation of *L. tarentolae* Mitochondria

L. tarentolae T7-TR strain promastigote was cultured as previously described with slight modification [20]. Briefly, parasites were aerobically cultured in 24 g/L yeast extract (YE) medium supplemented with 65 mM D-glucose, 71 mM K_2HPO_4, 17 mM KH_2PO_4, 50 U/mL penicillin, 50 μg/mL streptomycin, 10 μg/mL hemin, and 100 μg/mL hygromycin B. Glycerol stock of *L. tarentolae* was reactivated in 5 mL medium, and the culture was step-wisely scaled up at early-log phase to 10 L by mixing a 1 L of final pre-culture with 9 L of fresh YE medium in a BIOFLO 2000 Fermentor (New Brunswick Scientific, NJ, USA). After cultivation at 37 °C, 200 rpm, and maximum aeration, cells were harvested at late-log phase by centrifugation at $2,000 \times g$ for 60 min at 4 °C. The pellet was suspended in 20 mM sodium phosphate buffer (pH 8.0) containing 150 mM NaCl and 20 mM glucose, and centrifuged at $2,000 \times g$ for 60 min at 4 °C. Washed cells were resuspended to a density of 3×10^{12} cells/mL in lysis buffer [25 mM Tris-HCl (pH 7.6) containing 1 mM disodium malonate, 200 mM sodium orotate, 1 mM sodium EDTA, 0.25 M sucrose, and 1 bottle/L Protease Inhibitor Cocktail for general use (Sigma-Aldrich, USA)], and disrupted with a French pressure cell at 75 MPa. Homogenate was centrifuged at $50 \times g$ for 30 min at 4 °C to remove cell debris. Supernatant was centrifuged at $1,000 \times g$ for 30 min, $10,000 \times g$ for 40 min, and $33,000 \times g$ for 40 min at 4 °C. The pellet obtained after each centrifugation step was resuspended in a minimal volume of lysis buffer. For each fraction, SQR activity was measured before storage at −80 °C until required. The $1,000 \times g$ pellet possessed the highest SQR activity, hence, was taken as the mitochondrial fraction.

9.4 Preparation of *L. donovani* Mitochondria

Promastigotes of *L. donovani* D10 (provided by Institute of Tropical Medicine, Nagasaki University) were cultured in medium 199 (containing 25 mM HEPES, Life Technologies) supplemented with 10 % heat-inactivated fetal bovine serum, 100 U/mL penicillin, and 100 µg/mL of streptomycin at 25 °C. The culture was scaled up to 1 L and harvested at late-log phase. The subsequent steps were as described for *L. tarentolae* mitochondria preparation except that the cell density in the lysis buffer was 7.5×10^{11} cells/mL and the initial centrifugation at $50 \times g$ was for 15 min at 4 °C. The mitochondria-rich fraction was isolated by centrifugation at $1{,}000 \times g$ for 30 min at 4 °C and suspended in lysis buffer. This sample was frozen in liquid nitrogen and stored at −80 °C until needed.

9.5 Preparation of *T. b. brucei* Mitochondria

In vitro cultivation and isolation of the *T. b. brucei* procyclic form (GUTat 3.1) was performed as previously described [28]. The harvested parasite cells were kept on ice throughout the subsequent steps. Two milliliters of packed cells and an equal volume of glass beads (diameter 0.10–0.11 mm, B. Braun Biotech International GmbH, Germany) were suspended in 4 mL of 50 mM HEPES buffer (pH 7.2) containing 0.27 M sucrose, 1 mM EDTA, 1 mM $MgCl_2$, and 1 mM PMSF, then shaken vigorously with a Vortex mixer for 2 min. The resultant homogenate was centrifuged to remove the debris, and a cell-free extract was obtained by centrifugation at $900 \times g$ for 5 min, after which crude mitochondria were precipitated by centrifugation at $12{,}000 \times g$ for 10 min. The precipitate was suspended in 10 mM potassium phosphate buffer (pH 7.5) containing 0.25 M sucrose and 1 mM EDTA (wash buffer), and centrifuged at $12{,}000 \times g$ for 10 min. The mitochondrial fraction was suspended in wash buffer to a protein concentration of 3.4 mg/mL and stored at −80 °C until required.

9.6 Preparation of *T. cruzi* Mitochondria

In vitro cultivation and mitochondria isolation of *T. cruzi* Tulahuen strain epimastigote was performed as previously described [12]. Briefly, parasites were grown to a density of $6\text{–}8 \times 10^7$ cells/mL and washed by rounds of centrifugation and resuspension in 20 mM Tris-HCl (pH 7.2) buffer containing 10 mM NaH_2PO_4, 1 mM sodium EDTA, 1 mM dithiothreitol, 0.225 M sucrose, 20 mM KCl, and 5 mM $MgCl_2$. Cells were disrupted by grinding resuspended pellet with silicon carbide (Carborundum 440 mesh; Nacalai Tesque, Kyoto, Japan) in the presence of a minimum volume of the lysis buffer [25 mM Tris-HCl (pH 7.6), 1 mM dithiothreitol, 1 mM sodium EDTA, 0.25 M sucrose, and complete, EDTA-free Protease Inhibitor Cocktail (Roche Applied Science)]. The resultant homogenate was centrifuged at 500 × g for 5 min and 1,000 × g for 15 min to remove silicon carbide and nuclear fraction, respectively. The mitochondrial fraction was recovered upon centrifugation of the last supernatant at 10,000 × g for 15 min, washed 3 times in the lysis buffer, and stored at −80 °C until required.

9.7 Preparation of Porcine Mitochondria

The mitochondria fraction of porcine was prepared as previously described [29]. Briefly, 10 porcine hearts were sliced and a total of 2,324 g of the mince was suspended in 7 L of 20 mM sodium phosphate buffer (pH 7.4). The porcine heart suspension was homogenized by high-capacity, high-speed laboratory blender (Waring) for 90 s and adjusted to pH 7.2–7.4 with 6 M NaOH. The homogenate was centrifuged at 3,000 × g for 20 min at 4 °C and the resultant supernatant was re-centrifuged at 18,000 × g for 60 min at 4 °C. The pellet was suspended in 50 mM Tris-HCl (pH 8.0) and centrifuged at 120,000 × g for 40 min at 4 °C. The resultant mitochondria-rich fraction (pellet) was suspended in 100 mM borate-phosphate buffer pH 7.2 and stored at −80 °C until use.

9.8 Partial Purification of LtSQR

All purification steps were performed at 4 °C. Fifty milligrams of mitochondria fraction were suspended in a solubilization buffer composed of 10 mM sodium phosphate buffer (pH 8.0), 1 mM EDTA, 1 mM disodium malonate, 20 % (v/v) glycerol, 1 bottle/L Protease Inhibitor Cocktail for general use plus 2 % (w/v) sucrose monolaurate (SML; Mitsubishi Food, Japan) to a final protein concentration of 20 mg/mL and gently agitated for 60 min on Rotator RT-50 (TAITEC). The mixture was ultracentrifuged at 198,000 × g for 60 min. Supernatant was mixed with the equal volume of 10 % (w/v) polyethylene glycol (PEG) 3350 solution and incubated for 30 min at 4 °C before centrifugation at 9,000 × g for 15 min. The resulting pellet was suspended in 2 mL of solubilization buffer plus 0.1 % (w/v) SML and loaded onto DEAE Sepharose Fast Flow column (1 × 3 cm; GE Healthcare) using Econo Gradient Pump (Biorad), at a flow rate of 0.2 mL/min. After 10 columns volume wash at 1.0 mL/min with solubilization buffer plus 0.1 % (w/v) SML, proteins were eluted by 50 columns volume of solubilization buffer plus 0.1 % (w/v) n-dodecyl-D-maltoside (DDM; Dojindo). Flow though and eluted fractions were collected at 1 mL/tube, and absorbance was monitored at 280, 412, and 450 nm. SQR activity was measured in each fraction; active fractions were subjected to high resolution clear native electrophoresis (hrCNE). Fractions with similar hrCNE band pattern were pooled and concentrated by Amicon Ultra-15 (100 kDa cut off). The concentrated samples were stored in 50 % (v/v) glycerol at −20 °C.

9.9 High Resolution Clear Native Electrophoresis and LC-MS/MS

Protein solutions were mixed with 50 % (v/v) glycerol plus 0.1 % (w/v) ponceau S at the ratio 8:1, and subjected to hrCNE [30] on 4–16 % NativePAGE Novex Gel (Invitrogen) using hrCNE cathode buffer [50 mM tricine, 7.5 mM imidazole/HCl (pH 7.0), 0.02 % (w/v) DDM and 0.05 % (w/v) sodium deoxycholate] and anode buffer [25 mM imidazole/HCl (pH 7.0)]. SQR in the various samples was visualized by in-gel succinate dehydrogenase (SDH) activity staining [31] or Coomassie Brilliant Blue staining. Activity-stained bands in native gel were cut out and loaded onto 2-D 12.5 % tricine SDS-PAGE gel [32]. Each protein band in the tricine-PAGE gel was analyzed with a nano LC-MS/MS system (Wako, Japan) for detection of SQR subunits.

9.10 Enzyme Assay

Enzymatic activity measurements for quinone-dependent respiratory chain enzymes-type 2 NADH: quinone reductase (NDH2), succinate: quinone reductase (SQR; complex II), glycerol 3-phosphate dehydrogenase (G3PDH), and cytochrome c oxidase (complex IV) were performed for each mitochondrial preparation in a 1 mL quartz cuvette containing 1 mM $MgCl_2$ in a 50 mM potassium phosphate buffer (pH 7.5) as follows: NDH2; 50 μM oxidized cytochrome c and 1 mM KCN were added to the reaction buffer, reaction was initiated by addition of 50 μM NADH (Sigma), and the NADH-cytochrome c oxidase activity was measured by following NADH oxidation (ε_{340} = 6.3 mM^{-1} cm^{-1}). Complex II SQR activity was measured in the presence of 60 μM each of 2,4-dichlorophenolindophenol (DCIP) and ubiquinone-1 (UQ_1), and the rate of DCIP reduction was monitored (ε_{600} = 21 mM^{-1} cm^{-1}) after the addition of 10 mM sodium succinate. Complex II SDH activity was determined in 50 mM Tris-HCl buffer (pH 8.0) containing 120 μg/mL each of phenazinium methylsulfate (PMS) and 3-(4, 5-dimethyl-2-thiazolyl)-2, 5-diphenyl-2H-tetrazolium bromide (MTT) by following the rate of PMS-mediated reduction of MTT (ε_{570} = 17 mM^{-1} cm^{-1}), after the addition of 10 mM sodium succinate. For G3PDH, in a reaction mixture containing 20 mM glycerol 3-phosphate (G3P), 50 μM oxidized cytochrome c, and 1 mM KCN, the cytochrome c reduction (ε_{550} = 21 mM^{-1} cm^{-1}) was monitored. The reaction was initiated by addition of 10 mM of G3P. Each assay was performed after initial incubation at 25 °C for 300 s before addition of initiating substrate. Measurements were taken with V-650 UV-Vis Spectrophotometer (JASCO, Tokyo, Japan).

9.11 Identification of Trypanosomatids Complex II Inhibitors

Inhibitory effects of chemical compounds in our library of respiratory chain inhibitors were assessed on the SQR activity of trypanosomatid complex II. The screening assays were carried out using a 96-well plate SQR assay ($n = 3$). For determination of half maximal inhibitory concentration (IC_{50}), ~800 μg/mL of mitochondria-rich fraction (final concentration) from *T. cruzi* epimastigote or *T. b. brucei* procyclic form, *L. tarentolae* promastigote, *L. donovani* promastigote, and porcine heart mitochondria were suspended in 193 μL (per well) of reaction mixture containing 50 mM potassium phosphate buffer (pH 7.5), 60 μM DCIP, and 60 μM ubiquinone-1 (UQ_1). Serial concentrations of inhibitors (malonate, thenoyltrifluoroacetone [TTFA] [Sigma-Aldrich, USA], 2-heptyl-4-hydroxyquinoline 1-oxide [HQNO] [Wako, Japan]), and atpenin A5 (Enzo Life Science, USA) were dissolved in solvents, and 2 μL of inhibitor solution was added to each well. Alternatively, quinone-type respiratory chain inhibitors in the library (antimycin A [Sigma-Aldrich, USA], ascofuranone, atovaquone [USP, USA], ferulenol [Enzo Life Science, USA], fluopyram [Sigma-Aldrich Japan, Japan], Ku-55933 [Abcam, UK], nitazoxanide, nifurtimox, resveratrol, rotenone [Wako, Japan], or siccanin) was added into reaction mixtures containing *L. tarentolae* mitochondria to a final concentration of 0.025–50 μM. After mixing (750 rpm vortexing) for 60 s at room temperature, the SQR reaction was initiated by addition of 5 μL of 250 mM disodium succinate per well.

Of the inhibitors screened, siccanin was found to be most potent and its inhibitory mechanism on trypanosomatid complex II was determined. *L. tarentolae* mitochondrial membrane were suspended in 100 μL of reaction mixture containing 50 mM potassium phosphate buffer (pH 7.5), 60 μM DCIP, 0.5 % (v/v) dimethyl sulfoxide (DMSO) and 0, 0.25, 0.5, and 1 μM of siccanin. Serial concentrations of UQ_1 were prepared, covering a range of 8 points from 0.050 to 100 μM in a 96-well plate. After mixing with Mixmate (Eppendorf) at 750 rpm for 60 s, the plate was incubated for 300 s at 25 °C. The reaction was started by the addition of 100 μL of reaction mixture containing 20 mM disodium succinate, and enzymatic activity was measured as the time rate of DCIP reduction for 600 s.

9.12 In vitro Anti-leishmanial Assay of Siccanin

The effect of siccanin on the growth of promastigotes of *L. tarentolae*, *L. major*, and *L. donovani* was evaluated using the AlamarBlue method [33]. *L. tarentolae* was maintained in YE medium at 26 °C, and 50 μL of varying siccanin concentrations were prepared to cover a range of 10 points from 0.05 to 50 μM in a 96-well plate ($n = 3$). Fifty microliters of culture medium containing 2×10^6 promastigotes was mixed with the inhibitor dilutions and incubated at 26 °C for 72 h. To access the viability of *L. tarentolae* promastigotes, 5 μL AlamarBlue

(TREK Diagnostic Systems) was added to each well. The plate was incubated for 3 h and the absorbance difference was measured at 570 and 600 nm using SpectraMax Microplate Reader (Molecular Devices) DMSO and 50 µM pentamidine (Sigma-Aldrich) were used as negative and positive controls, respectively. Growth inhibition rate (%) was calculated from $[100 - (Abs_{570} - Abs_{600})/(Abs_{570DMSOcontrol} - Abs_{600DMSOcontrol}) \times 100]$.

For *L. major* and *L. donovani*, promastigotes were maintained in medium 199 (Life Technologies) supplemented with 20 and 10 % (v/v) respectively, of heat-inactivated fetal bovine serum (Thermo Fisher Scientific), and 100 U/mL penicillin and 100 µg/mL of streptomycin (Life Technologies), at 25 °C. Analysis for siccanin sensitivity was performed with a culture containing 2×10^5 cells in each well of the microplate and in the presence of siccanin (ranging from 0.05 to 100 µM) or amphotericin B (Sigma-Aldrich, ranging from 0.000139 to 11 µM). After 3 days, 10 µL of AlamarBlue was added to each well and the fluorescence intensity was measured using the SpectraMax Paradigm microplate reader (Molecular Devices) with an excitation wavelength of 555 nm and the emission wavelength of 595 nm, at the time of addition as well as after 4 h incubation. The growth inhibition rate (%) was calculated as $[100 - (FI_{test}$ at 4 h $- FI_{test}$ at 0 h$)/(FI_{control}$ at 4 h $- FI_{control}$ at 0 h$) \times 100]$. One percent DMSO was used as the control for siccanin, and medium without drug was used as the control for amphotericin B. IC_{50} values of drugs were calculated using GraphPad Prism5 (GraphPad Software).

9.13 In vitro Drug Assay for *T. b. brucei*

The blood stream form of *T. b. brucei* was propagated in HMI-9 medium at 37 °C under an atmosphere of 5 % CO_2. A stock of siccanin was dissolved in DMSO and 60 µL of five fold dilutions were transferred in triplicates to HMI-9 medium in a 96-well plate to arrive at final concentrations of 0.05 to 90 µM in 100 µL. Forty microliters of cell culture containing 5×10^4 parasites was added per well. The final concentration of DMSO was 0.5 % (w/v) in 100 µL of HMI-9. After 18 h incubation at 37 °C, 10 µL of AlamarBlue was added to each well and incubated for a further 6 h. Parasite growth after a 24 h incubation period was determined according to the change in color of the AlamarBlue. Measurement of absorbance was performed with Spectra-Max M2e-TUY Plate Reader (Molecular Devices, Japan) at wavelength values of 570 and 600 nm.

9.14 Inhibitory Effects of Siccanin on *T. cruzi* Infection and Growth in Mammalian Cells

In vitro infection of the mouse 3T3-SWISS Albino fibroblast cells (JCRB9019, Japanese Collection of Research Bioresources Cell Bank, National Institute of Biomedical Innovation, Japan) by trypomastigotes of *T. cruzi* Tulahuen stock was as described elsewhere [34]. A round coverslip was placed into the bottom of each well of a 24-well plate. Exponentially growing 3T3-SWISS Albino cells (5×10^3 cells in 1 mL per well) were transferred to 24-well plates, followed by incubation for 2 days, inoculation with 1×10^5 trypomastigotes, and the addition of siccanin in DMSO. For negative and positive controls, DMSO and 1 μM benznidazole (final concentration) were used, respectively. Three to four days later, the host cells attached to the coverslip were washed 3 times with PBS, pH 7.2, fixed with methanol, and stained with Diff-Quick Dyeing (Kokusai-Shiyaku Co.). The percentage of the infected cells and the average number of amastigotes per infected cell were determined microscopically. More than 200 host cells were analyzed per 6 randomly selected microscopic fields. Statistical analysis between the groups was performed using one-way ANOVA and Fisher's PLSD post hoc test.

9.15 Results

9.15.1 Purification and Characterization of LtSQR

As the purification procedure for TcSQR[12] was not compatible for purifying LtSQR, a new protocol for LtSQR purification was therefore developed. We first solubilized LtSQR from 50 mg of mitochondrial membranes with 2 % (w/v) of SML and added 2 % (w/v) of PEG3350 to remove hydrophilic proteins. Subsequently, concentration of PEG3350 in supernatant was increased to 10 % (w/v) and the resultant red precipitate was re-suspended in 0.1 % (w/v) SML. The suspension was applied to a Fractogel-DEAE column (Merk Millipore) and the detergent was exchanged from SML to DDM for specific elution of complex II. LtSQR activity was mainly eluted during the detergent exchange step. The fractions were collected and concentrated as the SQR peak fraction. Although the CBB-stained hrCNE gel of the SQR revealed an impurity band that migrated as a 140 kDa protein, with this purification protocol the complex II was remarkably purified (Fig. 9.1, lane A), and with the last step (ion exchange chromatography on DEAE-Sepharose Fast Flow)) furnishing a specific activity (2.65 U) that corresponded to 23-fold of purification (Table 9.1). The protein yield from 8 L of culture was 2.9 g, which is over 14-folds of that obtainable from 10 L of *T. cruzi* culture; the specific activity was also significantly higher (Table 9.2), indicating that *L. tarentolae* is a superior system for large-scale preparation of trypanosomatid SQR. Moreover, the estimated size, 520 kDa for LtSQR showed that it is comparable to that of TcSQR [12]. To assess the subunit composition and structure of LtSQR, the enzyme was subjected to Tricine SDS-PAGE after hrCNE. Results revealed that as was reported for *T. cruzi*, the LtSQR is made up of 12 protein subunits of sizes ranging from 9–69 kDa (Fig. 9.2). Each protein band from the SDS-PAGE was analyzed by LC-MS/MS and all of them were identified and annotated as SQR subunits (Fig. 9.2). Together, these data indicate that structure of SQR is also conserved between *L. tarentolae* and *T. cruzi* at the protein level.

To validate the SQR of non-pathogenic *L. tarentolae* as a model for the pathogenic trypanosomatid enzyme, we first analyzed the whole genome sequence of the *L. tarentolae* T7-TR strain and identified orthologs for all the 12 subunits of the SQR (Table 9.3). Identities of the gene sequence of LtSQR orthologs are high for SQR of *T. cruzi* (42–90 % [35]) and *T. brucei* (45–90 % [36]) and almost same as for those of *L. major* and *L. donovani* (81–98 % [37]). The di-carbonate-binding site and FAD-binding histidine in Fp, iron-sulfur cluster-binding motifs in Ip, "$PX_{11}SX_2HR$" quinone-binding motif, and helix I in CybL, Helix V, IV, and heme-binding histidine in CybS [12] are all conserved in LtSQR. These features indicate that based on genomic evaluations, the substrate binding sites of SQR are expected to be similar in both *L. tarentolae* and the pathogenic trypanosomatids. However, it is not often the case with the trypanosomatids because of their well-characterized polycistronic RNA editing machinery, which often results in non-correlated data between actual peptide sequences and those expected from the genomic information [38]. Although the subunit-structure of *T. cruzi* and *T. brucei* complex II has been well studied [12, 39], identification of SQR subunits in *Leishmania* spp. at the protein level has been unsuccessful to date [40], hence the need for the subunits peptide analysis.

9.15.2 Inhibitor Screening Against the 12-Subunit SQR

At first the susceptibility of LtSQR to the standard SQR inhibitors (malonate, atpenin A5, carboxin, TTFA, and HQNO) was evaluated. Judging from the calculated IC_{50} values of as high as 200 μM (Table 9.4), the LtSQR is not so sensitive to these established inhibitors that are known to inhibit the mammalian enzyme

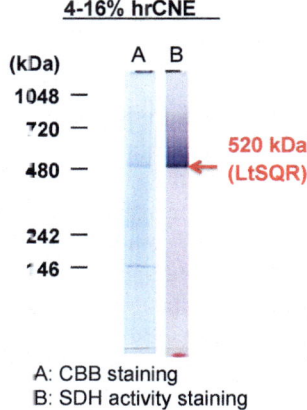

Fig. 9.1 High resolution clear native electrophoresis (hrCNE) of partially purified LtSQR; 20 μg of partially purified LtSQR was loaded onto 4–16 % Novex Native gel and visualized by CBB staining (*left*) or SDH-activity staining (*right*)

Table 9.1 Purification of LtSQR

Purification	Protein (mg)	Yield (%)	Specific activity (U/mg)
Mitochondria	50	100	0.113
PEG precipitation	6.5	50	0.356
DEAE-Sepharose fast flow	0.07	4.2	2.65 (23-fold)

Table 9.2 Preparation of mitochondria from 8 L culture of *L. tarentolae* T7-TR promastigote and 10 L culture of *T. cruzi* epimastigote. U: μmol/minute

Medium	Protein (g)	SQR total activity (U)	SQR specific activity (mU/mg)
T. cruzi 10 L LIT medium	0.31	27	85
L. tarentolae 8 L YE medium	2.9	340	113

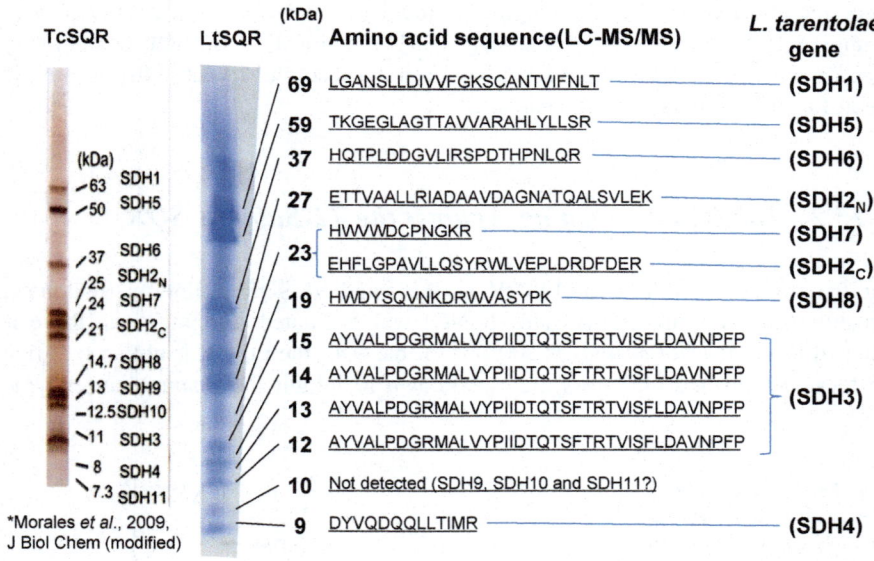

Fig. 9.2 Amino acid sequences of the peptides analyzed by mass spectrometry of subunits isolated from partially purified LtSQR

perfectly. For instance, atpenin A5 elicits an IC_{50} of 0.004 μM against mammalian SQR [12] as compared to 2.5 μM for LtSQR, accounting for a selectivity index of 0.0006 (Table 9.4). Expectedly, the lack of sensitivity of LtSQR for these inhibitors follows the pattern displayed by TcSQR [12] (Table 9.4), therefore, it could be inferred that these inhibitors cannot be drug candidates against the trypanosomatids and other organisms possessing the 12-subunit SQR. As a result, it is important to

Table 9.3 Comparison of genomic identity of SQR subunits between *L. tarentolae* and pathogenic trypanosomatids. The amino acid identity (%) compared to *L. tarentolae* gene is shown in parenthesis

Gene	*L. tarentolae*	*T. cruzi*	*T. brucei*	*L. major*	*L. donovani*
SDH1	67	67 (90)	67 (89)	67 (98)	67 (98)
SDH2n	27	32 (62)	27 (64)	27 (94)	27 (95)
SDH2c	22	21 (72)	21 (76)	22 (89)	22 (97)
SDH3	12	8 (73)	12 (61)	12 (94)	12 (95)
SDH4	16	14 (48)	14 (59)	16 (85)	16 (85)
SDH5	53	53 (44)	53 (42)	53 (91)	53 (91)
SDH6	37	36 (50)	37 (50)	37 (85)	37 (86)
SDH7	29	28 (53)	28 (55)	29 (89)	29 (89)
SDH8	17	16 (65)	17 (60)	17 (95)	17 (93)
SDH9	16	16 (58)	16 (79)	16 (81)	16 (81)
SDH10	16	16 (50)	14 (60)	16 (85)	16 (87)
SDH11	10	10 (50)	10 (77)	10 (85)	10 (89)
Complex II (kDa)	323	319	323	323	323

search for inhibitors possessing different scaffolds that are specific for the 12-subunit SQR.

Then we conducted a small-scale screening of SQR inhibitors using LtSQR. Because the difference of SQR IC_{50} between mammals and trypanosomatids were larger in quinone-binding site inhibitors (atpenin A5, carboxin, TTFA) than in the succinate-binding site inhibitor (malonate), we mainly focused on quinone-like compounds as candidates of specific inhibitors. We tested inhibitory effects of compounds in our library of quinone-site (Q-site) inhibitors, which are known to specifically target quinone-utilizing enzymes. With the exception of ferulenol and the old fungicide, siccanin, we found most of the compounds showed poor inhibition against LtSQR (IC_{50} > 50 µM). Intriguingly, siccanin very potently inhibited the enzyme with the IC_{50} of 0.19 µM (Fig. 9.3). Ferulenol showed only a milder inhibition (IC_{50} = 5.3 µM), coupled to being a well-known mammalian SQR inhibitor, ferulenol cannot be developed further as a selective trypanosomatid drug candidate. Siccanin being the first µM-order inhibitor of the 12-subunit SQR and possessing a novel structural scaffold was therefore selected for further studies.

Siccanin was originally known as a specific inhibitor of fungi SQR [43] and widely used for the treatment of tinea pedis. In our previous study [27], we identified siccanin as a specific inhibitor of *Pseudomonas aeruginosa* SQR (IC_{50} = 0.87 µM) showing the bactericidal effect in vitro. We also investigated the inhibitory effect of siccanin on various species and found its species-selective SQR inhibition property. Siccanin strongly inhibited not only *P. aeruginosa* SQR, but also those of *Trichophyton mentagrophytes* (IC_{50} = 0.09 µM), rat (IC_{50} = 9 µM), while showing low inhibition against SQR of *Escherichia coli* (IC_{50} = 210 µM) and porcine (IC_{50} = 860 µM). The inhibition mechanism of siccanin to *P. aeruginosa* SQR was

Table 9.4 Inhibitor sensitivity of LtSQR and TcSQR to classical SQR inhibitors

Inhibitor	Structure	SQR IC$_{50}$ (μM)			Selectivity	
		Mammal	Lt	Tc	Lt	Tc
Malonate	HO-CH$_2$-C(=O)-OH (malonic acid)	3.4[1]	200	40[4]	0.017	0.085
TTFA	F$_3$C-C(=O)-CH$_2$-C(=O)-thienyl	5.4[2]	>100	>100[4]	>0.054	>0.054
Atpenin A5	(atpenin A5 structure)	0.004[3]	2.5	6.4[4]	0.0016	0.0006

[1]Mogi et al. [27], J Biochem
[2]Grivennikova et al. [41], Biochim Biophys Acta
[3]Miyadera et al. [42], Proc Natl Acad Sci USA
[4]Morales et al. [12], J Biol Chem

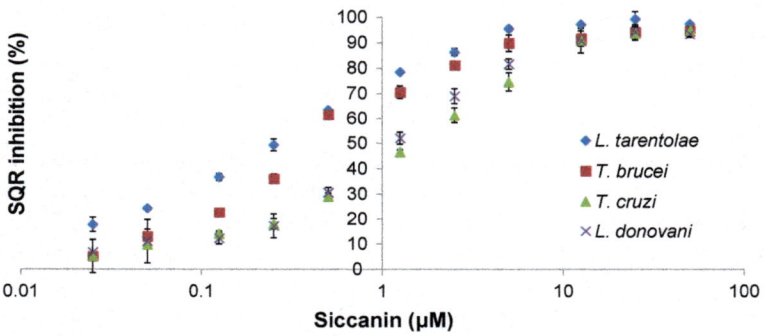

Fig. 9.3 Effect of siccanin on SQR activity of mitochondria from trypanosomatids

mixed type for ubiquinone 1 (UQ$_1$) and noncompetitive for succinate. These results indicated that siccanin binds to the vicinity of the quinone pocket. To investigate the inhibition mechanism of siccanin on LtSQR, we performed kinetic analysis at varying substrate and siccanin concentrations. *Lineweaver-Burk plot for the enzymatic activities (Fig. 9.4) indicated that* inhibition mode was the mixed type with 39 nM of K_{i1} and 102 nM of K_{i2}. Lower K_{i1} than K_{i2} means that the affinity of siccanin was reduced by the presence of quinone. Interestingly, siccanin's K_{i1} and K_{i2} for LtSQR were much lower than those of *P. aeruginosa* SQR despite their similar IC$_{50}$. This result may reflect the difference of the siccanin binding mode in quinone binding sites between 4-subunit and 12-subunit SQR, but further studies are required to elucidate details of the interactions.

9.15.3 Site and Target Specificity of Siccanin

To evaluate whether siccanin binds exclusively to the Q-site of trypanosomatids SQR and if it is specific for only the SQR and not other mitochondrial quinone-utilizing enzymes, we first measured the phenazine methosulfate (PMS)- methyl-thiazolyl-tetrazolium (MTT)-mediated succinate dehydrogenase (SDH) activity of mitochondria to investigate electron transfer in the Fp and Ip subunits; interestingly, the activity was not inhibited even up to 50 μM (not shown), suggesting that siccanin interacted with the quinone pocket but not with the hydrophilic portion of SQR composed of Fp and Ip, because SDH activity does not require the quinone-binding site for the enzymatic reaction. Next, we determined the inhibitory effect of siccanin against the other quinone-utilizing enzymes in *L. tarentolae* mitochondria (Fig. 9.5). Results revealed that siccanin did not inhibit NDH2, complex III (data not shown), glycerol-3-phosphate dehydrogenase up to 50 μM (Fig. 9.5), and we excluded these mitochondrial enzymes as targets of siccanin.

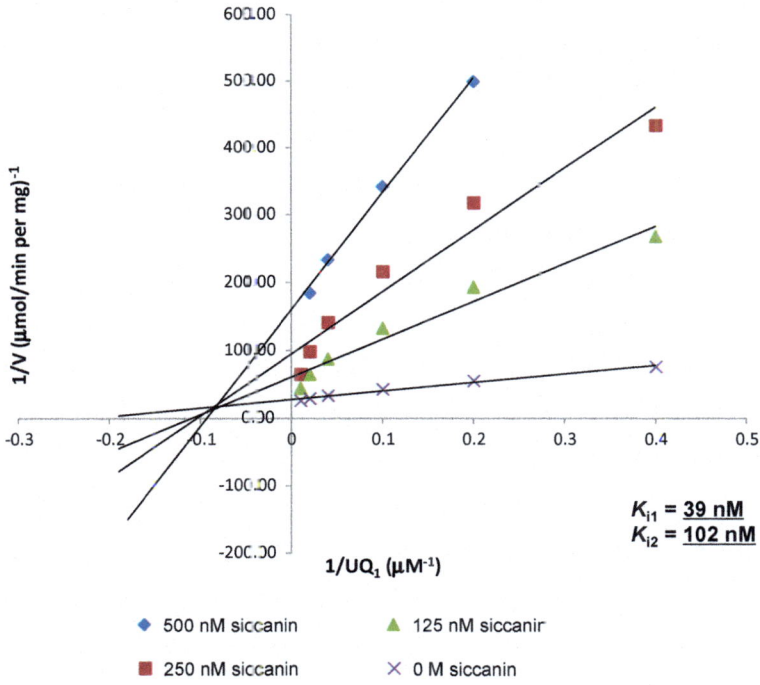

Fig. 9.4 Inhibition mechanism of siccanin to LtSQR (UQ_1). SQR activity was measured by the absence (0 nM) or presence of 125, 250, and 500 nM of siccanin under various concentrations of UQ_1 in 96-well plates. Activity was determined as time course of DCIP absorbance during 15-min incubation. Data were presented as double reciprocal plots taking UQ_1 concentration as x-axis and SQR activity as y-axis

9.15.4 Effects of Siccanin on Pathogenic Trypanosomatids

Reported herein, and to the best of our knowledge, siccanin is the only strong trypanosomatid SQR inhibitor discovered. To investigate its effect on the SQR of other pathogenic trypanosomatids, we determined IC_{50} of siccanin on the SQR of *T. cruzi*, *T. b. brucei* (same species as the African human trypanosomes, *T. b. gambiense* and *T. b. rhodensiense*), and *L. donovani* (Fig. 9.3). Siccanin strongly inhibited the enzymatic activities of all the parasites with the observed IC_{50} of 1.48, 0.368, and 1.17 µM, respectively; indicating its broad spectrum of activity across the genus of organisms in the *Trypanosomatidae* family (Fig. 9.3 and Table 9.5). Further, it is of great interest that siccanin has almost no effect on the activity of SQR from the mammalian host. Its IC_{50} against porcine SQR is 861 µM [27], accounting for a selectivity index of up to 4,532 when compared to those of the parasites (Table 9.5), meaning that it has potential to be a highly selective drug that will display no toxicity against the treated mammals.

Subsequently, we checked the trypanocidal effect of siccanin on parasite cells (Table 9.4). Because of ease of handling, we first used *L. tarentolae* promastigote as a drug screening model and checked the leishmanicidal effect of siccanin. The determined IC_{50} of siccanin on *L. tarentolae* was 4.4 µM by AlamarBlue assay (Table 9.6), a value comparable to that of pentamidine (6.3 µM, not shown). Then the effect of siccanin on *L. donovani* and *L. major*, the causative agents of visceral

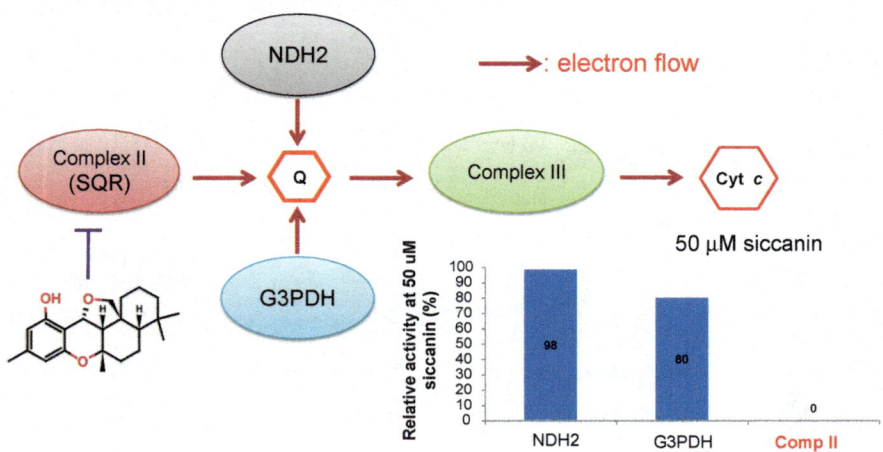

Fig. 9.5 Target specificity of siccanin. Electron transfer activity by enzyme complexes composed of NADH2-III, II-III, and G3PDH-complex III, respectively was monitored as reduction of cytochrome *c* in presence of *L. tarentolae* mitochondria after the addition of 100 µM NADH, 10 mM succinate, or 10 mM glycerol-3-phosphate, respectively

and cutaneous leishmaniasis, respectively, was determined. Similar to *L. tarentolae*, the promastigote of *L. donovani* and *L. major* are also sensitive to siccanin with IC_{50} of 0.7 and 13 µM, respectively. Notably, *L. donovani* was more sensitive to siccanin than were *L. major* and *L. tarentolae*. Because siccanin also inhibited the SQR of *T. cruzi* and *T. brucei*, we also evaluated the trypanocidal effect on *T. brucei* and *T. cruzi* as described in Materials and Methods. We found that siccanin killed the bloodstream form of *T. b. brucei* with an IC_{50} of 8.3 µM, and in the presence of 10 µM siccanin, the number of *T. cruzi* amastigote in mouse fibroblast infected with trypomastigotes decreased to 50.4 % of DMSO control (Table 9.6). Together, our results show that siccanin inhibited the SQR activity and cell viability of the major causative agents of trypanosomiasis and leishmaniasis, providing a chemical validation for their complex II as a specific and selective drug target candidate.

9.16 Discussion

This study is the first successful attempt at obtaining a drug candidate inhibitor against the mitochondrial complex II of organisms in the genus *Trypanosoma* and *Leishmania*, the pathogens of human and animal diseases known as trypanosomiasis and leishmaniasis. Hundreds of millions of humans and animals in endemic countries are infected [1–3], significantly contributing to the global burden of diseases. Because challenges such as high cost, toxicity, and treatment failures have long been associated with the currently available chemotherapy, it has become

Table 9.5 Effect of siccanin on SQR activity of mitochondria from porcine and trypanosomatids

Organisms	SQR IC_{50} (µM)	Selectivity
Porcine	861[1]	1
L. tarentolae	0.190	4,532
T. brucei	0.368	2,340
T. cruzi	1.48	582
L. donovani	1.17	736

[1]Mogi et al. [27], J Biochem

Table 9.6 Growth inhibition of pathogenic trypanosomatids by siccanin

Organism	IC_{50} (µM)		Life cycle stage
	SQR	Cell growth	
Porcine[1]	861	–	–
T. cruzi	1.4	10	Amastigote
T. b. brucei	0.37	8.3	Blood stream form
L. tarentolae	0.19	4.4	Promastigote
L. donovani	1.17	0.7	Promastigote
L. major	–	13	Promastigote

[1]Mogi et al. [27], J Biochem

necessary to develop new, safe, and more effective drugs. A major prerequisite towards development of effective and non-toxic drugs is the identification of molecular targets that are essential to the disease agents (*trypanosomes* and *leishmania*, in this case), and absent or different in the mammalian hosts (humans and animals). From the knowledge that mitochondrial complex II is pivotal to aerobic energy metabolism and to the adaptability of parasites to their host environments, and based on our previous finding that the complex II of *T. cruzi* (12-subunits enzyme) is distinct from that of mammals (4-subunits enzyme) [12], we rationalized that the protein is a good drug target candidate. Hence we sought to find inhibitors of the parasites complex II, for possible use in formulation of a new drug. However, because this approach requires a large amount of mitochondria and purified protein, for safety considerations during large-scale culture, *T. cruzi* is not an ideal source. Herein, we established that the non-pathogenic trypanosomatid, *L. tarentolae*, also possesses the 12-subunits complex II, and being safe and much easier to handle, it is validated as a model organism for obtaining large amount of the trypanosomatid complex II. Moreover, yield for the protein from *L. tarentolae* is approximately 14-folds that from *T. cruzi*, and in addition, the enzymatic activity is slightly higher (Table 9.2).

The newly established protocol using PEG precipitation and an ion exchange column chromatography resulted in purification of ∼50 %-pure LtSQR with 3 % activity yield, with a comparable efficiency to that of the TcSQR purification procedure. These data indicated that milligram-order of partially purified LtSQR is obtainable from the large cultivation of *L. tarentolae*. As inferred from genomic analysis (Table 9.1), 2-D native electrophoresis of the partially purified LtSQR detected 12 subunits, however, only 9 of the peptides were detectable by LC-MS/MS analysis. The 3 missing subunits (SDH9-11) could be attributed to their hydrophobic nature and not differences of subunit compositions between LtSQR and TcSQR, The molecular weights of these proteins in native states on hrCNE gels were almost identical (Fig. 9.1).

The insensitivity of TcSQR to standard SQR inhibitors, which is most likely owing to blocking of the inhibitor-binding sites by extra-subunits (that are absent in other SQRs), was also observed in LtSQR (Table 9.3). Our data implies that LtSQR shares common subunit- and catalytic site-structure with pathogenic trypanosomatids like *T. cruzi*. These unique properties of the trypanosomatids' SQR encouraged the screening for inhibitors targeting the 12-subunit SQRs. Remarkably, the current efforts have resulted in the rediscovery of the old fungicide, siccanin, as the first potent and selective SQR inhibitor of the trypanosomatid parasites. The high selectivity of siccanin against these 12-subunit SQRs could be that it interacts with some of the extra-SQR subunits that may be contributing to forming or maintaining the active site in the right conformation. The inhibition kinetics of siccanin on these SQRs is a non-competitive pattern (Fig. 9.4), suggesting that siccanin does not directly bind their catalytic site. However, the accurate elucidation of the mechanism of inhibition requires structural analysis of the protein in the presence and absence of siccanin. Nevertheless, the mixed type inhibition mode of siccanin against quinone revealed by the *Lineweaver-Burk plot*

(Fig. 9.4) is a desirable feature for medical application of the compounds, because this type of inhibitor can bind to both substrate-free enzymes and substrate-enzyme complexes, hence inhibiting functions of enzymes regardless of quinone concentration in parasite cells.

Considering that one of the major limitations of the current drugs for the treatment of trypanosomiasis and leishmaniasis is narrow potency [3] i.e., they are highly species specific; for instance, even the most recently introduced NECT is only effective against *T. b. gambiense*, and not *T. b. rhodesiense* [44]. Therefore, our finding that siccanin is effective at killing a wide-range of the trypanosomatid parasites (Table 9.6) is of remarkable clinical significance; siccanin will be helpful in treating patients that are infected by multiple parasite types (of different genus and species). Such cases of mixed infections have been widely reported in human and animals in endemic areas [45]. Furthermore, the present data indicated that while the potency of siccanin against trypanosomes is comparable, its effect against *L. donovani*, is significantly higher than that against *L. major*. These two *Leishmania* species cause distinct ailments: visceral leishmaniasis and cutaneous leishmaniasis, respectively. This feature of differential potency is another advantage of siccanin in chemotherapy, because cutaneous leishmaniasis is a skin infection and a higher dose of siccanin can be tolerable compared to the case of visceral leishmaniasis, in which parasites migrate into internal organs and the drug must be systemic.

To our knowledge, siccanin is the first and strongest specific inhibitor of trypanosomatids SQR and possesses killing effect on all the tested pathogens belonging to the group. Siccanin is a promising lead compound as the quality of the chemical starting points for drug discovery is a key factor in improving the likelihood of clinical success [46]. Animal experiments are ongoing to assess the side effects and in vivo trypanocidal efficacy of siccanin.

Acknowledgments This work has been financially supported by JSPS KAKENHI Grant Number 26253025 (to K. K.), and by JST/JICA Science and Technology Research Partnership for Sustainable Development (SATREPS; no. 10000284 to K. K.). We also acknowledge the support of the Program for Promotion of Basic and Applied Researches for Innovations in Bio-oriented Industry (BRAIN) and the Science and Technology Research Promotion Program for Agriculture, Forestry, Fisheries and Food Industry. We thank Daiichi-Sankyo Co., Ltd. and Dr. Kazuro Shiomi, Kitasato University for siccanin supply.

References

1. Brun R, Blum J, Chappuis F, Burri C. Human African trypanosomiasis. Lancet. 2010;375:148–59.
2. Rassi A, Marin-Neto JA. Chagas disease. Lancet. 2010;375:1388–402.
3. World Health Organization. Research priorities for Chagas disease, human African trypanosomiasis and leishmaniasis. World Health Organ Tech Rep Ser. 2012;v–xii:1–100.
4. Baker N, de Koning HP, Mäser P, Horn D. Drug resistance in African trypanosomiasis: the melarsoprol and pentamidine story. Trends Parasitol. 2013;29:110–8.

5. Chakravarty J, Sundar S. Drug resistance in leishmaniasis. J Glob Infect Dis. 2010;2:167–76.
6. Purkait B, Kumar A, Nandi N, Sardar AH, Das S, Kumar S, Pandey K, Ravidas V, Kumar M, De T, et al. Mechanism of amphotericin B resistance in clinical isolates of *Leishmania donovani*. Antimicrob Agents Chemother. 2012;56:1031–41.
7. Opperdoes FR, Coombs GH. Metabolism of *Leishmania*: proven and predicted. Trends Parasitol. 2007;23:149–58.
8. Bringaud F, Rivière L, Coustou V. Energy metabolism of trypanosomatids: adaptation to available carbon sources. Mol Biochem Parasitol. 2006;149:1–9.
9. van Hellemond JJ, Opperdoes FR, Tielens AG. The extraordinary mitochondrion and unusual citric acid cycle in *Trypanosoma brucei*. Biochem Soc Trans. 2005;33:967–71.
10. Escobar P, Matu S, Marques C, Croft SL. Sensitivities of *Leishmania* species to hexadecylphosphocholine (miltefosine), ET-18-OCH(3) (edelfosine) and amphotericin B. Acta Trop. 2002;81:151–7.
11. Carvalho L, Luque-Ortega JR, Manzano JI, Castanys S, Rivas L, Gamarro F. Tafenoquine, an antiplasmodial 8-aminoquinoline, targets leishmania respiratory complex III and induces apoptosis. Antimicrob Agents Chemother. 2010;54:5344–51.
12. Morales J, Mogi T, Mineki S, Takashima E, Mineki R, Hirawake H, Sakamoto K, Omura S, Kita K. Novel mitochondrial complex II isolated from *Trypanosoma cruzi* is composed of 12 peptides including a heterodimeric Ip subunit. J Biol Chem. 2009;284:7255–63.
13. Horsefield R, Iwata S, Byrne B. Complex II from a structural perspective. Curr Protein Pept Sci. 2004;5:107–18.
14. Harada S, Inaoka DK, Ohmori J, Kita K. Diversity of parasite complex II. Biochim Biophys Acta. 2013;1827:658–67.
15. Eubel H, Heinemeyer J, Sunderhaus S, Braun HP. Respiratory chain supercomplexes in plant mitochondria. Plant Physiol Biochem. 2004;42:937–42.
16. Huang S, Millar AH. Succinate dehydrogenase: the complex roles of a simple enzyme. Curr Opin Plant Biol. 2013;16:344–9.
17. Millar AH, Eubel H, Jänsch L, Kruft V, Heazlewood JL, Braun HP. Mitochondrial cytochrome *c* oxidase and succinate dehydrogenase complexes contain plant specific subunits. Plant Mol Biol. 2004;56:77–90.
18. Perez E, Lapaille M, Degand H, Cilibrasi L, Villavicencio-Queijeiro A, Morsomme P, González-Halphen D, Field MC, Remacle C, Baurain D, et al. The mitochondrial respiratory chain of the secondary green alga Euglena gracilis shares many additional subunits with parasitic Trypanosomatidae. Mitochondrion 2014;19(part B):338–349.
19. Tkacz ID, Gupta SK, Volkov V, Romano M, Haham T, Tulinski P, Lebenthal I, Michaeli S. Analysis of spliceosomal proteins in trypanosomatids reveals novel functions in mRNA processing. J Biol Chem. 2010;285:27982–99.
20. Fritsche C, Sitz M, Weiland N, Breitling R, Pohl HD. Characterization of the growth behavior of *Leishmania tarentolae*: a new expression system for recombinant proteins. J Basic Microbiol. 2007;47:384–93.
21. Taylor VM, Muñoz DL, Cedeño DL, Vélez ID, Jones MA, Robledo SM. *Leishmania tarentolae*: utility as an in vitro model for screening of antileishmanial agents. Exp Parasitol. 2010;126:471–5.
22. Turner TL, Nguyen VH, McLauchlan CC, Dymon Z, Dorsey BM, Hooker JD, Jones MA. Inhibitory effects of decavanadate on several enzymes and *Leishmania tarentolae* in vitro. J Inorg Biochem. 2012;108:96–104.
23. Graziose R, Rojas-Silva P, Rathinasabapathy T, Dekock C, Grace MH, Poulev A, Ann Lila M, Smith P, Raskin I. Antiparasitic compounds from *Cornus florida* L. with activities against *Plasmodium falciparum* and *Leishmania tarentolae*. J Ethnopharmacol. 2012;142:456–61.
24. Hooker JD, Nguyen VH, Taylor VM, Cedeño DL, Lash TD, Jones MA, Robledo SM, Vélez ID. New application for expanded porphyrins: sapphyrin and heterosapphyrins as inhibitors of *Leishmania* parasites. Photochem Photobiol. 2012;88:194–200.

25. Ishibashi K. Studies on antibiotics from *Helminthosporium* sp. fungi, 7. siccanin, a new antifungal antibiotic produced by *Helminthosporum siccanias*. J Antibiotics. 1962;15:161–7.
26. Arai M, Ishibashi K, Okazaki H. Siccanin, a new antifungal antibiotic. I. In vitro studies. Antimicrob Agents Chemother (Bethesda). 1969;9:247–52.
27. Mogi T, Kawakami T, Arai H, Igarashi Y, Matsushita K, Mori M, Shiomi K, Omura S, Harada S, Kita K. Siccanin rediscovered as a species-selective succinate dehydrogenase inhibitor. J Biochem. 2009;146:383–7.
28. Yabu Y, Yoshida A, Suzuki T, Nihei C, Kawai K, Minagawa N, Hosokawa T, Nagai K, Kita K, Ohta N. The efficacy of ascofuranone in a consecutive treatment on *Trypanosoma brucei brucei* in mice. Parasitol Int. 2003;52:155–64.
29. Sun F, Huo X, Zhai Y, Wang A, Xu J, Su D, Bartlam M, Rao Z. Crystal structure of mitochondrial respiratory membrane protein complex II. Cell. 2005;121:1043–57.
30. Wittig I, Karas M, Schägger H. High resolution clear native electrophoresis for in-gel functional assays and fluorescence studies of membrane protein complexes. Mol Cell Proteomics. 2007;6:1215–25.
31. Wittig I, Carrozzo R, Santorelli FM, Schägger H. Functional assays in high-resolution clear native gels to quantify mitochondrial complexes in human biopsies and cell lines. Electrophoresis. 2007;28:3811–20.
32. Schägger H. Tricine-SDS-PAGE. Nat Protoc. 2006;1:16–22.
33. Räz B, Iten M, Grether-Bühler Y, Kaminsky R, Brun R. The Alamar Blue assay to determine drug sensitivity of African trypanosomes (*T.b. rhodesiense* and *T.b. gambiense*) in vitro. Acta Trop. 1997;68:139–47.
34. Nara T, Kamei Y, Tsubouchi A, Annoura T, Hirota K, Iizumi K, Dohmoto Y, Ono T, Aoki T. Inhibitory action of marine algae extracts on the *Trypanosoma cruzi* dihydroorotate dehydrogenase activity and on the protozoan growth in mammalian cells. Parasitol Int. 2005;54:59–64.
35. El-Sayed NM, Myler PJ, Bartholomeu DC, Nilsson D, Aggarwal G, Tran AN, Ghedin E, Worthey EA, Delcher AL, Blandin G, et al. The genome sequence of *Trypanosoma cruzi*, etiologic agent of Chagas disease. Science. 2005;309:409–15.
36. Berriman M, Ghedin E, Hertz-Fowler C, Blandin G, Renauld H, Bartholomeu DC, Lennard NJ, Caler E, Hamlin NE, Haas B, et al. The genome of the African trypanosome *Trypanosoma brucei*. Science. 2005;309:416–22.
37. Ivens AC, Peacock CS, Worthey EA, Murphy L, Aggarwal G, Berriman M, Sisk E, Rajandream MA, Adlem E, Aert R, et al. The genome of the kinetoplastid parasite, *Leishmania major*. Science. 2005;309:436–42.
38. De Gaudenzi JG, Noé G, Campo VA, Frasch AC, Cassola A. Gene expression regulation in trypanosomatids. Essays Biochem. 2011;51:31–46.
39. Acestor N, Zíková A, Dalley RA, Anupama A, Panigrahi AK, Stuart KD. *Trypanosoma brucei* mitochondrial respiratome: composition and organization in procyclic form. Mol Cell Proteomics. 2011;10(M110):006908.
40. Verner Z, Cermáková P, Skodová I, Ková̌cová B, Lukeš J, Horváth A. Comparative analysis of respiratory chain and oxidative phosphorylation in *Leishmania tarentolae, Crithidia fasciculata, Phytomonas serpens* and procyclic stage of *Trypanosoma brucei*. Mol Biochem Parasitol. 2014;193:55–65.
41. Grivennikova VG, Gavrikova EV, Timoshin AA, Vinogradov AD. Fumarate reductase activity of bovine heart succinate-ubiquinone reductase. New assay system and overall properties of the reaction. Biochim Biophys Acta 1993;1140:282–92
42. Miyadera H, Shiomi K, Ui H, Yamaguchi Y, Masuma R, Tomoda H, Miyoshi H, Osanai A, Kita K, Omura S. Atpenins, potent and specific inhibitors of mitochondrial complex II (succinate-ubiquinone oxidoreductase). Proc Natl Acad Sci USA 2003;100:473–7
43. Nose K, Endo A. Mode of action of the antibiotic siccanin on intact cells and mitochondria of *Trichophyton mentagrophytes*. J Bacteriol 1971;105:176–84
44. WHO. 2016. http://www.who.int/mediacentre/factsheets/fs259/en/

45. Viol MA, Lima VM, Aquino MC, Gallo G, Alves IP, Generoso D, Perri SH, Lucheis SB, Langoni H, Nunes CM, Bresciani KD. Detection of cross infections by *Leishmania* spp. and *Trypanosoma* spp. in dogs using indirect immunoenzyme assay, indirect fluorescent antibody test and polymerase chain reaction. Parasitol Res 2012;111:1607–13
46. Katsuno K, Burrows JN, Duncan K, van Huijsduijnen RH, Kaneko T, Kita K, Mowbray CE, Schmatz D, Warner P, Slingsby BT. Hit and lead criteria in drug discovery for infectious diseases of the developing world. Nat Rev Drug Discov 2015;14:751–8

Part III
Diagnostic Strategy Enhancing Kala-Azar Elimination Program

Chapter 10
Challenges in the Diagnosis of Visceral Leishmaniasis on the Indian Subcontinent

Suman Rijal, François Chappuis, Jorge Alvar and Marleen Boelaert

Abstract Visceral leishmaniasis (VL) or kala-azar, an endemic vector-borne disease, affects populations in the lowest socioeconomic strata living in rural areas on the Indian subcontinent—a group that has limited access to proper health care. Untreated, kala-azar is almost always fatal, and the drugs currently in use are quite toxic. Confirmation of diagnosis before starting therapy is therefore crucial. Early diagnosis and treatment are a key strategy of the Kala-azar Elimination Programme launched in 2005 in Bangladesh, India, and Nepal. VL care must be decentralized to primary health centers to achieve this goal, which has become possible with the availability of rapid diagnostic tests. Parasitological diagnosis is limited to referral hospitals and specialized VL treatment centers. Two serological tests for field use— the direct agglutination test and the rK39 immunochromatographic test (ICT)— have both shown excellent performance on the Indian subcontinent, but the latter is preferred, because it is simpler to use. The proper implementation of these diagnostic strategies within the VL elimination programme involves not only the procurement, training, and supervision of staff, but also quality control both before and after deployment in the field. The logistical requirements are enormous, and therefore standardized guidelines for procurement and quality control must be established.

Keywords Kala-azar · Visceral leishmaniasis · Kala-azar elimination programme · Diagnostic test · Rapid diagnostic test · Access to health care · Rk39 immunochromatographic test · Direct agglutination test

S. Rijal (✉)
Drugs for Neglected Diseases Initiative, India Regional Office,
PHD House, Siri Institutional Area, New Delhi 110016, India
e-mail: srijal@dndi.org

10.1 Introduction

Visceral leishmaniasis (VL) or kala-azar is a major public health problem in Bangladesh, India, and Nepal, and the populations affected represent the poorest residents living in remote rural regions [1]. Of the 200,000–400,000 annual cases, 90 % occur in just 6 countries: Bangladesh, Brazil, Ethiopia, India, Nepal, and Sudan [2]. Around 200 million people in South Asia are estimated to be at risk. A regional elimination programme was jointly launched in Bangladesh, India, and Nepal in 2005 with the aim of eliminating kala-azar as a public health problem by 2018 [3]. The endemic regions within these 3 countries constitute a contiguous area. The causative agent is *Leishmania donovani*, and genetic studies of parasites from these countries have shown a very homogeneous population regardless of geographical origin [4].

Kala-azar is fatal if untreated, and even with treatment, particularly with pentavalent antimonials, up to 5 % of patients may die due to the toxicity of the drugs [5]. Increased fatality has also been found to be associated with advanced disease [6]. Increasing treatment failure of the anti-VL drug sodium stibogluconate on the Indian subcontinent (in particular in India and Nepal) has been well documented over the last two decade [7, 8]. A recent study from Nepal showed a significant association of treatment failure with delayed diagnosis [9]. There is, therefore, a need for a diagnostic test that is not only highly accurate but can also easily be used at the peripheral health facilities where the majority of kala-azar patients are seen.

In the kala-azar elimination programme, early diagnosis and appropriate treatment is a key strategy [10]. Decentralization to the peripheral level health facilities —e.g., primary health care centers—is making VL care more accessible. The availability of a rapid diagnostic test (RDT), the rK39 immunochromatography test (ICT) [11], has made it feasible to implement this strategy.

This chapter will present a review of the currently available diagnostic tools for VL for field use and their utility, the current limitations to VL care access, and issues in the implementation of the RDT in the kala-azar elimination programme on the Indian subcontinent.

10.2 Diagnostic Tests for VL

Accurate diagnosis is crucial before starting treatment with anti-VL drugs. The clinical manifestations of kala-azar (fever, anemia, splenomegaly) are nonspecific and are also commonly seen in other diseases that occur in this region. In a phase 3 validation study for the diagnosis of VL in Nepal, [12] the common differential diagnosis of VL included malaria, disseminated tuberculosis, hematological malignancies, and sepsis [13]. When applying the diagnostic test in suspect VL cases (fever of $\geq$ 2 weeks with palpable spleen) the proportion of VL cases ranged from 50 % at a district level hospital to 70 % at a referral hospital [14]. With a persistent decrease in the number of cases in the last few years, the proportion of kala-azar cases among the suspect VL cases should be further reduced. Currently there are no algorithms that have clear diagnostic pathways for VL suspect cases negative for kala-azar.

The currently available diagnostic tests for VL can be broadly divided into nonspecific VL tests, parasite demonstration, serological tests, and antigen detection tests.

10.2.1 Nonspecific Tests

Nonspecific tests for VL include the demonstration of pancytopenia and the aldehyde or formol gel test (FGT). The former was found in a study of suspected cases in Nepal to have excellent specificity, but the sensitivity was only 16 % [15]. The FGT, developed by Napier in the 1920s, is based on detecting the increased gamma globulin levels found in kala-azar patients [16]. Due to the simplicity of this test, the non-feasibility of performing microscopy on tissue aspirates, and the absence of an easily administered alternative, the FGT remained in use in health facilities on the Indian subcontinent until an RDT became readily available a decade ago. The sensitivity of FGT has been shown to be quite low (35 %), although the specificity is excellent [15].

10.2.2 Parasite Detection

Parasite detection may be broadly described as microscopy of tissue aspirates to demonstrate the presence of *Leishmania* amastigotes, along with culture and molecular biology techniques. Since the discovery of the parasite by Leishman and Donovan in 1903, microscopy has been considered the gold standard for confirmation of the diagnosis [17]. The sensitivity of microscopy depends on the tissue aspirate; for example, it is higher for spleen tissue (93–99 %) than for bone marrow (53–86 %). Splenic aspiration carries a 0.1 % risk of life-threatening hemorrhage even in centers specializing in kala-azar. It is also not feasible to perform this procedure at peripheral centers, as it requires technical expertise to perform and interpret, and ideally facilities for blood transfusion and surgery must be available. Molecular biology tools such as polymerase chain reaction (PCR) have demonstrated excellent sensitivity even when blood samples are used, but the specificity was as low as 62 % (95 % confidence interval [CI], 51–72) in a phase 3 diagnostic study from Nepal [18]. PCR also requires sophisticated tools, limiting its use to research centers and university hospitals in endemic countries [19], though efforts have been made to simplify the procedure by using less sophisticated technology [20].

10.2.3 Serological Tests

There are many test formats that detect antibodies to *Leishmania* from patients' serum, but most are not suitable for use in field conditions on the Indian subcontinent. Tests based on indirect fluorescence antibody, enzyme linked immunosorbent assay, and Western blot techniques show high diagnostic accuracy but are considered too complicated for field use. The direct agglutination test (DAT) and the RDT in ICT format are the 2 tests that are considered appropriate for field use. There are currently several ICTs available using different antigens: rK39, rK26, and rKE16. The most extensively validated point of care tests in the field are rk39 based ICT and the DAT.

The DAT, a semi-quantitative test, requires microtiter plates in which increasing dilutions of a patient's serum or blood are mixed with stained, killed *L. donovani* promastigotes [21]. The results are read with the naked eye after an overnight incubation. Although the DAT has been extensively validated in most endemic areas, there is still no consensus for recommending it, as not all the studies give the same results. A meta-analysis that included 30 studies showed sensitivity and specificity estimates of 94.8 % (95 % CI, 92.7–96.4) and 97.1 % (95 % CI, 93.9–98.7), respectively [22]. The performance of the DAT was quite similar in the different endemic foci of India, East Africa, and Brazil. The main drawbacks of the DAT are that it is not easily available commercially, as it is manufactured in only 2 institutes in Europe, and that it requires transportation through a cold chain. The latter obstacle has been overcome by the development of a freeze-dried version that delivers a performance similar to that of the liquid antigen [23]. Although the DAT is simpler than many other test formats, it still requires equipment (microplates, micropipettes), the extensive training of laboratory technicians, and regular quality control; it is also relatively expensive.

The rK39 dipstick contains a 39-amino acid repeat that is part of a kinesin-related protein of *L. chagasi*. The repeat is conserved within the *L. donovani* complex. This test developed in an ICT format (dipstick) is suitable for field use, as it is easy to administer, and results—which are reproducible—can be obtained in 10–15 min. However, as with the DAT, it has not been universally accepted, as the results from different studies vary. A recent Cochrane review on rapid tests for the diagnosis of VL in patients with suspected disease found a pooled sensitivity of 0.97 (95 % CI 0.90–1.00) and a specificity of 0.90 (95 % CI 0.76–0.98) with the rK39 ICT on the Indian subcontinent [24]. However, in studies from East Africa the rk39 ICT showed a lower sensitivity of 0.85 (95 % CI 0.75 to 0.93) and a specificity of 0.91 (95 % CI 0.80–0.97).

Recently, significantly lower anti-leishmania IgG responses have been observed among VL patients in Sudan when compared to VL patients from India, which may contribute to the lower sensitivity of the rK39-ICT in East Africa [25].

There are only limited studies that have validated the rK26 and rKE16 ICT. One study from India validating the rK26 ICT (InBios International, Seattle,

Washington) in a population of 352 patients with suspect VL observed a sensitivity of 21.3 % and a specificity 100 % [26].

A study validating rKE16-based ICT (Signal KA, Span Diagnostics, India) in Kenya observed a sensitivity of 77.1 % and a specificity of 95.5 % [27]. A recent multicentre evaluation of commercially available ICTs was coordinated by TDR-WHO [28]. The rKE16 ICT (Signal KA, Span Diagnostics, India) had a sensitivity of 92.8 % (88.9–95.4 %) and specificity of 99.2 % (97.1–99.8 %) in samples from the Indian subcontinent but the sensitivity was only 36.8 % (31.1–42.9 %) in samples from East Africa. This multicentre evaluation also looked at the thermal stability of the different ICTs, which was found to be good in the Indian subcontinent.

Most of the ICTs are recommended by manufacturers for use with serum. This requires venepuncture for the collection of blood and separation of the blood product. It has also been observed in the field that tests are being used on whole blood (obtained by finger prick). Thus there was a need to demonstrate that the diagnostic accuracy of these RDTs in whole blood was equivalent to that in serum; 2 recent evaluations in the Indian subcontinent have shown excellent performance of ICTs in blood [29, 30].

It is quite clear that the rk39 ICT is currently the best available VL diagnostic tool for use in the field; it has been appropriately recommended as a diagnostic test for use in the kala-azar elimination initiative on the Indian subcontinent.

However, there are some major limitations on the use of the serological tests. Because specific antibodies continue to be detected up to several years after cure in a high proportion of patients [31, 32], VL relapse cannot be diagnosed by serology. In addition, healthy people living in endemic areas with no history of VL may carry anti-leishmanial antibodies because of past or active asymptomatic infections. Seroprevalence in the healthy population in low transmission areas has been shown to vary between 5 and 15 % [33], while in high transmission areas it can be above 30 % [34]. Thus it is important to emphasize that antibody tests must always be used in combination with clinical case definitions for VL diagnosis.

There is a need to develop tests to diagnose relapse in the field. Currently most suspected relapse patients are referred to central referral hospitals for tissue aspiration and microscopy, but those who are unable to travel are started on treatment without confirmation of infection. A recent paper, measuring IgG subclass and IgG isotype antibody levels, observed elevated levels of specific IgG1 being associated with treatment failure and relapse of VL, whereas no or low levels of IgG1 were detected in cured VL patients [35]. This is being further evaluated using a lateral flow RDT, which could be a potentially useful test to confirm relapses in field conditions.

The elimination programme also emphasizes an active case-finding approach for the early detection of VL cases. This both benefits the individual with kala-azar and decreases transmission in the community. As the validation studies have used a passive case detection strategy, there is still a need to assess whether the performance of the diagnostic test is similar when it is employed using an active case detection strategy that assumes a shorter duration of the illness before diagnosis.

10.2.4 Antigen Detection Tests

In theory, antigen detection tests are more specific than serological tests, as they avoid cross-reactions and can distinguish active from past infections. A latex agglutination test (KAtex) detecting a heat-stable low molecular weight carbohydrate antigen in the urine of kala-azar patients showed promising initial results [36, 37]. Several studies conducted in East Africa and on the Indian subcontinent showed good specificity but low-to-moderate (48–87 %) sensitivity [38, 39]. As expected for an antigen detection test, KAtex turns negative in a very high proportion of patients (97–100 %) during anti-leishmanial treatment [40]. Apart from its low sensitivity, there are 2 practical limitations with KAtex. First, the urine must be boiled to avoid false-positive reactions, and second, it is difficult to distinguish weak positive from negative results, which affects the test's reproducibility [14, 38]. A simplified and improved form of KAtex is currently under development.

10.3 Access to VL Care

The determinants of access to VL care include knowledge of the disease, the inclination to seek health care, and the accessibility of diagnostic facilities.

The national control programmes that have been ongoing in these countries for many years include behavior change communication activities. Therefore, one might expect that the majority of the population in endemic areas would be knowledgeable about the disease and its transmission. However, a recent study showed that knowledge of the disease remained low in some regions. Fever, the most common symptom, was known to 32, 72, and 92 % of the interviewees from Bangladesh, Nepal, and India, respectively. In India and Nepal, almost all the respondents were aware that kala-azar was a curable disease, while in Bangladesh only 64 % were aware of this [41].

Numerous studies have shown that local unqualified village health care workers are the first-choice health care providers for many patients because of their easy accessibility [41–43]. Beyond the community, patients tend to prefer private providers, local pharmacists, or government health care facilities. A study from Bangladesh found that patients went through a median of 7 visits to different providers before starting VL treatment [42], and a similar situation has been found in Nepal [43, 44]. As a consequence, catastrophic expenditures are incurred by the families of these kala-azar patients despite the existence of a control programme providing free VL care. Over the last few years, efforts have been made to improve access to care by decentralizing VL treatment. This has been reflected by the decrease in the proportion of unreported cases in the national programme. The proportion of unreported cases from 4 to 8 times the government figures has come to be reduced to about 20 % of the total reported cases.

Because VL is a focal disease, active case detection strategies are being used to decrease the time to diagnosis and also prevent unnecessary visits to different care providers.

10.4 Issues in the Use of RDT in the Field

Issues to be addressed concerning the field use of the RDT within control programmes include the procurement and supply of the test kits, training and monitoring of their use by health care workers, and quality control.

After the validation of the rK39 ICT, this RDT was recommended as a diagnostic test and has been introduced into the control programmes up to the primary health center level in all 3 countries. This has encouraged many manufacturers to start producing RDTs, and currently many brands of RDT using the dipstick format are available, of which many are not sufficiently validated. Standard guidelines for the registration of tests similar to those for drugs have not yet been established. To address this, a multicenter evaluation of the commercially available brands was conducted, to help in making a selection for use in the programmes [28]. An additional major factor is the cost of the kit, which may discourage selection of the scientifically validated brands. The introduction of a new test requires training on a regular basis for the health personnel involved, which is a challenging task for the programme. The large number of health workers poses logistical challenges. Training manuals must also be developed that correspond to the skill level of individual health workers. In addition, the frequent transfer of health workers—which is quite common in these countries—and changes in the format of the RDT (dipstick or cassette versions) produce major difficulties.

Quality control of both the products procured and their use in the field is essential. Not only the performance of the different brands needs to be checked after the product reaches the country, but also that of different batches of the same brand. Most RDTs must be transported and stored within specified temperatures, avoiding extremes of heat. In these endemic countries on the Indian subcontinent, where both the temperature and humidity can be quite high, transportation under ideal conditions may not always be practically possible.

To avoid treating asymptomatic *Leishmania* infections, RDTs should be applied in combination with a clinical case definition, as many healthy people in the endemic region may also test positive for the disease [45, 46]. In these countries there is currently no set protocol for quality control after procurement of the RDT, or for monitoring its use in the field.

10.5 Conclusion

Diagnostic tests in the VL elimination programme will only have an impact if they are widely available to patients. The currently available RDT (rK39 ICT), though not ideal, is essential for the success of the kala-azar elimination initiative. There is a need for an RDT based on antigen detection to assess cure and relapse after VL treatment, and standards for the registration and quality control of RDTs must be established within each of the control programmes.

References

1. Alvar J, Vélez ID, Bern C, et al. Leishmaniasis worldwide and global estimates of its incidence. PLoS ONE. 2012;7:e35671.
2. Boelaert M, Meheus F, Sanchez A, et al. The poorest of the poor: a poverty appraisal of households affected by visceral leishmaniasis in Bihar, India. Trop Med Int Health. 2009;14:639–44.
3. http://www.searo.who.int/mediacentre/events/striving-for-elimination-kala-azar-story/en/.
4. Alam MZ, Kuhls K, Schweynoch C, Sundar S, Rijal S, et al. Multilocus microsatellite typing (MLMT) reveals genetic homogeneity of Leishmania donovani strains in the Indian subcontinent. Infection Genetics and Evolution. 2009;9:24–31.
5. Sundar S, More DK, Singh MK, et al. Failure of pentavalent antimony in visceral leishmaniasis in India: report from the centre of the Indian epidemic. Clin Infect Dis. 2000;31:1104–7.
6. Collin S, Davidson R, Ritmeijer K, et al. Conflict and kala-azar: determinants of adverse outcomes of kala-azar among patients in southern Sudan. Clin Infect Dis. 2004;38:612–9.
7. Sundar S. Drug resistance in Indian visceral leishmaniasis. Trop Med Int Health. 2001;6:849–54.
8. Rijal S, Chappuis F, Singh R, et al. Treatment of visceral leishmaniasis in south-eastern Nepal: decreasing efficacy of sodium stibogluconate and need for a policy to limit further decline. Trans R Soc Trop Med Hyg. 2003;97:350–4.
9. Rijal S, Bhandari S, Koirala S, et al. Clinical risk factors for therapeutic failure in Kala azar patients treated with pentavalent antimonials in Nepal. Trans R Soc Trop Med Hyg. 2010;104:225–9.
10. Bhattacharya SK, Sur D, Sinha PK, Karbwang J. Elimination of leishmaniasis (kala-azar) from the Indian subcontinent is technically feasible & operationally achievable. Indian J Med Res. 2006;123:195–6.
11. Sundar S, Reed SG, Singh VP, et al. Rapid accurate field diagnosis of Indian visceral leishmaniasis. Lancet. 1998;351:563–5.
12. Zhou XH, Obuchowski NA, McClish DK. The design of diagnostic accuracy studies in statistical methods in diagnostic medicine. In: Zhou XH, Obuchowski NA, & McClish DK, editors. Wiley, New York 2002, pp. 57–99.
13. Chappuis F, Rijal S, Singh R, Acharya P, Karki BMS, Das ML, et al. Prospective evaluation and comparison of the direct agglutination test and an rK39-antigen-based dipstick test for the diagnosis of suspected kala-azar in Nepal. Trop Med Int Health. 2003;8(3):277–85.

14. Chappuis F, Rijal S, Jha UK, Desjeux P, et al. Field validity reproducibility and feasibility of diagnostic tests for visceral leishmaniasis in rural Nepal. Trop Med Int Health. 2006;11:31–40.
15. Boelaert M, Rijal S, Regmi S, et al. A comparative study of the effectiveness of diagnostic tests for visceral leishmaniasis. Am J Trop Med Hyg. 2004;70:72–7.
16. Napier LE. A new serum test for kala-azar. Indian J Med Res. 1992;9:830–46.
17. WHO Technical Report Series 949. Control of the leishmaniasis: report of a WHO expert committee. World Health Organization, Geneva 2010.
18. Deborggraeve S, Boelaert M, Rijal S, et al. Diagnostic accuracy of a new Leishmania PCR for clinical visceral leishmaniasis in Nepal and its role in diagnosis of disease. Trop Med Int Health. 2008;13:1378–83.
19. Reithinger R, Dujardin JC. Molecular diagnosis of leishmaniasis: current status and future applications. J Clin Microbiol. 2007;45:21–5.
20. Deborggraeve S, Laurent T, Espinosa D, et al. A simplified and standardized polymerase chain reaction format for the diagnosis of leishmaniasis. J Infect Dis. 2008;198:1565–72.
21. El Harith A, Kolk AHJ, Kager PA, et al. A simple and economical direct agglutination test for serodiagnosis and sero-epidemiological studies of visceral leishmaniasis. Trans R Soc Trop Med Hyg. 1986;80:583–6.
22. Chappuis F, Rijal S, Soto A, et al. A meta-analysis of the diagnostic performance of the direct agglutination test and rK39 dipstick for visceral leishmaniasis. Br Med J. 2006;1(333):723–6.
23. Jacquet D, Boelaert M, Seaman J, Rijal S, et al. Comparative evaluation of freeze-dried and liquid antigens in the direct agglutination test for serodiagnosis of visceral leishmaniasis (ITMA-DAT/VL). Trop Med Int Health. 2006;11:1777–84.
24. Boelaert M, Verdonck K, Menten J, Sunyoto T, van Griensven J, Chappuis F, Rijal S. Rapid tests for the diagnosis of visceral leishmaniasis in patients with suspected disease. Cochrane Database Syst Rev. 2014;6:CD009135.
25. Bhattacharyya T, Bowes DE, et al. Significantly lower anti-Leishmania IgG responses in Sudanese versus Indian visceral leishmaniasis. PLoS Negl Trop Dis. 2014;8(2):e2675.
26. Mohapatra TM, Singh DP, Sen MR, Bharti K, Sundar S. Compararative evaluation of rK9, rK26 and rK39 antigens in the serodiagnosis of Indian visceral leishmaniasis. J Infect Dev Ctries. 2010;4(2):114–7.
27. Mbui J, Wasunna M, Balasegaram M, Laussermayer A, Juma R, et al. Validation of two rapid diagnostic tests for visceral leishmaniasis in Kenya. PLoS Negl Trop Dis. 2013;7(9):e2441.
28. Cunningham J, Hasker E, Das P, et al. WHO/TDR Visceral Leishmaniasis Laboratory Network. A global comparative evaluation of commercial immunochromatographic rapid diagnostic tests for visceral leishmaniasis. Clin Infect Dis. 2012;55(10):1312–9.
29. Kumar D, Khanal B, Tiwary P, et al. Comparative evaluation of blood and serum samples in rapid immunochromatographic tests for visceral leishmaniasis. J Clin Microbiol. 2013;51(12):3955–9.
30. Ghosh P, Hasnain MG, Ghosh D, et al. A comparative evaluation of the performance of commercially available rapid immunochromatographic tests for the diagnosis of visceral leishmaniasis in Bangladesh. Parasit Vectors. 2015;16(8):331.
31. Bern C, Hightower AW, Chowdhury R, et al. Risk factors for kala-azar in Bangladesh. Emerg Infect Dis. 2005;11(5):655–62.
32. De Almeida Silva L, Romero HD, Prata A, et al. Immunologic tests in patients after clinical cure of visceral leishmaniasis. Am J Trop Med Hyg. 2006;75:739–43.
33. Rijal S, Uranw S, Chappuis F, et al. The epidemiology of Leishmania donovani infection in high transmission foci in Nepal. Trop Med Int Health. 2009;14:60.
34. Sundar S, Gidwani K, Anderson E, et al. The epidemiology of Leishmania donovani infection in high transmission foci in India. Trop Med Int Health. 2009 Sep; 14:186.
35. Bhattacharyya T, Ayandeh A, Falconar et al. IgG1 as a potential biomarker of post-chemotherapeutic relapse in visceral leishmaniasis, and adaptation to a rapid diagnostic test. PLoS Negl Trop Dis. 2014 Oct 23;8(10):e3273.

36. Attar ZJ, Chance ML, el-Safi S, et al. Latex agglutination test for the detection of urinary antigens in visceral leishmaniasis. Acta Tropica 2001;78:11–6.
37. Sarkari B, Chance M, Hommel M. Antigenuria in visceral leishmaniasis: detection and partial characterisation of a carbohydrate antigen. Acta Trop. 2002;82:339–48.
38. Rijal S, Boelaert M, Regmi S, et al. Evaluation of urinary antigen-based latex agglutination test in the diagnosis of kala-azar in eastern Nepal. Trop Med Int Health. 2004;9:724–9.
39. Sundar S, Singh RK, Bimal SK, et al. Comparative evaluation of parasitology and serological tests in the diagnosis of visceral leishmaniasis in India: a phase III diagnostic accuracy study. Trop Med Int Health. 2007;12:284–9.
40. El-Safi SH, Abdel-Haleem A, Hammad A, et al. Field evaluation of latex agglutination test for detecting urinary antigens in visceral leishmaniasis in Sudan. Eastern Mediterr Health J. 2003;9:844–55.
41. Mondal D, Singh SP, Kumar N, et al. Visceral leishmaniasis elimination programme in India, Bangladesh, and Nepal: reshaping the case finding/case management strategy. PLoS Neglected Trop Dis. 2009;3:e355.
42. Sharma AD, Bern C, Varghese B, et al. The economic impact of visceral leishmaniasis on households in Bangladesh. Trop Med Int Health. 2006;11:757–64.
43. Rijal S, Koirala S, Van der stuyft P, Boelaert M. The economic burden of visceral leishmaniasis for households in Nepal. Trans R Soc Trop Med Hyg. 2006;100:838–841.
44. Adhikari AR, Maskey NM, Sharma BP. Paying for hospital-based care of kala-azar in Nepal: assessing catastrophic, impoverishment and economic consequences. Health Policy and Planning. 2008;24:129–39.
45. Sundar S, Maurya R, Singh RK, et al. Rapid, noninvasive diagnosis of visceral leishmaniasis in India: comparison of two immunochromatographic strip tests for detection of anti-K39 antibody. J Clin Microbiol. 2006;44:251–3.
46. Sundar S, Singh RK, Maurya R, et al. Serological diagnosis of Indian visceral leishmaniasis: direct agglutination test versus rK39 strip test. Trans R Soc Trop Med Hyg. 2006;100:533–7.

Chapter 11
Changes of *Leishmania* Antigens in Kala-Azar Patients' Urine After Treatment

Sharmina Deloer, Sohel Mohammad Samad, Hidekazu Takagi, Chatanun Eamudomkarn, Kazi M. Jamil, Eisei Noiri and Makoto Itoh

Abstract A sandwich enzyme-linked immunosorbent assay (ELISA) to detect *Leishmania* circulating antigens (CA) in urine samples was applied to monitor the antigen levels after treatment of visceral leishmaniasis, also known as kala-azar. Fifty VL cases were followed up for 2 months after treatment with sodium stibogluconate. The CA levels decreased rapidly after the treatment, demonstrating a 76, 92, and 97 % reduction after 1, 2, and 3 weeks, respectively, and became negative after 8 weeks. The CA was detected in 40 of 50 urine samples (80 %). The positive rate of CA with sodium stibogluconate was slightly lower than that with nested PCR (88 %), but higher than that with KAtex (48 %) before the treatment, and it reduced more slowly than that with the others after the treatment. As the urine ELISA gives CA levels quantitatively, it will be useful to evaluate the efficacy of the treatments used and for surveillance of drug resistance.

Keywords Kala-azar · Urine · ELISA · *Leishmania donovani* · Circulating antigen · Bangladesh

11.1 Introduction

Visceral leishmaniasis (VL) remains endemic in 57 countries and 200,000–400,000 cases occur each year. More than 90 % of VL cases occur in just 6 countries: India, Bangladesh, Sudan, South Sudan, Brazil, and Ethiopia [1]. It is a vector-born disease caused by the protozoan *Leishmania* species. In different countries different species of *Leishmania* are responsible for different types of the disease. In 2005, a

---

S. Deloer (✉)
Department of Microbiology and Immunology, Aichi Medical University School of Medicine, Nagakute 480-1195, Japan; Department of Infection and Immunology, Aichi Medical University, Aichi, Nagakute, Japan
e-mail: kakon_sd@yahoo.com

memorandum of understanding was signed between Bangladesh, India, and Nepal to eliminate kala-azar or VL by 2015, aiming to reduce the incidence to less than 1 in 10,000 population.

Although detection of parasites in biopsy samples from the spleen or bone marrow is the gold standard for diagnosis of VL, it is invasive and trained personnel are required. Serological tests to detect antibodies to the parasite are now commonly used and are sensitive, and a specific immunochromatographic strip test is commercially available (rK39 dip stick), but such tests cannot distinguish present and past infections. PCR with peripheral blood to detect the parasite DNA is now commonly used, as well as loop-mediated isothermal amplification (LAMP), which can detect the DNA without thermal cyclers [2]. Detection of the parasite-derived antigens in circulation (circulating antigens, [CA]) is another way to confirm the presence of infection. A latex agglutination test kit, KAtex, which detects antigens in urine samples is commercially available but its sensitivity varied study to study [3]. We also developed an ELISA to detect CA in urine samples that are collected safely and easily with good compliance [4]. In this study, we examined CA in urine samples from VL patients before and after treatment and compared the changes with those of DNA in blood detected by nested PCR.

11.2 Detection of Leishmanial Derived Circulating Antigens (CA) in Urine Samples

Sandwich ELISA to detect CA in urine was carried out according to the method described previously [4]. Briefly, 96-well microtiter plates (MaxiSorp^{TM}, Nunc, Denmark) were coated with 10 µg/mL (100 µL/well) rabbit anti-*L. donovani* IgG and blocked with a casein buffer (1 % casein in 0.05 M Tris-HCL buffer with 0.15 M NaCl, pH 7.6) for 2 h at room temperature, then the wells were loaded with 100 µL of urine (1:2 dilution in casein buffer) and incubated at 25 °C overnight. After washing with PBS (pH 7.4) containing 0.05 % Tween 20, the plate was incubated with biotinylated rabbit anti-*L. donovani* IgG (10 µg/mL in the casein buffer) at 37 °C for 1 h. After being washed, the plates were incubated with horseradish peroxidase conjugated streptavidin (Vector Laboratories Inc., CA, USA) (1:1000 dilution in the casein buffer) at 37 °C for 1 h and then with substrate ABTS (KPL Inc., Gaithersburg, MD, USA) for 1 h at room temperature. The optical density was measured at 415 and at 492 nm as reference. Each sample was assayed in duplicate. Antigen levels were expressed as absolute concentrations estimated from the standard curve constructed with 3 folds of serially diluted *L. donovani* promastigote acetone treated with crude antigens of known concentrations (10,000–14 ng/mL): CA levels were indicated as nanogram equivalents to the crude antigens (ng eq/mL). As the crude antigens were used as the standard, CA detected in this ELISA might be overestimated because the crude antigens contain less CA. The cutoff was calculated from the values of 5.1 ng eq/mL.

A commercially available urinary antigen detection kit, KAtex, was also used according to the manual.

11.3 Detection of *Leishmania* Derived DNA by Nested PCR

DNA was extracted from blood samples by QIAamp DNA Blood Mini Kit (QIAGEN) according to the manufacturer's protocol. They were used for LAMP and nested PCR for the detection of parasite DNA. The nested PCR was carried out according to reports [5]. Briefly, primers 5'-AAATCGGCTCCGAGGCGGGAAAC-3' and 5'-GGTACACTCTATCAGTAGCAC-3' were used for the first PCR followed by the nested PCR with primers 5'-TCGGACGTGTGTGGATATGGC-3' and 5'-CCFATAATATAGTATCTCCCG-3'. The nested PCR amplified a 385 bps fragment internal to the 592 bps product of the first PCR. The nested PCR used 1 µL of the diluted (1:10) product from the first PCR under the same conditions as the first PCR except primers.

11.4 Changes After Treatment

11.4.1 Levels of CA in Urine

CA levels in urine detected by this ELISA are shown in Fig. 11.1. CA were detected in 40 among the 50 urine samples (80 %) obtained before treatment and no CA was detected in 40 negative samples (nonendemic healthy sample). The levels varied from 28 to 661 ng eq/mL; the average was 478 ng eq/mL. The CA levels decreased rapidly after initiation of the treatment with sodium stibogluconate, then gradually decreased and were diminished at 8 weeks: 115 ng eq/mL after 1 week and 4.5 ng eq/mL after 4 weeks.

11.4.2 CA Positive Rate

Compared to the rapid decrease of the CA levels, the positive rates decreased slowly: 64 % after 1 week, 30 % after 2 weeks, 12 % after 3 weeks, and 4 % after 4 weeks (Fig. 11.2). Those of KAtex were lower than those of others: 48 % before treatment, and 6 and 2 % at 1 and 2 weeks, respectively.

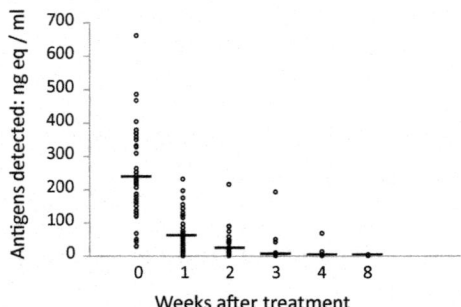

Fig. 11.1 Changes in levels of *L. donovani* derived circulating antigens (*CA*) in urine of patients with visceral leishmaniasis (*VL*) after treatment. Forty patients with VL whose urine samples were CA positive before treatment were followed up after the treatment. The *horizontal bars* in each week shows the averages. Urine samples were collected as follows; before treatment (0), 1, 2, 3, 4 and 8 weeks after treatment

11.4.3 Nested PCR Positive Rate

Forty-six of the 50 blood samples (92 %) were positive with the nested PCR before the treatment. The positive rate rapidly decreased to 20 and 2 % at 1 and 2 weeks after the treatment, respectively, then all became negative at 3 weeks.

11.5 Discussion

Fifty VL patients were followed up with testing for CA in urine by ELISA and KAtex, and testing for parasite DNA in blood by nested PCR after treatment with sodium stibogluconate. The nested PCR was the most sensitive (92 %); ELISA

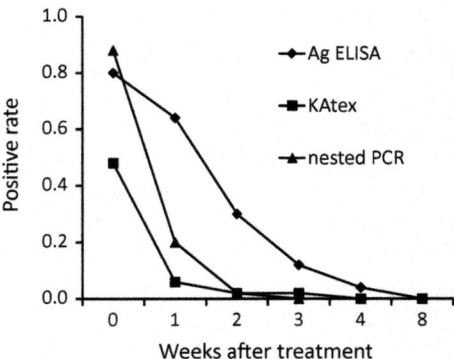

Fig. 11.2 Changes in the positive rates of ELISA, nested PCR, and KAtex after treatment for visceral leishmaniasis (*VL*)

(80 %) and KAtex (48 %) followed it. Among the 10 ELISA negatives, 3 were also negative with the nested PCR. One of the 3 nested PCR negatives was positive with the ELISA. Although the samples and the target to be detected were different, 84 % of the ELISA results were identical to those of the nested PCR. The simple kit, KAtex, is an excellent tool for antigen detection, but the sensitivity was low compared with the other 2 methods used.

The positive rate of the parasite DNA in blood decreased rapidly after initiation of the treatment; 92 % decreased to 20 % (1 week), 2 % (2 weeks), and 0 % (3 weeks). The positive rates of the CA in urine changed gradually; 64, 30, 12, and 4 % at 1, 2, 3, and 4 weeks, respectively. Pourabbas et al. [6] also reported rapid reduction of the parasite DNA in blood from *L. infantum* infected patients using real time PCR. The mechanism of the clearance of DNA and CA is not clear, but the DNA seems to be more rapidly cleared than CA from circulation.

Unresponsiveness of VL to anti-*Leishmania* drugs have been reported and has severely compromised disease control [7–11]. Surveillance of appearance of the resistance is essential and changes in levels of the parasite DNA and/or CA will provide this information. Both methods can be used for monitoring the efficiency of the treatments.

Two diagnostic methods with urine are now available for visceral leishmaniasis: anti-*Leishmania* antibody detection [12, 13] and CA detection. As control programs proceed, good compliance from residents of the *Leishmania* control areas becomes difficult to obtain. Urine samples that are safely and noninvasively collected will be useful in such situations.

11.6 Conclusion

Utilizing ELISA to detect CA quantitatively in urine is useful, not only to monitor the response to treatments, but also to conduct active surveys of visceral leishmaniasis by detecting anti-*Leishmania* antibodies in urine.

Acknowledgments The research underlying this chapter was conducted by the authors, and was partially supported by JST/JICA, SATREPS, and also by a Grant-in-Aid for Scientific Research (B) No. 18406013 from the Japan Society for the Promotion of Science.

References

1. Alvar J, Vélez ID, Bern C, et al. Leishmaniasis worldwide and global estimates of its incidence. PLoS ONE. 2012;7(5):e35671.
2. Takagi H, Itoh M, Islam MZ, et al. Sensitive, specific, and rapid detection of *Leishmania donovani* DNA by loop-mediated isothermal amplification. Am J Trop Med Hyg. 2009;81(4):578–82.

3. Abeijon C, Campos-Neto A. Potential non-invasive urine-based antigen (protein) detection assay to diagnose active visceral leishmaniasis. PLoS Negl Trop Dis. 2013;7(5):e2161.
4. Islam MZ, Itoh M, Ekram AR, et al. Detection of *Leishmania (Leishmania) donovani* antigen in urine by antigen capture enzyme-linked immunosorbent assay. In: Hashiguchi Y, editor. Studies on New and Old World leishmaniases and their transmission, with particular reference to ecuador, Peru, Argentina and Pakistan. Kochi, Japan; 2004. p. 78–84.
5. Sreenivas G, Ansari NA, Kataria J, et al. Nested PCR assay for detection of *Leishmania donovani* in slit aspirates from post-kala-azar dermal leishmaniasis lesions. J Clin Microbiol. 2004;42:1777–8.
6. Pourabbas B, Ghadimi Moghadam A, Pouladfar G, et al. Quantification of Leishmania infantum kinetoplast DNA for monitoring the response to meglumine antimoniate therapy in visceral leishmaniasis. Am J Trop Med Hyg. 2013;88(5):868–71.
7. Sundar S, More DK, Singh MK, et al. Failure of pentavalent antimony in visceral leishmaniasis in India: report from the center of the Indian epidemic. Clin Infect Dis. 2000;31(4):1104–7.
8. Olliaro PL, Guerin PJ, Gerstl S, et al. Treatment options for visceral leishmaniasis: a systematic review of clinical studies done in India, 1980–2004. Lancet Infect Dis. 2005;5(12):763–74.
9. Jha TK. Epidemiology of drug-resistant kala-azar in India and neighboring countries. Kala azar in south Asia. Jha TK, Noiri E, editors. New York: Springer; 2011. p. 21–34.
10. Stauch A, Duerr HP, Dujardin JC, et al. Treatment of visceral leishmaniasis: model-based analyses on the spread of antimony-resistant *L. donovani* in Bihar, India. PLoS Negl Trop Dis. 2012;6(12):e1973.
11. Rijal S, Ostyn B, Uranw S. Increasing failure of miltefosine in the treatment of Kala-azar in Nepal and the potential role of parasite drug resistance, reinfection, or noncompliance. Clin Infect Dis. 2013;56(11):1530–8.
12. Islam MZ, Itoh M, Takagi H, et al. Enzyme-linked immunosorbent assay to detect urinary antibody against recombinant rKRP42 antigen made from *Leishmania donovani* for the diagnosis of visceral leishmaniasis. Am J Trop Med Hyg. 2008;79(4):599–604.
13. Islam MZ, Itoh M, Islam MA, et al. ELISA with recombinant rKRP42 antigen using urine samples: a tool for predicting clinical visceral leishmaniasis cases and its outbreak. Am J Trop Med Hyg. 2012;87(4):658–62.

Chapter 12
Potentiality of Urinary L-FABP Tests to Kala-Azar Disease Management

Eisei Noiri, Yoshifumi Hamasaki, Bumpei Tojo, Kazi M. Jamil, Kent Doi and Takeshi Sugaya

Abstract The diagnostic process of disease detection and disease management is different in every disease. Recently, the diagnostic procedure for visceral leishmaniasis (VL), also known as kala-azar, was simplified from a time-consuming pathological examination into a simple blood test. But the monitoring of disease activity during therapy still relies on clinical findings and classical laboratory data in endemic areas. In this chapter, we introduce and examine the utility of a urinary biomarker, fatty acid-binding protein 1 (FABP1) or alternatively L-type fatty acid-binding protein (L-FABP), for monitoring VL disease activities and drug-induced side effects. A FABP1 analysis developed as an enzyme-linked immunosorbent assay was transformed into a urinary immuno-chromatography dipstick test for point of care use in endemic areas. We expect that a FABP1 dipstick test will serve as a triage marker in disease monitoring of VL. In addition, this chapter introduces the impact of FABP1 on predicting survival in septic acute kidney injury and the clinical interpretation of FABP1 measurements in VL.

Keywords Dipstick urine test · Fatty-acid-binding protein 1 · Leishmaniasis · Urinary biomarker · Pharmacovigilance · Renal toxicity · Kidney injury

12.1 Introduction

The diagnosis of visceral leishmaniasis (VL), also known as kala-azar, has been dependent on parasite detection in organ aspirates by expert pathologists, but now the rK39 test is a simpler alternative [1]. Though the diagnostic procedure is standardized, the monitoring procedure for disease activity after starting VL therapy remains dependent on the clinical symptoms, such as body temperature and physical examination of the spleen size. Blood tests such as the erythrocyte sedi-

E. Noiri (✉)
Hemodialysis and Apheresis, Nephrology 107 Lab, The University of Tokyo Hospital, Tokyo, Japan
e-mail: noiri-tky@umin.ac.jp

© Springer International Publishing 2016
E. Noiri and T.K. Jha (eds.), *Kala Azar in South Asia*,
DOI 10.1007/978-3-319-47101-3_12

mentation test, complete blood count, and prothrombin time are not suitable for frequent evaluation. Because antibodies persist for a long period after the onset of disease, antibody-based immune-detection tools are not useful for monitoring the disease activity of VL [2]. Furthermore, diagnostic techniques that do not require electricity are necessary for use in the frontier locations of endemic areas. For this reason, we focused on urinary tests to evaluate the disease activity of VL.

Proteinuria has been reported empirically by physicians in endemic areas and is well known to occur in tropical diseases. However, specific attention has not been paid to the value of this symptom as a marker for VL disease activity. Recently, urinary biomarkers have been a focus in the field of acute kidney injury (AKI) [3], for which the definition is broader than that for acute renal failure [4]. These biomarkers have been developed to detect AKI much faster than is possible by measuring serum creatinine (SCr) or blood urea nitrogen (BUN) levels. For instance, an increase of SCr levels may be detected 24–48 h after cardio-pulmonary bypass surgery, but new urinary biomarkers, such as neutrophil gelatinase-associated lipocalin (NGAL), L-type fatty-acid binding protein (L-FABP, FABP1), kidney injury molecule-1 (KIM-1), etc., can detect AKI within 2 h after surgery [5–7]. Because the kidney is sensitive to blood circulation and mirrors the body condition through factors in the urine, sensitive urinary biomarkers in AKI are potential candidates to sense subtle changes in the health of patients with tropical diseases. Among urinary biomarkers, NGAL is partly derived from neutrophils and therefore NGAL blood and urine levels often increase in various infectious diseases [8, 9]. Careful consideration should be necessary to use NGAL in endemic area as a biomarker. In addition, enormous gender differences are seen in the normal range of children, often victim of tropical diseases [10]. There is accumulating evidence demonstrating FABP levels may indicate ischemic conditions in various organs. FABP1 is a potential predictive marker detectable in urine [11]. In addition, FABP1 can detect cisplatinum-induced nephrotoxicity [12] and contrast-medium-induced nephropathy [13] earlier than can SCr and BUN levels. KIM-1 levels in the urine increase in AKI at a slightly delayed time compared to those of NGAL and FABP1, but faster than do SCr levels. Recent reports have clarified that KIM-1 is an efficient reporter molecule to sense various drug-induced nephrotoxicities [14]. Highly sensitive urinary biomarkers may contribute not only to the monitoring of disease activities in tropical diseases including VL, but also drug-induced nephrotoxicities, which are more serious issues for VL therapy. Unlike therapies targeting bacteria and viruses, therapies against eukaryotic organisms such as parasites are complex because their cellular structures are similar to those of human beings. FABP1 is one of the best studied urinary biomarkers in AKI and has been introduced for the purpose of disease control in VL. Herein, we review the basic aspects of FABP1 in the human kidney, demonstrate our recent investigation into a VL cohort using newly developed urine dipstick tests driven by immunochromatography and the ELISA assay, and discuss the specific applicability of FABP1 for control of VL.

12.2 Physiological and Pathophysiological Attributes of FABP1 in the Human Kidney

FABPs are known as intracellular lipid chaperones that transport lipids to specific components in the cell; however, little is known about their exact biological functions and mechanisms of action [15]. FABPs are found in many different species including Drosophila melanogaster, Caenorhabditis elegans, mice, rats, and humans and their tissue distribution of FABP is ubiquitous [16]. Nine different FABPs have been reported including liver (L-), intestinal (I-), muscle and heart (H-), adipocyte (A-), epidermal (E-), ileal (Il-), brain (B-), myelin (M-), and testis (T-). Through an analysis of the human kidney, Veerkamp and co-workers found 2 types of FABPs in renal tubular cells; FABP1 and heart and muscle type (FABP3 or H-FABP; H-type fatty acid binding protein) [17]. They investigated the characteristics of renal FABP1 and FABP3 in detail. Renal FABP1 and FABP3 showed the same Kd values for oleic acid derived from liver and heart. Compared with FABP1 expressed in the liver, FABP1 in the kidney showed a more neutral isoelectric point (pI 5.8) and had 2 additional tryptophan residues. Therefore, renal FABP1 is seemingly kidney specific and could be a new subtype of FABP1. FABPs in the human kidney are predominantly expressed in epithelial cells. FABP1 is specifically localized around proximal tubular cells and FABP3 is in distal tubular cells. Immunohistochemical analysis of FABP1 may also show a weak positive staining in parietal glomerular epithelial cells. FABP1 is a 14 kDa protein with a beta barrel structure of beta sheets shaped like a clam shell (Fig. 12.1). Helix-turn-helix motif by 2 alpha-helix plays a role of hatch to the inside and stabilize the molecular structure. The promoter region of FABP1 contains binding sites for hepatocyte nuclear factor (HNF), hypoxia inducible factor-1 (HIF-1), and peroxisome proliferator activated receptor (PPARs). Because of the anatomical structure of the kidney, the outer medullary region can be easily injured by hypoxia resulting from decreased peritubular capillary blood flow and subsequent oxidative stress in the renal IR model [18–20]. Proximal tubules are rather more susceptible to hypoxic stress than are distal tubules. Under hypoxic conditions, proximal tubules are apt to undergo necrosis, but distal tubules under the same level of hypoxic stress instead show apoptosis [21]. The FABP1 gene is responsive to hypoxic stress because the promoter region of the FABP1 gene has a hypoxia responsive element (HRE).

A simultaneous decrease of peritubular capillary blood flow should provoke FABP1 excretion into urine. As a proof of this concept, we examined the peritubular capillary flow in transplanted kidneys and compared that level with urinary FABP1 concentrations derived from the ureter of transplanted kidneys. Among urinary indicators measured, such as NAG, $\alpha 1MG$, and $\beta 2MG$, urinary FABP1 levels showed a significant correlation with the degree of peritubular capillary blood

Fig. 12.1 Crystal structure of human FABP1. FABP1 binds free fatty acids and their coenzyme A derivatives, bilirubin, and some other small molecules including HNE in cytoplasm, and therefore are considered to be involved in intracellular lipid and lipid peroxidation product transport. The image was obtained from Protein Data Base

flow. Ischemic time, defined as beginning at the harvest point from the donor to the first urine draining from the ureter of the transplanted kidney, correlated extremely well with urinary FABP1 levels (Fig. 12.2). A 1-h biopsy sample demonstrated that the urinary FABP1 was derived from the proximal epithelial tubular cells shed into the tubular lumen. One possible mechanism for FABP1 projection from the proximal cellular cytoplasm to the tubular lumen is proposed. The physiological role of FABP1 is related to the metabolism of fatty acids often bound with albumin. Fatty acids are filtered together with albumin through the glomerular basement membrane and re-absorbed from proximal tubular epithelial cells together with albumin. FABP1 captures fatty acids and relocates them to cellular organelles such as mitochondria, lysosomes, and peroxisomes for their energy metabolism (Fig. 12.3, right side). When proximal epithelial cells are subjected to injury, presumably related to the decrease of peritubular blood flow, intracellular oxidative stress will be increased by the breakdown of cytoplasmic membrane structures, of which the majority are lipid peroxidation products. Because of the flexibility of the recognition site of FABP1, highly cytotoxic aldehydes (HNE, HHE, etc.) can be bound to FABP1 and presumably excreted into the tubular lumen in the role of a chaperone for fatty acids and related products (Fig. 12.3, left side). The latter part of

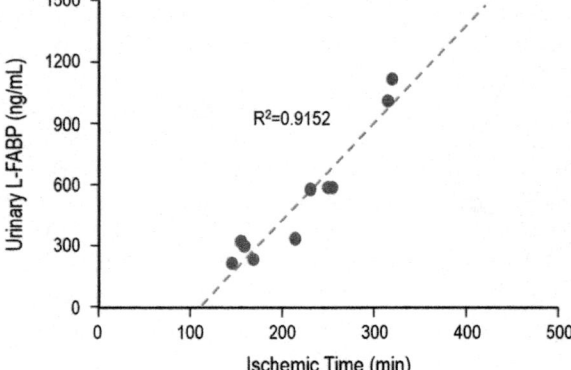

Fig. 12.2 Correlation between ischemic time and urinary FABP1. The ischemic time in living-related renal transplantation was defined as the period between the time point of clamping the donor's renal artery and the time point of appearance of virgin urine from the recipient's ureter. "Reproduced with permission from the American Society for Nephrology, [22]"

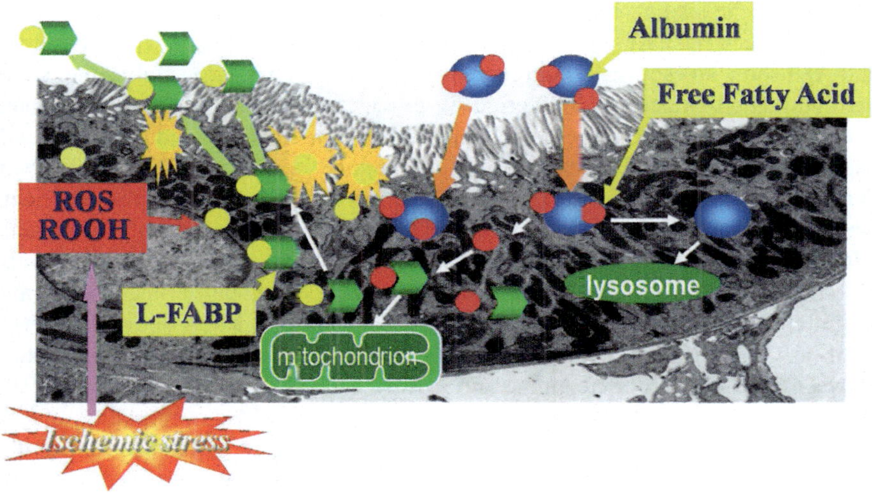

Fig. 12.3 Conceptual schema for the renal FABP1 mechanism. In the kidney, albumin filtered from glomeruli is reabsorbed predominantly in proximal tubules together with free fatty acids (FFA) under physiologic conditions. After reabsorption, cytosolic albumin transfer to lysosome and fatty acid was released and received by L-FABP (FABP1) during this process. Fatty acid-bound L-FABP will usually be relocated to cytosolic peroxisome for size reduction of fatty acids. Under ischemic conditions lipid peroxidation products will accumulate in proximal tubules and damage proximal tubules (*left*). L-FABP is presumably capable of binding these noxious lipid peroxidative products and transferring them to urinary spaces. L-FABP is excreted from the proximal tubules into urine by binding cytotoxic lipids. ROOH denotes as hydroperoxide radicals. "Reproduced with permission from the American Society for Nephrology, [22]"

explanation is presumably accepted in purposiveness but the details of the mechanism are not clarified.

We reported that in the human pathophysiological condition urinary FABP1 was useful for early detection of AKI after pediatric cardio-pulmonary bypass surgery (CPB-AKI) [6], in which urinary FABP1 levels were examined at 4 and 12 h after surgery. Receiver operating characteristic (ROC) curve analysis of urinary FABP1 levels for CPB-AKI diagnosis was performed. The area under the ROC curve of urinary FABP1 at 4 h after surgery was 0.810, which is an acceptable level for a single predictive biomarker. Univariate logistic regression analyses showed that both bypass time and urinary FABP1 levels were significant independent risk indicators for AKI. Pediatric cardiac surgery is recognized as an ideal clinical setting for the initial studies in biomarker development in terms of minimal comorbidity and known timing of renal injury [23]. Likewise, early detection of AKI after adult CPB surgery was reported even immediately after surgery showing an area under the ROC of 0.86, at 3 h after surgery of 0.85, and at 6 h after surgery of 0.83 [7].

Sepsis is the leading cause of death in critically ill patients, and the incidence of sepsis has been increasing [24, 25]. Sepsis causes AKI and patients with both sepsis and AKI show an especially high mortality rate [26]. In one large multicenter study involving approximately 30,000 critically ill patients, 50 % of AKI was associated with septic shock [27]. Therefore, early prediction of sepsis-induced AKI will enable us to improve patient survival. Recently urinary FABP1 levels were examined in patients with septic shock. Urinary FABP1 levels in these patients were significantly higher than those in healthy subjects. In a cohort of 145 patients with septic shock, a logistic regression analysis including urinary FABP1 levels, blood endotoxin level, CRP, peripheral white blood cell count, urinary NAG, and serum creatinine levels revealed that only urinary FABP1 levels showed a significant association with patient survival (Fig. 12.4). There was no correlation between levels of serum creatinine and urinary FABP1. Likewise, urinary FABP1 testing in a mixed ICU population was able to predict 14-day mortality, one of important major adverse kidney events, with an accuracy of 0.9 area under the curve of ROC for the independent cohort of a single-center prospective study ($n = 339$) [29].

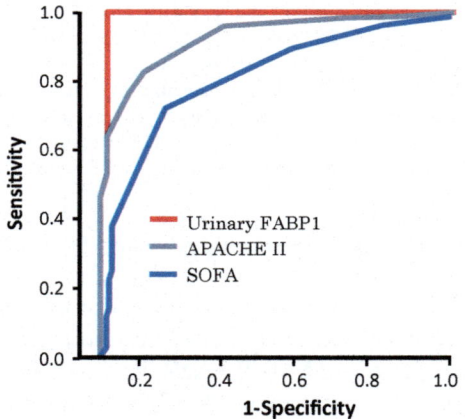

Fig. 12.4 Receiver operating characteristic curve analysis for 14-days mortality in the mixed ICU. Receiver operating characteristic curve analysis for deaths was performed by using urinary fatty acid-binding protein 1 (*FABP1*) measured at admission. Acute Physiology and Chronic Health Evaluation II (*APACHE II*) and Sepsis-related Organ Failure Assessment (*SOFA*) scores at the first day at the intensive care unit with 145 septic shock patients. The area under the receiver operating characteristic curves (95 % confidence interval) of each parameter was as follows: urinary FABP1 0.993 (0.956–0.999), APACHE II 0.927 (0.873–0.959), and SOFA 0.813 (0.733–0.872). "Reproduced with permission from Wolters Kluwer, [28]"

Polymyxin B-immobilized fiber (PMX-F) hemoperfusion has been performed to treat severe sepsis in more than 30,000 patients in Japan since 1994, and a meta-analysis demonstrated its efficacy in septic shock treatment [30]. Of a cohort of 40 patients with sepsis treated with PMX-F, 28 survived and 12 died [31]. Among the survivors, urinary FABP1 levels were reduced by treatment. However, the non-survivors showed higher urinary FABP1 levels, with a smaller decrease after the treatment, compared with those of the survivors. These results suggested that urinary FABP1 levels could reflect the severity of sepsis, and also monitor the effectiveness of treatment [31].

Renal toxicity is one of the major concerns in VL therapy with both traditional and newly developed drugs. The potential for urinary FABP1 to detect the early onset of nephrotoxicity was recently demonstrated in animal experiments. We have shown that urinary FABP1 levels can reflect interstitial changes after the administration of cyclooxygenase 2 (COX2) inhibitors under low-sodium diet conditions [32]. The low sodium levels enhanced renin-angiotensin-aldosterone activity and reduced peritubular capillary blood flow. The administration of the COX2 selective inhibitor meloxicam further decreased peritubular capillary blood flow significantly. A mild interstitial fibrotic region and partial cell infiltration was confirmed by histological examination. A more suggestive finding of this study is the concept of responder and non-responder for the evaluation of drug toxicology. Celecoxib, another COX2 inhibitor, induced low peritubular capillary blood flow in a certain fraction of animals (about 50 %), which were categorized as responders. Animals of the responder group showed higher urinary FABP1 levels compared with that seen in the non-responder group. When kidneys were harvested 4 weeks after celecoxib administration, kidneys derived from responders showed more interstitial fibrotic regions compared with those of non-responders (Fig. 12.5). This kind of individual variation in phenotypes frequently occurs in human clinical studies. It is crucial to distinguish between drug responders and non-responders to improve treatment strategies and reduce adverse effects. Urinary FABP1 levels may be able to distinguish responders at risk for drug renal toxicity.

The carry-over of renal toxicity to chronic kidney diseases (CKD), or that of AKI to CKD, is not well studied in the field of nephrology. However, there is a potential for urinary FABP1 levels to be used to monitor that progression, given its proven utility in various types of CKD [33]. When mice expressing human FABP1 were exposed to low doses of cis-platinum (CDDP; 10 mg/kg once weekly for 3 weeks), urinary FABP1 levels increased. Kidneys that received CDDP multiple times showed increasing fibrotic changes (Fig. 12.6). These histological changes were also associated with increasing levels of urinary FABP1 comparing 0 and 3 week values (bar graph, Fig. 12.6). Results from this preclinical study suggest that urinary FABP1 levels may predict the progression of AKI to CKD. The target population for intensive treatment may have the features of lower blood flow in the peritubular capillaries, subsequent hypoxic conditions, and the presence of interstitial fibrosis, all of which can be monitored by urinary FABP1 levels.

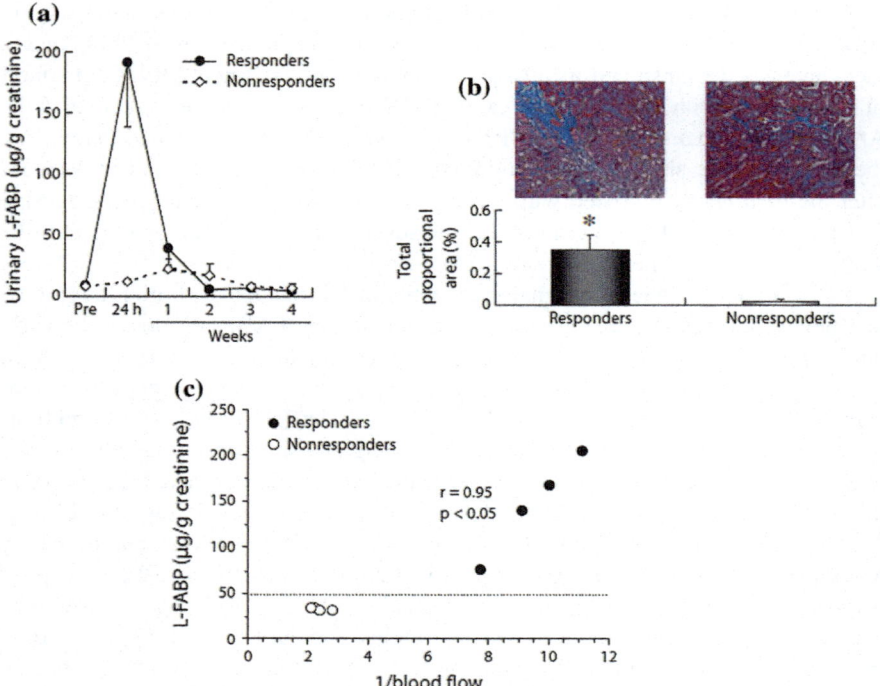

Fig. 12.5 Analyses of urinary FABP1 responders and non-responders after starting celecoxib. **a** Time course of urinary FABP1 (L-FABP) levels. Animals excreting urinary FABP1 >50 μg/g creatinine during the follow-up period were defined as responders ($n = 4$). Those excreting lower levels were defined as non-responders or low responders ($n = 4$). **b** Representative images obtained from responder and non-responder kidneys. Quantitative analyses were performed for interstitial fibrosis. Asterisk depicts $p < 0.05$. **c** Correlations between urinary FABP1 and peritubular capillary blood flow 2 days after starting celecoxib. "Reproduced with permission from S. Karger AG, [32]"

12.3 Potential of FABP1 as a Urinary Marker for VL

In a cross-sectional study, urinary FABP1 levels were examined in both 162 patients with VL (median age 18.0 years; range 5–60; male:female = 89:73) and 46 healthy endemic control individuals (median age 24.5 years; range 7–50; male:female = 30:16). VL was diagnosed by rK39 and spleen biopsy examination and only cases with clear diagnosis of VL were allowed entry into the study. Urine from the patients was obtained at the initial clinical visit and measured for urinary creatinine, protein, NAG, and FABP1 to determine any associations with disease severity and kidney stress. The summary data in Table 12.1 demonstrate that both urinary FABP1 and protein levels were increased significantly in the VL cohort compared with findings seen in the control. When urinary concentrations were adjusted for urinary creatinine, those values showed double. This is presumably

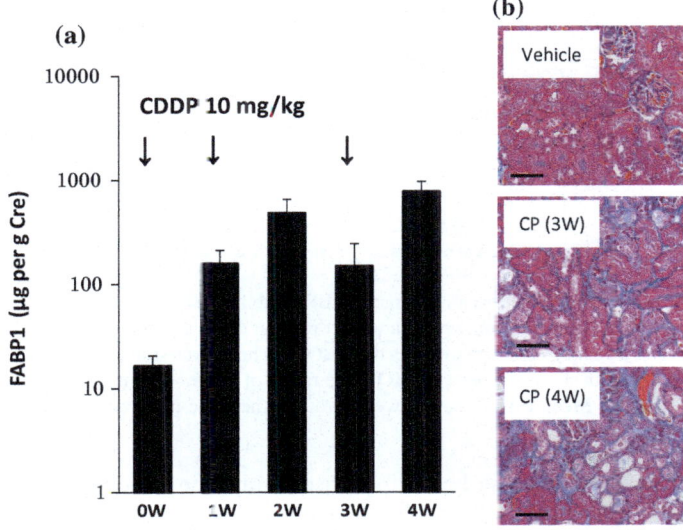

Fig. 12.6 Response to multiple CDDP treatments in the mice with CP-induced tubulointerstitial injury. **a** Animals were administered 10 mg/kg CDDP 3 times over 4 weeks (0, 1, and 3 weeks). After exposure to CDDP, levels of BUN and urinary FABP1 were significantly higher at 4 weeks (4 W) ($n = 7$–8). FABP1 levels at 3 W were decreased according to the lack of CDDP injection at 2 W. **b** To assess the cumulative CDDP induced toxicity for this model, some mice ($n = 6$) were killed at 3 W after first CDDP injection. Moderate renal interstitial fibrosis was already seen at 3 W after two CDDP injections, this tissue significantly deteriorated at 4 W after the third CDDP injection (size bar = 50 μm)

Table 12.1 Urinary L-FABP (uL-FABP) and protein (uProtein) levels in both visceral leishmaniasis (VL) and endemic control cohorts in Mymensingh, Bangladesh

		N	Mean	SD	SE	<95 %CI	>95 %CI	Prob > |t|
uL-FABP (ng/mL)	VL	162	113.41	223.30	17.54	78.77	148.06	
	Endemic control	46	0.94	1.95	0.29	0.36	1.52	<0.0001
uProtein (mg/dL)	VL	162	35.83	46.01	3.62	28.69	42.97	
	Endemic control	46	3.41	2.22	0.33	2.76	4.07	<0.0001
uL-FABP (μg/g Cre)	VL	162	153.20	326.59	25.66	102.53	203.87	
	Endemic control	46	1.93	3.85	0.57	0.78	3.07	<0.0001
uProtein (mg/g Cre)	VL	162	486.56	510.66	40.12	407.33	565.79	
	Endemic control	46	125.48	164.69	24.28	76.57	174.38	<0.0001

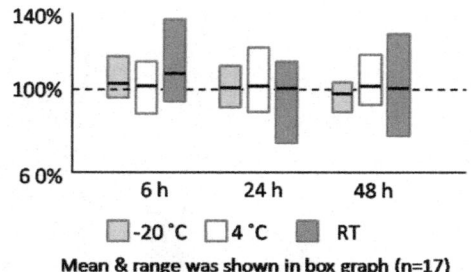

Fig. 12.7 Urine storage conditions and urinary FABP1 stability. Box graphs demonstrate the range of FABP1 (maximum and minimum value) and the thick bar depicts mean value of each group. The reproducibility of urinary FABP1 results was the best when urine was stored at −20 °C. When urine samples were stored in room temperature (RT), the range of values of 6 h samples was similar to that of 48 h samples. Mean FABP1 levels were virtually the same in all conditions

because the serum creatinine levels for those residing in endemic areas are lower than those of Western individuals and Japanese residents and may reflect the basal malnutrition condition of the residents in the endemic areas [34]. Urinary NAG often showed a value under the detection limit in both VL and the control cohort. Though samples were stored at under −20 °C, high external temperatures and the transfer time between sentinel sites and the laboratory affected the outcome of the urinary indicators. Similar sample handling effects are expected in case of NGAL, which requires more strict storage conditions after securing the samples [35]. FABP1 is a stable protein in urine; Fig. 12.7 compares values measured immediately after sampling with those measured when urine samples were stored at −20, 4 °C, or room temperature for either 6, 24, or 48 h. Urine concentrations stored at room temperature for 48 h were within 20 % variation of the initial measurements. In addition, the FABP1 levels did not show a sex difference or a daily fluctuation. The cut-off value of FABP1 in the immuno-chromatography assay, based on this information, was placed at 100 ng/mL. Proteinuria was measured by the Bradford method (BioRad) with the standardization of bovine serum albumin, but the potential cut-off value was too low to be distinguishable by protein dipstick tests.

We followed-up 50 patients in the VL cohort (median age 17.5 years; range 5–45; male:female = 29:21) by retaking the urine and serum samples for 3 months, including a hospital admission period of 4 weeks with sodium stibogluconate (SSG) treatment; this was a cooperative investigation by the International Center for Diarrhoeal Disease Research, Bangladesh (icddr,b; Dr. Kazi M. Jamil and colleagues). The clinical study was approved by the ethics committee of the icddr,b. Duration of fever was 85.8 ± 32.7 days, 58 % of patients showed abdominal swelling, especially on the left side, and 94 % had a history of weight loss. Family history of VL was present in 82 %. Inclusion criteria were as follows, considering the risks of spleen biopsy: (1) hemoglobin levels >6.0 g/dL; (2) platelet count >5 × 10^4 cells/μL; (3) prothrombin time <4 s. All patients had confirmed diagnoses of VL by both spleen biopsy and rK39. All patients showed complete

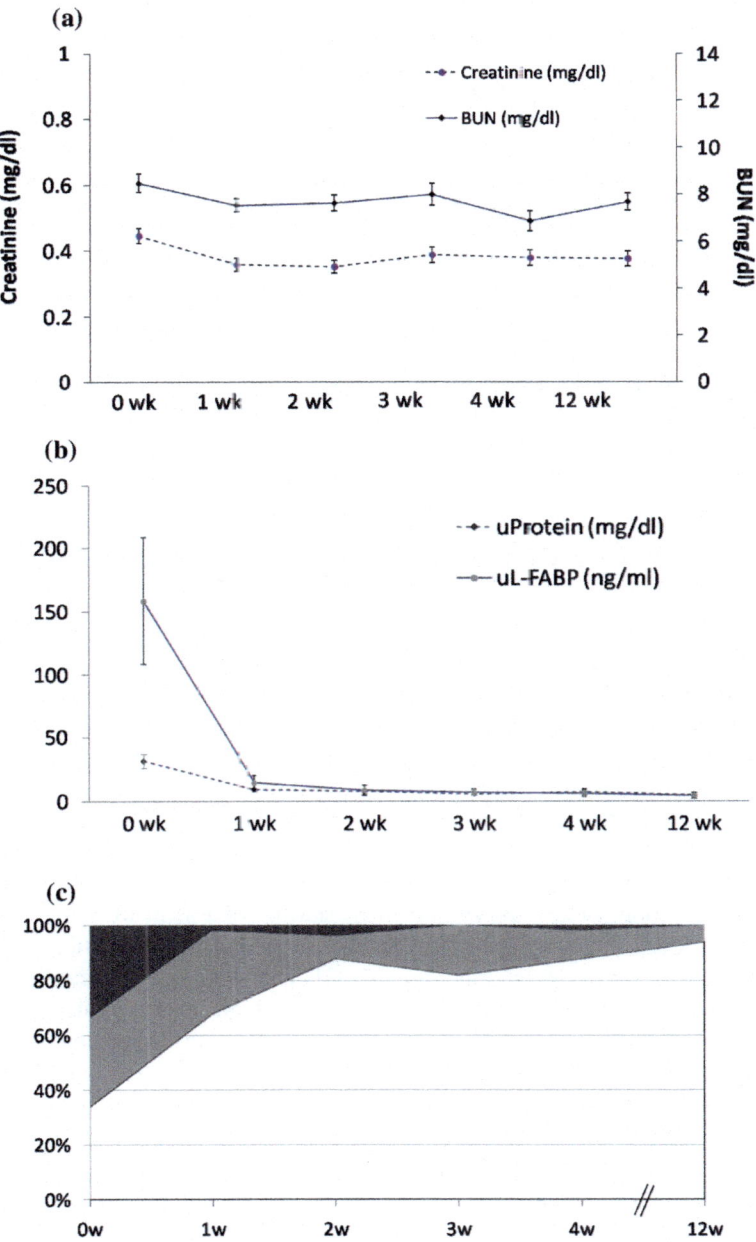

Fig. 12.8 The follow-up data of 50 patients with visceral leishmaniasis (VL). All 50 patients received a diagnosis of VL confirmed by spleen biopsy and rK39. Sample correction was performed immediately before treatment and 1, 2, 3, 4, 12 weeks after treatment by SSG. **a** SCr and serum urea nitrogen (BUN), **b** Urinary FABP1 and protein. **c** Follow-up data was stratified by urinary FABP1 level in each sampling point

remission clinically by 4 weeks after starting therapy. SCr and BUN levels were steady during the follow-up period (Fig. 12.8a). In contrast, the increase of urinary FABP1 levels seen at the initial clinic visit decreased by 1 week after starting therapy, which initially developed by dehydration-induced prerenal and amastigote involvement induced renal damage were improved and presumably reflects the efficacy of the prescription (Fig. 12.8b). A similar trend was seen in proteinuria. When the 50 cases were stratified by urinary FABP1 levels, the cut-off value of 10 ng/mL was considered to more sensitively reflect the disease activity and the efficacy of the prescription (Fig. 12.8c).

12.4 Efficacy to Detect Renal Toxicity by Urinary FABP1

The potential of urinary FABP1levels to be used to detect renal toxicity has been reported in an animal CDDP-AKI model [12] and in human contrast-medium-induced AKI [13]. In this chapter, the potential utility of FABP1 levels in monitoring amphotericin-B induced kidney injury will be discussed. Amphotericin-B binds to the sterols in the cell membrane of renal tubular epithelial cells and endothelial cells and modifies membrane permeability. Amphotericin-B, acting as a pseudo-phospholipid, interacts with sterol molecules to cause formation of aqueous pores. These pores consist of an annulus of polyene and sterol in which the hydrophilic region of the drug molecule faces the interior of the pore. This mechanism is more or less inevitable when starting treatment of amphotericin-B and related drugs, but acute spasms of renal vasculature may be induced by doses toxic enough to induce severe renal injury [36]. Therefore, we expected that urine FABP1 levels may identify more susceptible individuals among treated patients.

To examine the toxicity of amphotericin-B and liposome-conjugated amphotericin-B (AmBisome®), we injected different dosages as shown in Fig. 12.9 into humanized FABP1 mice every 24 h [37]. When animals were subjected to amphotericin-B 50 mg/kg, all were dead within 24 h. AmBisome® 50 mg/kg was tolerable to the animals, but urine volume was increased to three fold within 24 h. When animals were subjected to amphotericin-B 10 mg/kg, urine volume increased day by day and reached 3-fold levels in a week. AmBisome® 10 mg/kg increased urine volume more moderately compared with that seen with amphotericin-B 10 mg/kg. Urinary FABP1 levels within 24 h after starting those treatments increased, clearly reflecting the level of injury shown by increased urine volumes often seen in clinical situation. But urinary FABP1 levels started to decrease after 24 h although those injections were continued. The reason for this

Fig. 12.9 Amphotericin-B and AmBisome® toxicity were examined using humanized FABP1 transgenic mice. Both drugs were injected to each animal every day at the indicated dose (mg/kg). Urine samples were serially collected in glass-made metabolic cages. Mean and SE values are shown in the graph. The doses used in this experiment are higher compared with those used in previous studies. When animals were subjected to amphotericin-B 50 mg/kg, all were dead within 24 h. **a** Urinary volume. **b** Urinary FABP1 level

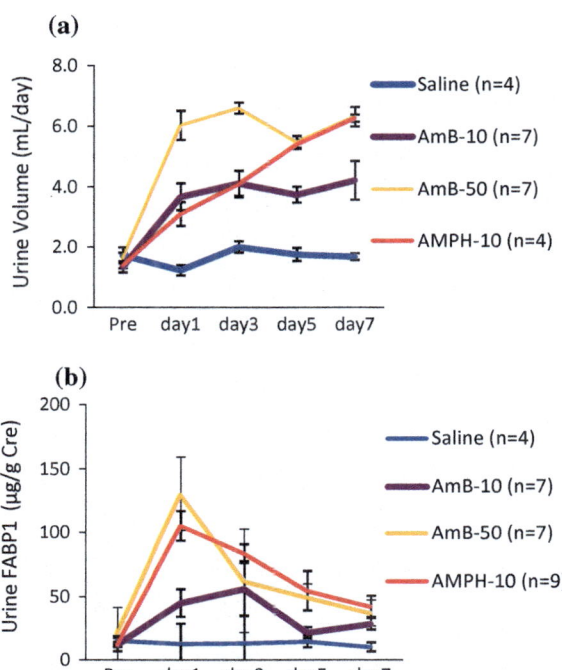

partial response of FABP1 is not clear. But because renal localization of FABP1 is limited to proximal tubular cells [37], and the effect of amphotericin-B in proximal tubules is presumably limited in higher dosages because peritubular blood flow is affected by the initiation of therapy. Thus FABP1 response may be lessened with time. An alternative explanation is that epithelial membrane disruption precludes holding cytosolic components including FABP1. Recently, higher doses of AmBisome® and other therapies are being considered to shorten the duration of therapy and attain better patient compliance. Single-dose therapy of AmBisome® has been started in Suruya-Kanta Kala-azar Research Center (SKKRC), Mymensingh, Bangladesh. SKKRC is the referral center for intractable cases such as treatment failure or resistance in Bangladesh. SKKRC was established by SATREPS (joint program of Japan Scientific Agency and Japan International Cooperation Agency), DNDi. icddr,b, and the government of Bangladesh. Urine samples were collected at three time points between May and December 2014; before starting AmBisome®, one day after, and at discharge from SKKRC. Each case was plotted in Fig. 12.10. Almost all patients had urinary FABP1 levels higher than 7 µg per g creatinine (µg/gCre) before starting treatment. When treatment was successful, the level of FABP1 fell to below 10 µg/gCre. Interestingly, FABP1 levels higher than 10 µg/gCre in urine 24 h after AmBisome® administration

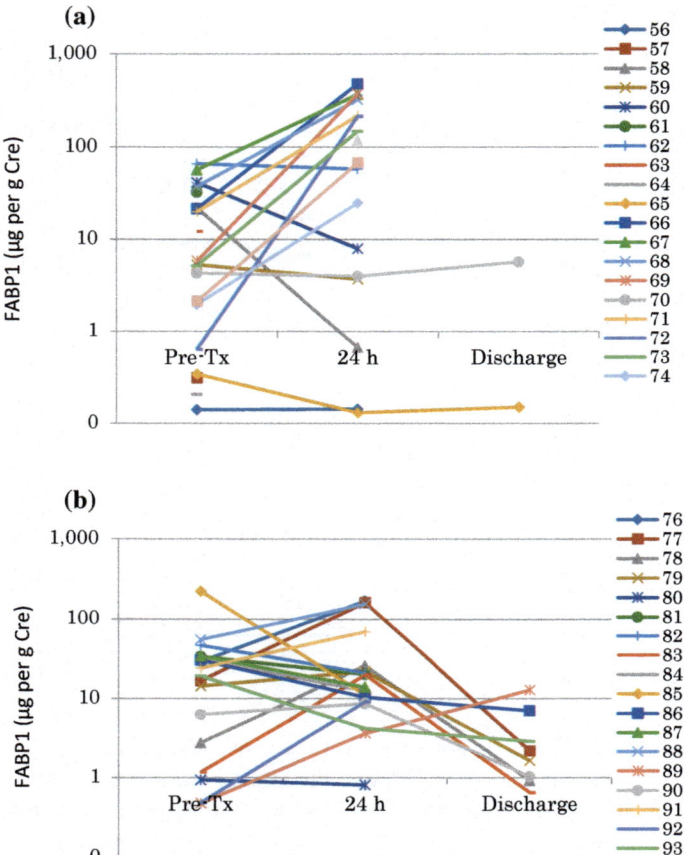

Fig. 12.10 Change of urinary FABP1 before and after single injection protocol of AmBisome® Urinary FABP1 levels are plotted for 39 cases. Almost all the patients show higher urine FABP1 values. Similar to results shown in Fig. 12.9, the increase of FABP1 levels was often seen 24 h after the administration of AmBisome®. These increases were seemingly transient and FABP1 levels had almost normalized at discharge

reflects AKI, similarly to findings for humanized FABP1 transgenic mice data in Fig. 12.9.

These data convinced us that urinary FABP1 levels are a promising indicator for VL disease activity and potential drug-induced kidney injury. In this way, urinary FABP1, especially that level before and 24 h after initiation of therapy, will be useful as the triage biomarker to promote individualized patient care.

12.5 Fabrication the of FABP1 Immuno-chromatography Dipstick

A more realistic approach is necessary for VL diagnosis at the frontier of the endemic area, the so-called ground level. A urine dipstick test will be much more suitable in these primitive conditions than the ELISA method, even for qualitative study. We developed a dipstick test using immuno-chromatography methods; the mechanistic explanation is summarized in Fig. 12.11. This strategy enables the original sandwich ELISA method for FABP1 to be incorporated into a convenient urine immuno-dipstick. Approximately 90 µL of urine is applied by a tiny dropper and the run-in period is 10 min. The cut-off value of the current FABP1 dipstick is 100 ng/mL. The intensity of the band is similar to that of the control band, which is a sign of a successful run of the urine sample.

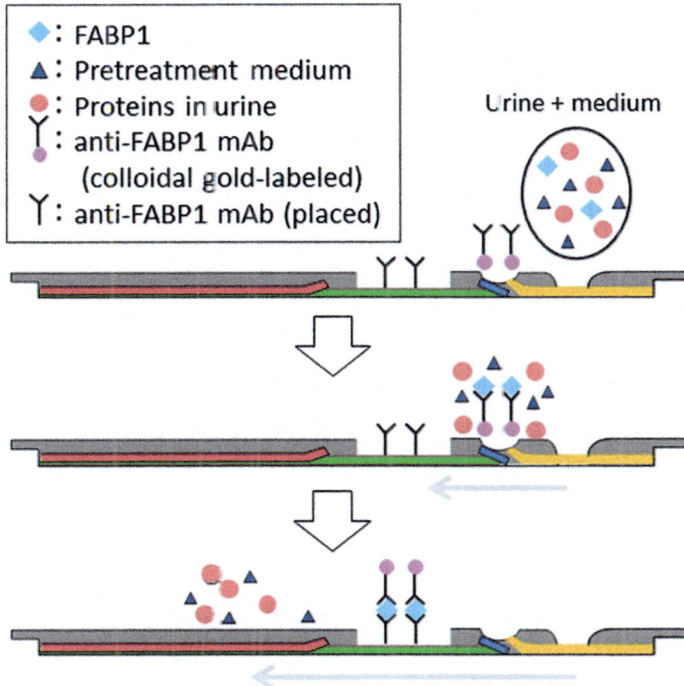

Fig. 12.11 The principle of the immunochromatography test for urinary FABP1. For the purpose of VL screening, the cut-off value was tentatively assigned to 100 ng/mL and the siliconized gold-colloid filter (pink circle with antibody mark) was optimized for endemic urine. From top to bottom schema, applied urine will run to left side and the colorimetric reaction will appear as a single band. Light blue rectangle, blue triangle, and salmon pink circle are FABP1, pretreatment agent, contaminant proteins, respectively. FABP1 is captured by fixed anti-FABP1 antibody and color will appear by the gold-colloid effect

Fig. 12.12 Tone level list of the urinary FABP1 immunochromatography test (FABP1 dipstick). Dip-score was determined by tone level list

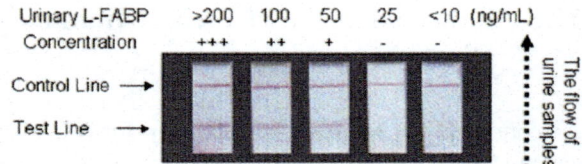

We tried the dipstick method at the Kalazar Research Center in Bihar with Dr T. K. Jha and his team in 2008. We had a chance to examine 25 patients. As seen in Fig. 12.12, the tone level list was made to read the sample concentration from (−) to (3+). And then the level of tone list assigned the digits from 0 to 4. The actual distribution of urine sample results is plotted in Fig. 12.13a with the comparison between ELISA and dipstick data (dip-score), and results are further summarized in the dip-score average seen in Fig. 12.13b. Both exhibit extremely good correlation, as expected. The within-run reproducibility was examined three times at different dates using pre-measured samples and standard FABP1-dependent positive and negative controls. Reproducibility was excellent (Fig. 12.14). It was necessary to optimize the gold-colloid filter in the dipstick cassette maximize performance in endemic areas and a different type of siliconized filter was used to reduce oozing and to increase contrast.

Quantification of dipstick results was examined using a touch panel operative immuno-reader (Fig. 12.15). ROC analysis for detection was examined for

Fig. 12.13 Comparison of urinary FABP1 (L-FABP) Dip-score versus ELISA. Of note that y-axis of ELISA uses logarithmic scale. **a** The distribution of each sample data. **b** The average of each dip-score was summarized in line graph

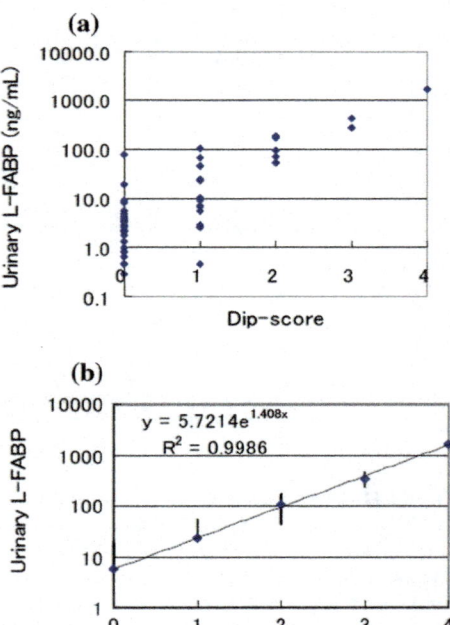

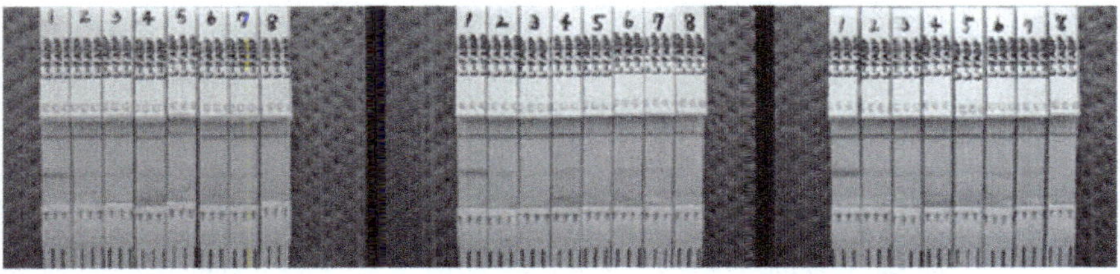

	Urine sample	$n=1$	$n=2$	$n=3$
1	Standard FABP1 (400 ng/mL)	++	++	++
2	Positive High urine (125.1 ng/mL)	+	+	+
3	Positive Medium urine (68.0 ng/mL)	+	+	+
4	Positive Low urine (17.8 ng/mL)	+	+	+
5	Negative urine 1	-	-	-
6	Negative urine 2 (false positive)	+	+	+
7	102-1 (105.8 ng/mL)	+	+	+
8	3-1 (0.5 ng/mL)	-	-	-

Fig. 12.14 Within-run reproducibility test. Tests were examined 3 times at different dates by using pre-measured samples and standard FABP1 (L-FABP)-dependent positive and negative controls

Fig. 12.15 Touch panel operative immuno-reader

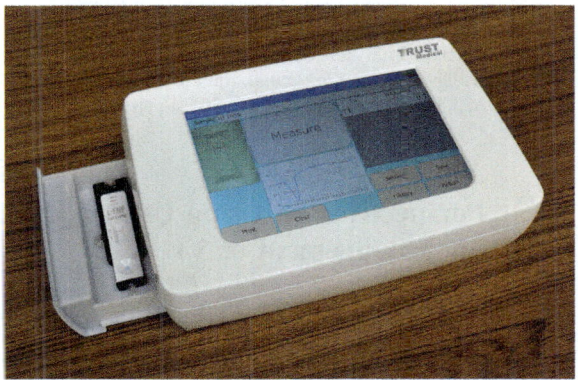

dip-score or immuno-reader results compared with ELISA. Urine samples from 99 VL patients were determined using FABP1 cut-off value 30, 50, 100 ng/mL, and compared with the performance of dip-score or that of immuno-reader. Area under ROC curve of dip-score values detected 0.77, 0.87, and 0.89, and that of the immuno-reader showed 0.81, 0.88, and 0.90 accuracy for the positive detection with the cut-off value of 30, 50, and 100 ng/mL (Fig. 12.16).

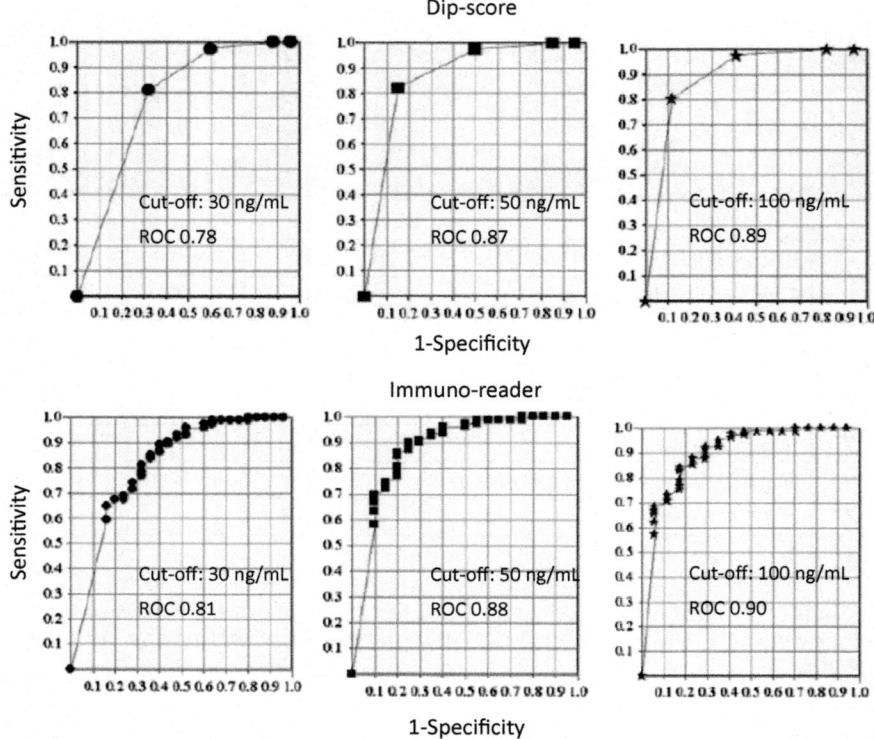

Fig. 12.16 ROC analyses determined by ELISA were compared to those for Dip-score or immuno-reader. From the *left panel*, the cut-off values were assigned 30 (positive individuals 25, negative 74), 50 (positive individuals 20, negative 79), and 100 ng/mL (positive individuals 17, negative 82)

12.6 Perspectives on Using Urinary Biomarkers for Management of VL

We had a chance to follow-up 50 individuals with VL for 3 months. Almost all patients responded well to SSG treatment within 2 weeks after starting the drug. Urinary biomarkers data have demonstrated that AmBisome® has a potential for kidney injury just after treatment. Therefore, careful monitoring of these patients is necessary during and shortly after this treatment. Initial disease activity, which should be immediately resolved with the appropriate prescription, occasionally is sustained with clinical symptoms. Disease activity remaining in patients with treatment failure will be easily and rapidly detectable by the use of urinary biomarker dipsticks. Further study will be necessary with other types of tropical diseases which are often co-infected in residents living in endemic areas.

Acknowledgments Part of this study was supported by Science and Technology Research Partnership for Sustainable Development from Japan Agency for Medical Research and Development (SATREPS, JST/JICA) (EN).

References

1. Sundar S, Singh RK, Maurya R, Kumar B, Chhabra A, Singh V, et al. Serological diagnosis of Indian visceral leishmaniasis: direct agglutination test versus rK39 strip test. Trans R Soc Trop Med Hyg. 2006;100(6):533–7.
2. Singh DP, Sundar S, Mohapatra TM. The rK39 strip test is non-predictor of clinical status for kala-azar. BMC Res Notes. 2009;2:187.
3. Vaidya VS, Ferguson MA, Bonventre JV. Biomarkers of acute kidney injury. Annu Rev Pharmacol Toxicol. 2008;48:463–93.
4. Hoste EA, Clermont G, Kersten A, Venkataraman R, Angus DC, De Bacquer D, et al. RIFLE criteria for acute kidney injury are associated with hospital mortality in critically ill patients: a cohort analysis. Crit Care. 2006;10(3):R73.
5. Mishra J, Dent C, Tarabishi R, Mitsnefes MM, Ma Q, Kelly C, et al. Neutrophil gelatinase-associated lipocalin (NGAL) as a biomarker for acute renal injury after cardiac surgery. Lancet. 2005;365(9466):1231–8.
6. Portilla D, Dent C, Sugaya T, Nagothu KK, Kundi I, Moore P, et al. Liver fatty acid-binding protein as a biomarker of acute kidney injury after cardiac surgery. Kidney Int. 2008;73(4):465–72.
7. Matsui K, Kamijo-Ikemori A, Sugaya T, Yasuda T, Kimura K. Usefulness of urinary biomarkers in early detection of acute kidney injury after cardiac surgery in adults. Circ J Off J Jpn Circ Soc. 2012;76(1):213–20.
8. Ichino M, Kuroyanagi Y, Kusaka M, Mori T, Ishikawa K, Shiroki R, et al. Increased urinary neutrophil gelatinase associated lipocalin levels in a rat model of upper urinary tract infection. J Urol. 2009;181(5):2326–31.
9. Alpizar-Alpizar W, Laerum OD, Illemann M, Ramirez JA, Arias A, Malespin-Bendana W, et al. Neutrophil gelatinase-associated lipocalin (NGAL/Lcn2) is upregulated in gastric mucosa infected with helicobacter pylori. Virchows Arch. 2009;455(3):225–33.
10. Bennett MR, Nehus E, Haffner C, Ma Q, Devarajan P. Pediatric reference ranges for acute kidney injury biomarkers. Pediatr Nephrol. 2015;30(4):677–85.
11. Noiri E, Doi K, Negishi K, Tanaka T, Hamasaki Y, Fujita T, et al. Urinary fatty acid-binding protein 1: an early predictive biomarker of kidney injury. Am J Physiol Renal Physiol. 2009;296(4):F669–79.
12. Negishi K, Noiri E, Doi K, Maeda-Mamiya R, Sugaya T, Portilla D, et al. Monitoring of urinary L-type fatty acid-binding protein predicts histological severity of acute kidney injury. Am J Pathol. 2009;174(4):1154–9.
13. Nakamura T, Sugaya T, Node K, Ueda Y, Koide H. Urinary excretion of liver-type fatty acid-binding protein in contrast medium-induced nephropathy. Am J Kidney Dis. 2006;47(3):439–44.
14. Dieterle F, Marrer E, Suzuki E, Grenet O, Cordier A, Vonderscher J. Monitoring kidney safety in drug development: emerging technologies and their implications. Curr Opin Drug Discov Devel. 2008;11(1):60–71.
15. Furuhashi M, Hotamisligil GS. Fatty acid-binding proteins: role in metabolic diseases and potential as drug targets. Nat Rev Drug Discov. 2008;7(6):489–503.
16. Veerkamp JH, Paulussen RJ, Peeters RA, Maatman RG, van Moerkerk HT, van Kuppevelt TH. Detection, tissue distribution and (sub)cellular localization of fatty acid-binding protein types. Mol Cell Biochem. 1990;98(1–2):11–8.

17. Maatman RG, Van Kuppevelt TH, Veerkamp JH. Two types of fatty acid-binding protein in human kidney. Isolation, characterization and localization. Biochem J. 1991;273(Pt 3):759–66.
18. Doi K, Suzuki Y, Nakao A, Fujita T, Noiri E. Radical scavenger edaravone developed for clinical use ameliorates ischemia/reperfusion injury in rat kidney. Kidney Int. 2004;65(5):1714–23.
19. Yamamoto T, Tada T, Brodsky SV, Tanaka H, Noiri E, Kajiya F, et al. Intravital videomicroscopy of peritubular capillaries in renal ischemia. Am J Physiol Renal Physiol. 2002;282(6):F1150–5.
20. Noiri E, Nakao A, Uchida K, Tsukahara H, Ohno M, Fujita T, et al. Oxidative and nitrosative stress in acute renal ischemia. Am J Physiol Renal Physiol. 2001;281(5):F948–57.
21. Bonventre JV, Weinberg JM. Recent advances in the pathophysiology of ischemic acute renal failure. J Am Soc Nephrol. 2003;14(8):2199–210.
22. Yamamoto T, Noiri E, Ono Y, Doi K, Negishi K, Kamijo A, et al. Renal L-type fatty acid-binding protein in acute ischemic injury. J Am Soc Nephrol. 2007;18(11):2894–902.
23. Thurman JM, Parikh CR. Peeking into the black box: new biomarkers for acute kidney injury. Kidney Int. 2008;73(4):379–81.
24. Angus DC, Linde-Zwirble WT, Lidicker J, Clermont G, Carcillo J, Pinsky MR. Epidemiology of severe sepsis in the United States: analysis of incidence, outcome, and associated costs of care. Crit Care Med. 2001;29(7):1303–10.
25. Martin GS, Mannino DM, Eaton S, Moss M. The epidemiology of sepsis in the United States from 1979 through 2000. N Engl J Med. 2003;348(16):1546–54.
26. Russell JA, Singer J, Bernard GR, Wheeler A, Fulkerson W, Hudson L, et al. Changing pattern of organ dysfunction in early human sepsis is related to mortality. Crit Care Med. 2000;28(10):3405–11.
27. Uchino S, Kellum JA, Bellomo R, Doig GS, Morimatsu H, Morgera S, et al. Acute renal failure in critically ill patients: a multinational, multicenter study. JAMA. 2005;294(7):813–8.
28. Doi K, Noiri E, Maeda-Mamiya R, Ishii T, Negishi K, Hamasaki Y, et al. Urinary L-type fatty acid-binding protein as a new biomarker of sepsis complicated with acute kidney injury. Crit Care Med. 2010;38(10):2037–42.
29. Doi K, Negishi K, Ishizu T, Katagiri D, Fujita T, Matsubara T, et al. Evaluation of new acute kidney injury biomarkers in a mixed intensive care unit. Crit Care Med. 2011;39(11):2464–9.
30. Cruz DN, Perazella MA, Bellomo R, de Cal M, Polanco N, Corradi V, et al. Effectiveness of polymyxin B-immobilized fiber column in sepsis: a systematic review. Crit Care. 2007;11(2):R47.
31. Nakamura T, Sugaya T, Koide H. Urinary liver-type fatty acid-binding protein in septic shock: effect of polymyxin B-immobilized fiber hemoperfusion. Shock; 2008.
32. Tanaka T, Noiri E, Yamamoto T, Sugaya T, Negishi K, Maeda R, et al. Urinary human L-FABP is a potential biomarker to predict COX-inhibitor-induced renal injury. Nephron Exp Nephrol. 2008;108(1):e19–26.
33. Kamijo-Ikemori A, Sugaya T, Kimura K. Urinary fatty acid binding protein in renal disease. Clin Chim Acta. 2006;374(1–2):1–7.
34. Hossain MI, Dodd NS, Ahmed T, Miah GM, Jamil KM, Nahar B, et al. Experience in managing severe malnutrition in a government tertiary treatment facility in Bangladesh. J Health Popul Nutr. 2009;27(1):72–9.
35. Haase-Fielitz A, Haase M, Bellomo R. Instability of urinary NGAL during long-term storage. Am J Kidney Dis. 2009;53(3):564–5; author reply 6.
36. Zgheib NK, Capitano B, Branch RA. Amphotericin B. In: DeBroe ME, Porter GA, Bennett WM, editors. Clinical nephrotoxins: renal injury from drugs and chemicals. Dordrecht: Springer; 2008. p. 323–52.
37. Yamamoto T, Noiri E, Ono Y, Doi K, Negishi K, Kamijo A, et al. Renal L-type fatty acid binding protein in acute ischemic injury. J Am Soc Nephrol. 2007;18(11):2894–902.

Chapter 13
Antibody Capture Direct Agglutination Test (abcDAT) for Diagnosis of Visceral Leishmaniasis with Urine

Fumiaki Nagaoka, Hidekazu Takagi, Eisei Noiri and Makoto Itoh

Abstract A sensitive diagnostic method to detect anti-*Leishmania* antibodies in urine samples has been developed. This simple method uses plastic plates coated with anti-human IgG. Urine samples are applied to the plate and then colored *Leishmania donovani* promastigotes are added serving as direct agglutination test (DAT) antigens. If the urine contains IgG to *Leishmania*, the DAT antigen attaches to the plate; sediment will be observed if there is no specific IgG. Among 56 urine samples from patients with visceral leishmaniasis (VL) 55 (98.2 %) were positive with this method. However, 13 endemic normals and 24 Japanese heathy controls were negative. As the results can be observed by the naked eye, no detection equipment is required, and only the antibody coated plate and DAT antigen which are stable at least for one year under ambient temperature, needed for the test. This method is considered useful for active surveys of visceral leishmaniasis at late stages of the control program.

Keywords Kala-azar · Urine · *Leishmania donovani* · DAT · Diagnosis

13.1 Introduction

Visceral leishmaniasis (VL) is caused by a protozoan parasite of the *Leishmania* species that is transmitted by sandflies. Poverty is one of the risk factors of the disease [1, 2], and the disease is potentially fatal without appropriate therapy. The WHO targeted VL, which is one of the neglected tropical diseases and a public health problem on the Indian subcontinent, for elimination by 2017. Sensitive, easy

---

M. Itoh (✉)
Department of Microbiology and Immunology, Aichi Medical University School of Medicine, Nagakute, Aichi 480-1195, Japan
e-mail: macitoh@aichi-med-u.ac.jp

© Springer International Publishing 2016
E. Noiri and T.K. Jha (eds.), *Kala Azar in South Asia*,
DOI 10.1007/978-3-319-47101-3_13

to use, and low cost diagnostic tools suitable for use in endemic regions are required for the control program. Detection of the parasite in amastigote form in spleen or bone marrow aspirate is the most accurate method [3]. But application of this invasive method is limited to hospitals with trained doctors. As the control program progresses, not only will the diagnosis in suspected patients be required, but also surveys of asymptomatic carriers. Detection of parasite-specific antibodies in the blood by the rK39 dipstick test is now commonly used. The antigen-specific antibodies are also detected in urine by enzyme-linked immunosorbent assay (ELISA) and the development of VL is predicted with the rise in antibody levels [4, 5].

The direct agglutination test (DAT) uses blue colored and freeze dried *L. donovani* promastigotes to detect anti-*Leishmania* antibodies in serum samples. The simple and sensitive DAT doesn't require any sophisticated equipment, and the results, agglutination of the colored promastigote, can be seen with the naked eye [6]. Nawa et al. developed an IgM-capture ELISA method to detect antigen-specific immunoglobulins for serodiagnosis of dengue [7, 8].

In this chapter, we report a new diagnostic tool that can detect anti-*Leishmania* IgG in urine samples using the DAT antigen and the immunoglobulin capture method.

13.2 Antibody Capture Direct Agglutination Test (abcDAT)

Diagram of the abcDAT is shown in Fig. 13.1. A 96-well microtiter plate with V-bottom (MICROLON®, Greiner Bio-One Co. Ltd., Tokyo, Japan) was coated with 0.5 μg/mL goat anti-human IgG (Protos Immunoresearch, Burlingame, CA) at 4 °C overnight and blocked with a solution, StabilCoat® (SurModics, Inc., MN, USA), then dried according to the manufacture's protocol. The dried plates were sealed in shading bags with deoxidizer until use. Urine samples were directly applied to each well (100 μl/well) and incubated at room temperature for 1 h. If the urine samples contained anti-*Leishmania* IgG, the immunoglobulins were captured by the anti-human IgG attached on the plate at this step. After the plates were washed 3 times with PBS-tween, 100 μl of 10 times diluted DAT antigen (KIT-Biomedical Research, Amsterdam, Netherlands, diluted by PBS-tween) was added and the plates were incubated overnight at room temperature. If the samples were positive, the blue colored DAT antigen bound to the captured anti-*Leishmania* IgG is visible as a small blue dot, the size of which is dependent on the antibody levels (Fig. 13.2 1–6). If it is negative, the unbound DAT antigen forms a sediment, of which size is the same as negative control applied to each plate, at the bottom of the V-shaped well (Fig. 13.2 1 and 2). The results of the test were able to be judged visually (Fig. 13.3).

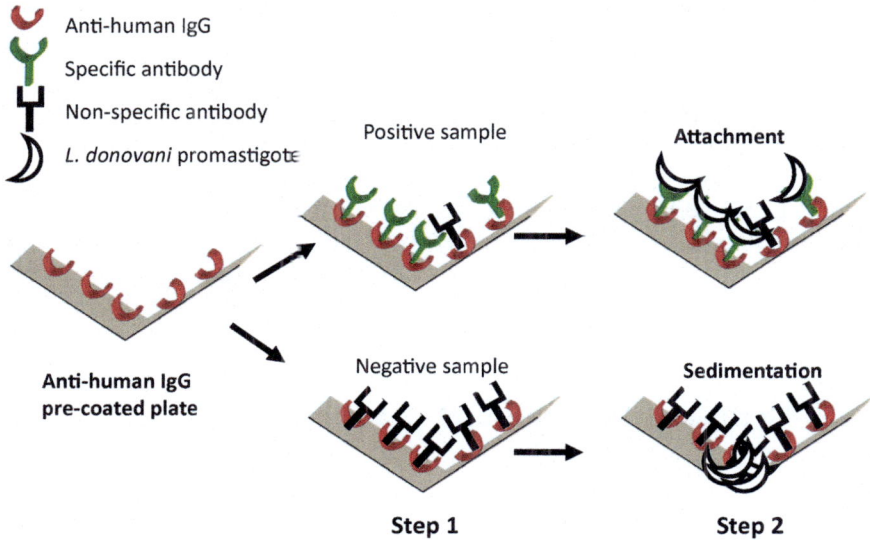

Fig. 13.1 Diagram of the abcDAT method. *Step 1* Application of urine samples to the anti-human IgG pre-coated plate. *Step 2* Application of DAT antigen to the plate

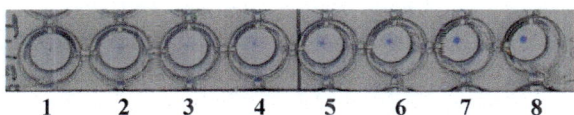

Fig. 13.2 Sensitivity of abcDAT. A series of serially diluted (3 times) pooled patient sera were used, from 20,000 times (#1) to 14,580,000 times (#7) dilution. Positive results were shown up to 4,860,000 times dilution. #8 is a blank

Fig. 13.3 An example of the abcDAT results with urine samples. Positive and negative results are indicated

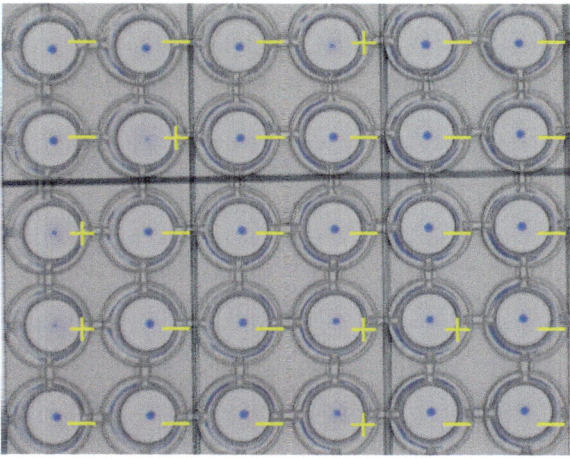

13.3 Sensitivity and Specificity of the abcDAT

A series of serially diluted pooled patient sera (20,000–14,580,000 times dilution) showed positive results at up to 4,860,000 times dilution (Fig. 13.2), which was approximately the same sensitivity as that seen with the ELISA system [4].

Among 56 urine samples collected from patients with parasitologically confirmed VL, 55 (98 %) were positive with abcDAT. However, urine samples from 13 endemic normal and 24 Japanese heathy controls were all negative with the abcDAT. Compared with the direct agglutination test (DAT) [6] and the new DAT [9], the abcDAT was 10 times and 3 times more sensitive, respectively.

13.4 Merits and Demerits of abcDAT

Although spleen or bone marrow biopsy is the definitive test for the diagnosis of VL, their use is limited to hospitals with skilled doctors. The number of new VL cases is now decreasing rapidly in many areas due to VL control programs. At this stage actively surveying to find new endemic foci and to detect resurgences of VL for prompt targeted measures will be important, and benign surveys that are accepted by residents with good compliance are needed. Because urine samples are collected simply, safely, and noninvasively, the abcDAT that detects *Leishmania* specific IgG in urine is convenient for these surveys. Such active surveys have been proven useful for control programs of lymphatic filariasis [10–12].

When we detect antigen specific IgG by ELISA, the immunoglobulin specificity is determined by the second antibody type used, such as anti-human IgG. The effect of other immunoglobulin classes already attached to antigens on the plate cannot be excluded by the ELISA system; however, only IgG attach to the plate made for the abcDAT evaluation. If the anti-human IgG used was changed to anti-human IgM, antigen-specific IgM can be detected.

The StabilCoat® treated plates are stable for at least 1 year at room temperature. Furthermore, the freeze dried DAT antigen is also stable for that same period at room temperature. No cold storage or transportation is required for the abcDAT.

By titration of the positive urine samples, prediction of development of VL will be possible as was observed with the urine ELISA [5].

Although present and past infections cannot be distinguished by detection of antigen-specific antibodies, antibody detection is more sensitive than antigen detection, especially at early stages and in cases with a mild infection. Changes of the *Leishmania* species, from *L. donovani* to other species, may make this new diagnostic tool useful for the leishmaniasis caused by the other *Leishmania* species.

13.5 Conclusion

We developed a highly sensitive and specific new urine VL diagnosis system, the abcDAT. The test system consists of an anti-human IgG coated plate and freeze dried DAT antigen that can be kept at room temperature for a long period. It enables the active survey of VL in endemic fields where the infrastructure is poor.

Acknowledgments Research underlying this chapter, conducted by the authors, was partially supported by JST/JICA, SATREPS, and also by a Grant-in-Aid for Scientific Research (B) No. 18406013 from the Japan Society for the Promotion of Science.

References

1. Boelaert M, Meheus F, Sanchez A, et al. The poorest of the poor: a poverty appraisal of households affected by visceral leishmaniasis in Bihar, India. Trop Med Int Health. 2009;14(6):639–44.
2. Argaw D, Mulugeta A, Herrero M, et al. Risk factors for visceral Leishmaniasis among residents and migrants in Kafta-Humera, Ethiopia. PLoS Negl Trop Dis. 2013;7(11):e2543.
3. Zijlstra EE, Ali MS, el-Hassan AM, et al. Kala-azar: a comparative study of parasitological methods and the direct agglutination test in diagnosis. Trans R Soc Trop Med Hyg. 1992;86(5):505–7.
4. Islam MZ, Itoh M, Takagi H, et al. Enzyme-linked immunosorbent assay to detect urinary antibody against recombinant rKRP42 antigen made from *Leishmania donovani* for the diagnosis of visceral leishmaniasis. Am J Trop Med Hyg. 2008;79:599–604.
5. Islam MZ, Itoh M, Ul Islam MA, et al. ELISA with recombinant rKRP42 antigen using urine samples: a tool for predicting clinical visceral leishmaniasis cases and its outbreak. Am J Trop Med Hyg. 2012;87(4):658–62.
6. Harith AE, Kolk AH, Kager PA, et al. A simple and economical direct agglutination test for serodiagnosis and sero-epidemiological studies of visceral leishmaniasis. Trans R Soc Trop Med Hyg. 1986;80:583–6.
7. Nawa M, Takasaki T, Yamada K, et al. Serotype-cross-reactive immunoglobulin M responses in dengue virus infections determined by enzyme-linked immunosorbent assay. Clin Diagn Lab Immunol. 2000;7(5):774–7.
8. Nawa M, Takasaki T, Yamada K, et al. Development of dengue IgM-capture enzyme-linked immunosorbent assay with higher sensitivity using monoclonal detection antibody. J Virol Methods. 2001;92(1):65–70.
9. Islam MZ, Itoh M, Mirza R, et al. Direct agglutination test with urine samples for the diagnosis of visceral leishmaniasis. Am J Trop Med Hyg. 2004;70:78–82.
10. Itoh M, Wu W, Sun D, et al. Confirmation of elimination of lymphatic filariasis by an IgG4 enzyme-linked immunosorbent assay with urine samples in Yongjia, Zhejiang Province and Gaoan, Jiangxi Province, People's Republic of China. Am J Trop Med Hyg. 2007;77(2):330–3.
11. Itoh M, Weerasooriya MV, Yahathugoda TC, et al. Effects of 5 rounds of mass drug administration with diethylcarbamazine and albendazole on filaria-specific IgG4 titers in urine: 6-year follow-up study in Sri Lanka. Parasitol Int. 2011;60:393–7.
12. Samad MS, Itoh M, Moji K, et al. Enzyme-linked immunosorbent assay for the diagnosis of *Wuchereria bancrofti* infection using urine samples and its application in Bangladesh. Parasitol Int. 2013;62(6):564–7.

Chapter 14
Loop-Mediated Isothermal Amplification (LAMP): Molecular Diagnosis for the Field Survey of Visceral Leishmaniasis

Hidekazu Takagi, Makoto Itoh and Eisei Noiri

Abstract Loop-mediated isothermal amplification (LAMP), a DNA amplification method, has been applied to the detection of various pathogenic organisms because it features rapid reaction time, accuracy, cost-effect and results that can be determined with the naked eye. The most advantage of this method is that DNA amplification occurs at a constant temperature, usually between 60 and 65 °C, therefore sophisticated equipments are unnecessary. It can be easy that this method is employed in resource-limited laboratories and flied. We have designed a LAMP primer set to detect the kinetoplast minicircle DNA of *Leishmania donovani*. The LAMP was sensitive and could detect 1 fg of *L. donovani* DNA, and was not cross-reacted with DNA of 5 other *Leishmania* species, *Plasmodium falciparum* and human. The LAMP sensitivity and specificity is equal to nested PCR in DNA samples form patients with visceral leishmaniasis (VL). Therefore, the LAMP can be a better alternative to nested PCR, especially under field conditions, and will be a powerful tool for VL mass screening in combination with urine ELISA.

Keywords Visceral leishmaniasis · Molecular diagnosis · LAMP

14.1 Introduction

Leishmaniasis is one of the neglected tropical diseases and consists of 4 main clinical syndromes: cutaneous leishmaniasis, mucocutaneous leishmaniasis, visceral leishmaniasis (VL; also known as kala-azar), and post-kala-azar dermal leishmaniasis (PKDL). VL is usually fatal if left untreated. It is caused by the *Leishmania donovani* complex protozoan and is transmitted by the bite of the infected female

---

H. Takagi (✉)
Department of Microbiology and Immunology, Aichi Medical University School of Medicine, Nagakute, Aichi 480-1195, Japan
e-mail: htakagi@aichi-med-u.ac.jp

phlebotomine sandfly [1]. In WHO's South-East Asia Region, about 147 million people in Bangladesh, India, and Nepal are at risk of VL [2].

To eliminate VL, it is essential that there are available therapeutic medications, accurate and simple diagnostic methods, effective surveillance systems, and vector controls. There are various diagnostic tools for VL, each having its own advantages and disadvantages. The microscopic examination of the parasites in aspirates from the spleen or tissues (such as bone marrow, lymph nodes, or skin) is the classical confirmatory test, but these techniques are invasive and require skilled personnel and proper facilities, and the sensitivity with bone-marrow aspirate was reported to be variable [3, 4]. There is an immunological diagnosis test to detect the *Leishmania*-specific antibody in blood (a dipstick test kit with recombinant antigen, rK39) and urine (an enzyme-linked immunosorbent assay [ELISA] with recombinant antigen, rKRP42) [5, 6]. Although these methods are simple, highly sensitive, and specific, they are unable to distinguish past and current infection. There is a test known as KATex that detects the parasite-derived antigen in urine by a latex agglutination reaction [7], but the sensitivity varies [8].

A DNA detection method by polymerase chain reaction (PCR) has also been applied to the diagnosis of VL [9, 10]. Peripheral blood is often used to avoid more invasive procedures, and the reported sensitivity of PCR with blood ranged from 70 to 96 % [4, 8–11]. Recently, loop-mediated isothermal amplification (LAMP) was developed as a novel method to amplify DNA rapidly with high specificity under an isothermal condition [12, 13]. We have developed a LAMP technique to detect *L. donovani* DNA and applied it to detect the parasite DNA in blood from patients with VL [14].

14.2 Detection of *L. donovani* DNA in Blood Samples from Patients with VL

We designed a LAMP primer set to detect the kinetoplast minicircle DNA of *L. donovani*. This LAMP could detect 1 fg of *L. donovani* DNA; equivalent to approximately 0.1 parasite [15]. However, testing was negative for 100 ng each of DNA from 5 other *Leishmania* species (*L. mexicana*, *L. major*, *L. infantum*, *L. tropica*, and *L. braziliensis*), 100 ng of *P. falciparum* DNA, or 500 ng of human genomic DNA. These results showed that this LAMP is highly sensitive and specific for the detection of *L. donovani* DNA.

The sensitivity of the LAMP was compared with that of nested PCR. DNA samples were extracted from a venous blood sample (1 mL) from patients with VL confirmed by spleen biopsy at Surya Kanta Kala-azar Research Center (SKKRC). Among the 50 samples, 45 (90 %) were positive with LAMP, and 44 (88 %) were positive with nested PCR (Table 14.1). With this very high sensitivity, LAMP would also be useful to detect *L. donovani* DNA in skin lesions of PKDL patients.

Table 14.1 Comparisons between loop-mediated isothermal amplification (LAMP) and nested PCR by examination of 50 samples from patients with visceral leishmaniasis

		LAMP		
		Pos	Neg	Total
Nested PCR	Pos	44	0	44
	Neg	1	5	6
	Total	45	5	50

14.3 LAMP as a Tool for Mass Survey

Although early diagnosis and complete treatment are efficacious for symptomatic patients with VL, they are not sufficient for the elimination of VL. There are many asymptomatic carriers of *L. donovani* in endemic areas [16, 17], patients with PKDL, and untreated patients with VL because of poverty and socio-cultural constraints. Those carriers could be parasite reservoirs for the sand fly vector. Diagnostic tests for these types of patients could reveal transmission hot spots in endemic areas. Urine ELISA to detect antibodies in urine samples is a simple and sensitive tool and suitable for mass survey in endemic countries [6]. Urine sample collections are easy to perform, noninvasive, have little chance of accidental infection, and the compliance of residents is expected to be good, especially among those with no symptoms. The combination LAMP with urine ELISA can be a practical mass survey method to identify VL carriers.

After the ELISA assay is completed with urine samples, blood samples for molecular diagnosis will be collected from patients with ELISA positive results to identify active infections. LAMP has many advantages over PCR. PCR requires a well-established laboratory, expensive equipment such as a thermal cycler, and a system to detect and analyze amplicons. However, LAMP consists of the following simple steps: (1) mix a sample of DNA, 4 different primers, *Bst* DNA polymerase, and substrates, and (2) incubate them for 1 h at 65 °C using only basic equipment like a heat block or a water bath. Judgment of positive or negative results is easily made by the naked eye (Fig. 14.1). In addition, LAMP testing is not very expensive [18]. LAMP reagents that are dried and stabilized by adding a cryoprotectant can be stored at ambient temperature [19]. LAMP was less affected by the contamination of serum, plasma, or other inhibitory components in DNA samples compared with findings for PCR [20, 21]. This means that the cost of the DNA purification step can be reduced. Therefore, we propose a strategy for finding endemic foci of VL and asymptomatic carriers with a combination of mass screening using urine ELISA followed by LAMP testing for confirmation of infection. This procedure is also useful to monitor the recurrence of VL and is expected to contribute to the control of VL.

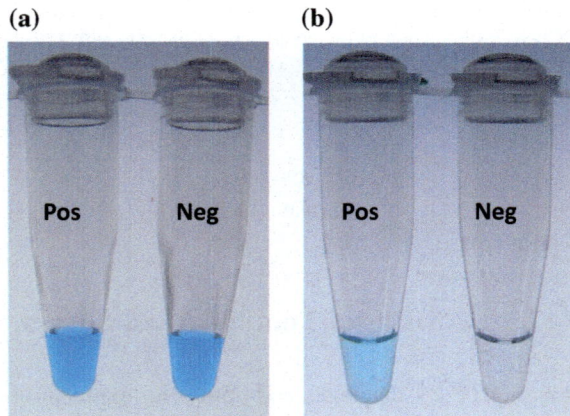

Fig. 14.1 Easy judgment of loop-mediated isothermal amplification (LAMP) results by colorimetric visual detection. **a** Before incubation. **b** After incubation at 65 °C for 60 min. Malachite green dye was added to the LAMP reaction mixture so that the mixtures show brilliant blue before incubation. After incubation, a positive sample (Pos) turns to light *blue*, negative (Neg) is clear [22]

14.4 Conclusion

VL has recently received public attention as a neglected tropical disease. The WHO has advocated an elimination target of reducing the annual incidence of VL to 1 case in 10,000 by 2015 in Southeast Asia. To reach this goal, healthcare workers must ensure early and accurate diagnosis and complete treatment. In addition, surveillance systems must appropriately determine the prevalence of infection in the endemic areas. The urine ELISA for mass survey of VL and the LAMP method for confirmation of *L. donovani* infection will be useful to detect foci of VL transmission and help to reach this goal.

Acknowledgments Research underlying this chapter, conducted by the authors, was partially supported by JST/JICA, SATREPS, and also by a Grant-in-Aid for Scientific Research (B) No. 18406013 from the Japan Society for the Promotion of Science.

References

1. Murray HW, Berman JD, Davies CR, Saravia NG. Advances in leishmaniasis. Lancet 2005;366:1561–77.
2. World Health Organization. Regional Strategic Framework for Elimination of Kala-azar from the Southeast Asia Region (2011–2015). New Delhi: WHO Regional Office for Southeast Asia 2004. Accessed August 2005, at http://www.searo.who.int/entity/world_health_day/2014/KA_CD239.pdf.
3. Sundar S, Rai M. Laboratory diagnosis of visceral leishmaniasis. Clin Diagn Lab Immunol. 2002;9:951–8.
4. Chappuis F, Sundar S, Hailu A, Ghalib H, Rijal S, Peeling RW, Alvar J, Boelaert M. Visceral leishmaniasis: what are the needs for diagnosis, treatment and control? Nat Rev Microbiol. 2007;5:873–82.
5. Sundar S, Pai K, Sahu M, Kumar V, Murray HW. Immunochromatographic strip-test detection of anti-K39 antibody in Indian visceral leishmaniasis. Ann Trop Med Parasitol. 2002;96:19–23.

6. Islam MZ, Itoh M, Takagi H, Islam AU, Ekram AR, Rahman A, Takesue A, Hashiguchi Y, Kimura E. Enzyme-linked immunosorbent assay to detect urinary antibody against recombinant rKRP42 antigen made from *Leishmania donovani* for the diagnosis of visceral leishmaniasis. Am J Trop Med Hyg. 2008;79:599–604.
7. Attar ZJ, Chance ML, el-Safi S, et al. Latex agglutination test for the detection of urinary antigens in visceral leishmaniasis. Acta Trop. 2001;78:11–6.
8. Singh OP, Sundar S. Developments in diagnosis of visceral leishmaniasis in the elimination era. J Parasitol Res. 2015;239469.
9. Antinori S, Calattini S, Longhi E, Bestetti G, Piolini R, Magni C, Orlando G, Gramiccia M, Acquaviva V, Foschi A, Corvasce S, Colomba C, Titone L, Parravicini C, Cascio A, Corbellino M. Clinical use of polymerase chain reaction performed on peripheral blood and bone marrow samples for the diagnosis and monitoring of visceral leishmaniasis in HIV-infected and HIV-uninfected patients: a single-center, 8-year experience in Italy and review of the literature. Clin Infect Dis. 2007;44:1602–10.
10. Reithinger R, Dujardin JC. Molecular diagnosis of leishmaniasis: current status and future applications. J Clin Microbiol. 2007;45:21–5.
11. Singh RK, Pandey HP, Sundar S. Visceral leishmaniasis (kala-azar): challenges ahead. Indian J Med Res. 2006;123:331–44.
12. Notomi T, Okayama H, Masubuchi H, Yonekawa T, Watanabe K, Amino N, Hase T. Loop-mediated isothermal amplification of DNA. Nucleic Acids Res. 2000;28:E63.
13. Nagamine K, Hase T, Notomi T. Accelerated reaction by loop-mediated isothermal amplification using loop primers. Mol Cell Probes. 2002;16:223–9.
14. Takagi H, Itoh M, Islam MZ, et al. Sensitive, specific, and rapid detection of *Leishmania donovani* DNA by loop-mediated isothermal amplification. Am J Trop Med Hyg. 2009;81:578–82.
15. Salotra P, Sreenivas G, Pogue GP, Lee N, Nakhasi HL, Ramesh V, Negi NS. Development of a species-specific PCR assay for detection of *Leishmania donovani* in clinical samples from patients with kala-azar and post-kala-azar dermal leishmaniasis. J Clin Microbiol. 2001;39:849–54.
16. Bern C, Haque R, Chowdhury R, Ali M, Kurkjian KM, Vaz L, Amann J, Wahed MA, Wagatsuma Y. The epidemiology of visceral leishmaniasis and asymptomatic leishmanial infection in a highly endemic Bangladeshi village. Am J Trop Med Hyg. 2007;76(5):909–14.
17. Ostyn B, Gidwani K, Khanal B, Picado A, Chappuis F, Singh SP, Rijal S, Sundar S, Boelaert M. Incidence of symptomatic and asymptomatic *Leishmania donovani* infections in high-endemic foci in India and Nepal: a prospective study. PLoS Negl Trop Dis. 2011;5(10): e1284.
18. Poon LL, Wong BW, Ma EH, Chan KH, Chow LM, Abeyewickreme W, Tangpukdee N, Yuen KY, Guan Y. Sensitive and inexpensive molecular test for falciparum malaria: detecting *Plasmodium falciparum* DNA directly from heat-treated blood by loop-mediated isothermal amplification. Clin Chem. 2006;52(2):303–6.
19. Hayashida K, Kajino K, Hachaambwa L, Namangala B, Sugimoto C. Direct blood dry LAMP: a rapid, stable, and easy diagnostic tool for Human African Trypanosomiasis. PLoS Negl Trop Dis. 2015;9(3):e0003578.
20. Wilson IG. Inhibition and facilitation of nucleic acid amplification. Appl Environ Microbiol. 1997;63:3741–51.
21. Kaneko H, Kawana T, Fukushima E, Suzutani T. Tolerance of loop-mediated isothermal amplification to a culture medium and biological substances J Biochem Biophys Methods. 2007;70:499–501.
22. Nzelu CO, Gomez EA, Caceres AG, Sakurai T, Martini-Robles L, Uezato H, Mimori T, Katakura K, Hashiguchi Y, Kato H. Development of a loop-mediated isothermal amplification method for rapid mass-screening of sand flies for Leishmania infection. Acta Trop. 2014;132:1–6.

Chapter 15
Applicability of Multiplex Real-Time PCR to Visceral Leishmaniasis

Yoshifumi Hamasaki, Hirofumi Aruga, Chizu Sanjoba,
Hidekazu Takagi, Shyamal Paul, Yoshitsugu Matsumoto
and Eisei Noiri

Abstract Visceral leishmaniasis (VL), also known as kala-azar caused by several species of *Leishmania*, is a life-threatening and disseminated parasite disease and considered one of the most neglected tropical diseases in the Old World. Rapid and accurate diagnosis is needed to provide appropriate treatment to VL patients. Identification of the species that cause VL is useful to develop an epidemiology of leishmaniasis and reveal the biology of each species of *Leishmania*. Herein, we describe a new multiplex real-time PCR assay to distinguish *L. donovani* from *L. infantum* with high sensitivity; both these organisms are major causes of VL. The real-time PCR assay targeting kinetoplast minicircle DNA can identify the infection of *Leishmania* species with high sensitivity in the DNA extracted from the peripheral blood of patients. Two assays were designed to distinguish the difference in the cysteine protease B gene copies between *L. donovani* and *L. infantum* species. The endogenous control assay targeting mammalian ribonuclease P RNA component H1 (RPPH1) was used to confirm whether the PCR reaction progressed precisely and to quantify the clinical sample input amount. Multiplex real-time PCR with these 4 assays successfully detected *Leishmania* DNA with high sensitivity and distinguished *L. donovani* from *L. infantum* with high specificity using DNA samples extracted from cultured parasites or the peripheral blood of patients. Multiplex real-time PCR can contribute to the therapeutic strategy for patients with kala-azar and to the research of *Leishmania* epidemiology and biology.

Keywords Multiplex real-time PCR · Kinetoplast DNA · cpb gene · RNase P · *L. donovani* · *L. infantum*

Y. Hamasaki (✉)
22nd Century Medical and Research Center, The University of Tokyo Hospital, Tokyo, Japan
e-mail: yhamasaki-tky@umin.ac.jp

15.1 Introduction

Kala-azar, known as visceral leishmaniasis (VL), is a life-threatening disseminated disease of leishmaniasis caused by the *Leishmania donovani* complex. VL is considered to be one of the most neglected tropical diseases [1] with an estimated incidence of new cases of VL each year of 0.2–0.4 million and causing 20,000–40,000 deaths annually [2]. Ninety percent of VL cases occur in just 5 countries: Bangladesh, India, Nepal, Sudan, and Brazil [2]. The etiological agent of human VL in the Old World is caused by 2 closely related parasites that belong to the *L. donovani* complex: *L. donovani* and *L. infantum*.

A rapid and accurate method to diagnosis of VL is desirable to provide appropriate treatment for individuals and to understand the epidemiology of *Leishmania* precisely. Conventional diagnosis of VL is done by one or more of the following tests: detecting parasite DNA in tissue and blood samples; detecting parasites in the smear of an aspiration specimen taken from the involved organ, such as the spleen; or serological tests using blood or urine [3–5]. Although the demonstration of parasites is the most specific diagnostic method and antibody testing is convenient, these techniques impose several challenges: spleen aspiration is invasive, serological tests require skilled personnel and proper facilities, and the sensitivity of these tests was reported to be variable [3]. To overcome these challenges, healthcare workers sought to develop a new diagnostic method with increased speed, accuracy, and reliability. In the past 2 decades, polymerase chain reaction (PCR) based techniques have been utilized for the detection of *Leishmania* DNA [6]. PCR is superior in terms of reducing time from sample-to-result and optimizes sensitivity and specificity. In addition, the test results are easy to interpret compared with those from other conventional traditional diagnostic methods for leishmaniasis [7–9]. Although there are many publications of different variants of PCRs for VL, real-time PCR is advantageous because it is faster, less labor intensive, reduces risk of contamination, and increases sensitivity and specificity by using well-designed probes compared with procedures for conventional PCR [10, 11].

The typical target of PCR for detection of *Leishmania* species is the kinetoplast DNA (kDNA) mini-circle due to its abundance and repetitive nature that enhances sensitivity and specificity [9]. PCR systems targeted for kDNA are genus specific; therefore they cannot differentiate leishmania species. The discrimination between *L. donovani* and *L. infantum* is important because these 2 species are the major cause of VL in humans. Although they are morphologically indistinguishable, they are associated with different epidemiology, ecology, and pathology [12, 13]. We deployed the PCR assay to detect *Leishmania* genus and an additional assay to distinguish *L. donovani* from *L. infantum*. Several PCR methods have been demonstrated for *L. donovani/L. infantum* discrimination [12, 13]. Multiplex PCR, the method of PCR that can amplify several different DNA sequences simultaneously, enables the detection of *Leishmania* species and has an internal control in the same reaction. Several investigators have reported the strategy of using multiplex PCR to diagnose leishmaniasis [14, 15]. There has been no report of multiplex real-time PCR assays to diagnose VL and distinguish *L. donovani* from *L. infantum*. We demonstrated a rapid and accurate molecular method to diagnose VL and to distinguish *L. donovani* from *L. Infantum* simultaneously by deploying new multiplex real-time PCR assays.

15.2 Materials and Methods

15.2.1 DNA Extraction

Parasite DNA: We used promastigotes of *L. donovani* strain BD38 and *L. infantum* strain EP56 in this study. Parasite strains were obtained from the faculty of agriculture at The University of Tokyo, Japan. DNA of each species was extracted by the classical phenol-chloroform extraction method, and then suspended in TE (Tris–EDTA) buffer. The DNA obtained from parasite cultures was adjusted to a concentration of 100 ng/μL and stored at 4 °C.

Clinical samples: This study was reviewed and approved by the Ethical Committee of Tokyo University, Japan, and the Ethical Review Committee of the Bangladesh Medical Research Council. Blood samples were collected from patients with VL at The Surya Kanta Kala-azar Research Centre (SKKRC) in Bangladesh after obtaining informed consent. Withdrawn peripheral whole blood was collected in the collection tube containing EDTA-2Na and centrifuged at 3,000 rpm for 10 min to obtain a buffy coat. DNA was extracted from the buffy coat using QIAamp DNA Blood Mini Kit (QIAGEN, Venlo, Netherlands) according to the manufacturer's protocol. The concentration of DNA samples was measured using NanoDrop 2000 (Thermo Fisher Scientific, Inc., Waltham, MA). All DNA samples were stored at 4 °C.

15.2.2 Primer Design and Multiplex Quantitative PCR

To design species-specific primers and probes for the discrimination of *L. donovani* and *L. infantum*, we targeted cysteine protease B (CPB) CPB enzymes play a role in destruction of the host protein and evasion of the host immune response. They are encoded by a tandem array located in a single locus [12, 13]. The last repeat of the cpb cluster, named cpbF of *L. donovani* and cpbE of *L. infantum*, was used as the target of discriminative real-time PCR [12, 13]. Sequences of the cpbEF gene of *L. infantum* and *L. donovani* were obtained from NCBI and specific primers and probes were designed to differentiate cpbF of *L. donovani* and cpbE of *L. infantum*.

In our preliminary experiments, we designed several sets of candidate assays targeting the cpbEF gene for discriminative PCR and optimized PCR conditions such as annealing temperature. We selected the most robust set of primers/probe to distinguish *L. donovani* from *L. infantum* with maximum sensitivity and specificity.

We used kDNA as a target because it is present in high copy numbers (about 10,000 copies per parasite) and is a more sensitive target than genomic DNA or ribosomal DNA [16, 17]. To increase the detection sensitivity of *Leishmania* DNA, we included the assay for the minicircle region of kDNA to multiplex real-time PCR.

Mammalian ribonuclease P RNA component H1 (RPPH1) was used to test the quality of DNA samples obtained from patients and as an endogenous control.

Reporter dyes for *Leishmania* (kDNA), RNase P, *L. donovani*, and *L. infantum* probes were FAM™, VIC®, ABY®, and JUN®, respectively; all primers and probes were synthesized by Thermo Fisher Scientific (Table 15.1).

The multiplex real-time PCR assay was performed in a final total volume of 20 μL containing 10 μL of 2x TaqMan® Multiplex Master Mix (Thermo Fisher Scientific). The final concentrations for the set of primers and probes were optimized as follows: 900 nM of each primer and 250 nM for probes for *L. donovani* assays, 200 nM of each primer and 125 nM for probes for *L. infantum* assays, 200 nM of each primer and 125 nM for probes for kDNA assays, and 200 nM of each primer and 125 nM for probes for RNase P (RPPH1) assays. The amount of DNA samples for each reaction was 10 pg parasite DNA and 1 μL patient DNA. A ViiA7™ Real-Time PCR System (Thermo Fisher Scientific) was used as the real-time PCR system. Thermal cycling conditions were as follows: holding step at 95 °C for 10 min for enzyme activation, followed by 40 cycles of the amplification step that consisted of 2 stages of denaturation (at 98 °C for 15 s) and annealing (at 70 °C for 1 min). The amplification reactions were analyzed by ViiA 7™ RUO Software (Thermo Fisher Scientific).

Table 15.1 Sets of primers and probes designed for the multiplex real-time PCR

Small scale part number	Assay name	Target	Dye label	Probe type
CCU001S	Black Fever_Leishmania	kinetoplast DNA	FAM	MGB
CCU001SNR	Black Fever_RNaseP-hiTm	RNase P (RPPH1)	VIC	QSY
CCU001SNR	Black Fever_*L. donovani*	cpb F gene	ABY	QSY
CCU001SNR	Black Fever_*L. infantum*	cpb E gene	JUN	QSY

15.3 Results

15.3.1 Parasite Samples

Amplification plots of multiplex real-time PCR are shown in Fig. 15.1. From the results of multiplex real-time PCR using parasite DNA, the combination of 2 assays could distinguish *L. donivani* from *L. infantum*. The assay for kDNA was positive in both *L. donovani* and *L. infantum* DNA samples.

15.3.2 Clinical Samples

The result of multiplex real-time PCR is shown in Fig. 15.2. When DNA samples from patients given clinical diagnoses of leishmaniasis were used, assays for RNase P (RPPH1) and kDNA were positive. Because the pathogen infecting the patient was *L. donovani*, the assay for *L. donovani* (cpbF gene) was positive, while the assay for *L. infantum* (cpbE gene) was negative.

15.4 Discussion

We developed a novel, multiplex, real-time PCR assay to diagnose leishmaniasis and distinguish *L. donovani* from *L. infantum*. To our knowledge, there has not been any report describing a multiplex real-time PCR method to differentiate *L. donovani* and *L. infantum*, which are the dominant causative microorganisms of VL. We demonstrated *Leishmania* detection with high sensitivity by targeting kDNA. Our method enabled us to characterize and distinguish 2 leishmanial species using 2 assays targeting the cpb E and F gene.

The gold standard for the diagnosis of VL has been the detection of protozoa by microscopy or culture. However, these methods have limitations such as insufficient sensitivity, operator dependence, and long process time from sample collection to detection. Real-time PCR methods offer a sensitive and specific alternative approach, especially for cases inconclusive with conventional diagnostic tests, thus enabling early and adequate treatment [18]. The utility of real-time PCR for the detection of protozoa in VL has been reported with good sensitivity and specificity [19–21]. Wortmann et al. [22] demonstrated that real-time PCR assays using designed primer/probe sets were retaining highly complex-specific diagnostic accuracy for cultured organisms and patient clinical samples. However, sets of primers and probes for the discrimination of *L. donovani* and *L. infantum* were not used in that study.

There are several reports describing PCR methods to distinguish *L. donovani* from *L. infantum*. Haralambous et al. [23] developed a PCR method for

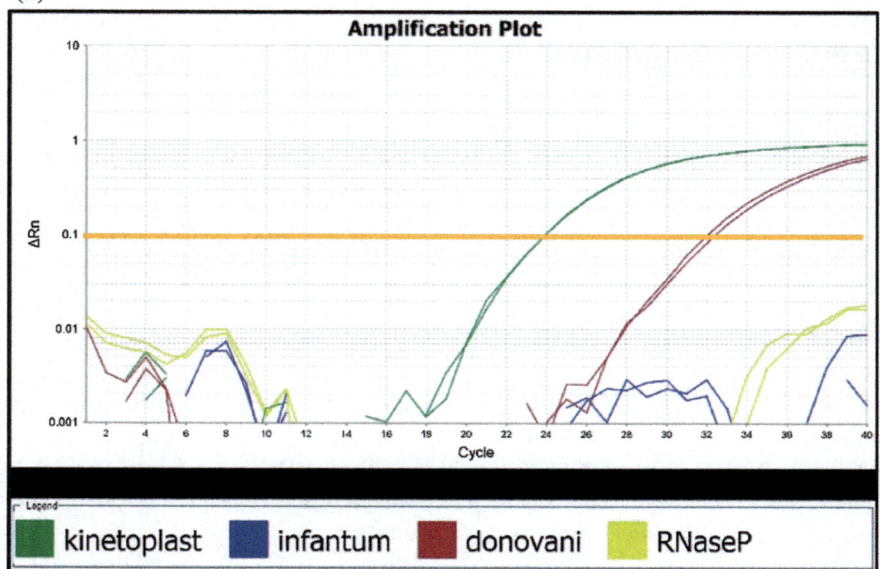
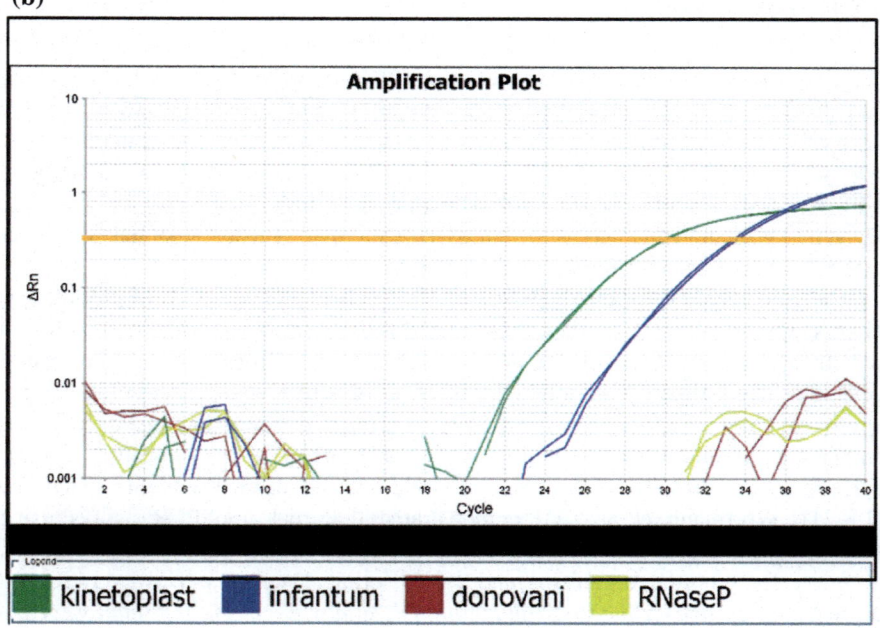

Fig. 15.1 Results of multiplex, real-time PCR using DNA samples extracted from cultured *L. donovani* and *L. infantum*. Amplification plot of the kDNA assay that became positive in samples of both *L. donovani* and *L. infantum* (**a**, **b**). When the sample contains *L. donovani* DNA, the assay targeting the cpb F gene of *L. donovani* was positive and the assay targeting the cpb E gene of *L. infantum* was negative (**a**). When the sample contained *L. infantum* DNA, the assay for *L. infantum* was positive and the assay for *L. donovani* was negative (**b**). DNA samples: 10 pg per each reaction

discrimination of *L. donovani* complex based on the amplicon size of the K26 gene. This method was more useful to distinguish geographical strains instead of species because there are some identical or very close PCR products for both *L. donovani* and *L. infantum*. Oshaghi et al. [13] showed 2 molecular tools, including restriction fragment length polymorphisms of amplified DNA (PCR-RFLP) and PCR. They found only a PCR primer pair for *L. donovani* but not for *L. infantum*, and did not test PCR for clinical samples. Hide et al. [12] demonstrated a PCR method that can discriminate between *L. donovani* and *L. infantum* based on different lengths of the product of the cpb gene and did not generate amplification for other trypanosomatids. However, this PCR was not sensitive enough for the diagnosis of clinical samples. We developed a new approach to distinguish *L. donovani* from *L. infantum* using a multiplex real-time PCR assay with TaqMan® primer/probe sets that were more specific than SYBR green chemistry. We demonstrated the performance of these assays using clinical samples from patients.

Multiplex PCR allows the simultaneous detection and discrimination of several parasite species with similar morphology from a single clinical specimen [24]. The highly sensitive and specific molecular detection techniques enable high-throughput and cost-effective screening without increasing the amount of samples. The implementation of such an approach is especially important in a setting of low parasite prevalence. Recently, several authors have highlighted the strategy of multiplex PCR to detect leishmania species [18]. Mohammadiha et al. [11] performed TaqMan® probe-based multiplex PCR with a sensitivity of 93.9–100 % in human and dog samples. To increase the diagnostic accuracy of leishmaniasis, Gonçalves-de-Albuquerque et al. [14] developed a triplex PCR assay targeting the parasite's DNA as the external control and glyceraldehyde 3-phosphate dehydrogenase as an internal control to clarify sample quality. In our TaqMan® probe-based multiplex real-time PCR, we not only diagnosed leishmaniasis by using an assay for kDNA with high sensitivity, but also distinguished *L. donovani* from *L. infantum* by deploying assays for the cpbEF gene. Also, our method included the assay for mammalian RNase P (RPPH1) to confirm the quality of the clinical samples.

There are several limitations in our multiplex real-time PCR assay for leishmania. Identification of species may be difficult when there is a very small amount of *Leishmania* DNA in the clinical samples. Due to the difference in Cq values between assays for *L. donovani/infantum* and assays for kDNA, it is difficult to detect *Leishmania* species in samples with a high (more than 30) Cq value for the kDNA assay. This is due to the difference in the number of gene copies in the

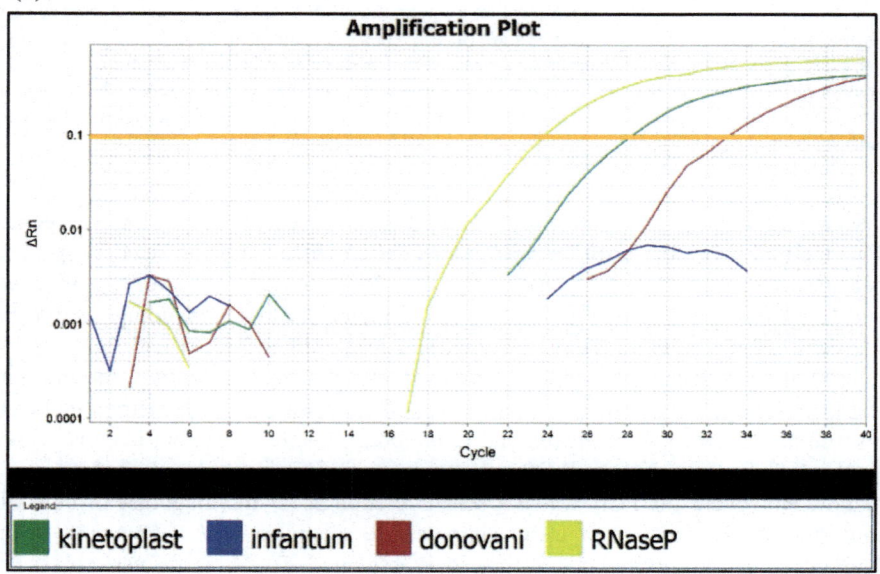

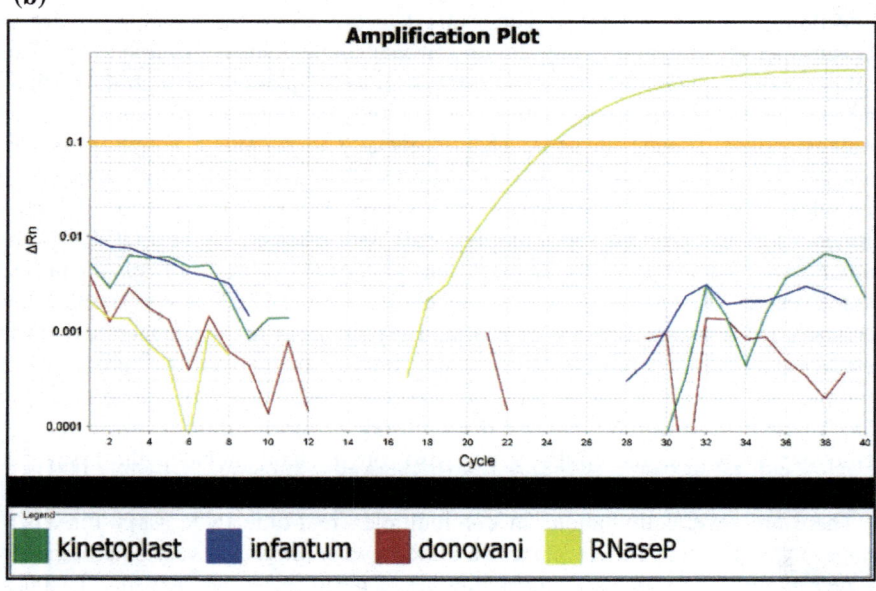

Fig. 15.2 Examples of the multiplex, real-time PCR using DNA samples from patients. DNA samples from patients with visceral leishmaniasis (VL) clinically diagnosed by symptoms and/or the detection of *Leishmania* by microscopy were used in (**a**). DNA samples from a control patient was used in (**b**). The amplification plots of the kDNA assay was positive in VL patients (**a**), whereas it was negative in a control patient (**b**). The amplification plot of the assay for mammalian RNase P (RPPH1) was positive (**a**, **b**). DNA samples: 1 μL per each reaction

extracted DNA and the sensitivity limit of *L. donovani*/*L. infantum* assays without deteriorating specificity. Polymorphisms in the cpb gene have also been reported in the *L. donovani* complex [25]. Further experiments using samples from various strains are needed to confirm the applicability of our assay for all *L. donovani* and *L. infantum* species. The real-time PCR method for the diagnosis of leishmaniasis is currently under development, therefore standardization is lacking among laboratories.

Multiplex real-time PCR has the potential to be established as a gold standard diagnostic technique for leishmaniasis, especially in the field setting of highly endemic areas of leishmaniasis, and due to its high sensitivity and specificity could be useful in many other situations. Multiplex real-time PCR will be useful for early and precise diagnosis of symptomatic patients, screening of asymptomatic patients suspected of leishmaniasis infection after physical examination, and patient monitoring that may enable early pre-emptive treatment before the systemic burden of the parasite.

15.5 Conclusion

We developed a new multiplex real-time PCR method to diagnose leishmaniasis and to distinguish between *L. donovani* and *L. infantum*. It enables rapid and accurate diagnosis of VL and will support informed decisions on the treatment strategy for patients. Our multiplex real-time PCR for the discrimination between *L. donovani* and *L. infantum* can also contribute to revealing the epidemiology of leishmaniasis and the biology of *Leishmania*.

References

1. Trouiller P, Olliaro P, Torreele E, Orbinski J, Laing R, Ford N. Drug development for neglected diseases: a deficient market and a public-health policy failure. Lancet. 2002;359(9324):2188–94.
2. Alvar J, Velez ID, Bern C, et al. Leishmaniasis worldwide and global estimates of its incidence. PLoS ONE. 2012;7(5):e35671.
3. Sundar S, Rai M. Laboratory diagnosis of visceral leishmaniasis. Clin Diagn Lab Immunol. 2002;9(5):951–8.

4. Chappuis F, Sundar S, Hailu A, et al. Visceral leishmaniasis: what are the needs for diagnosis, treatment and control? Nat Rev Microbiol. 2007;5(11):873–82.
5. Takagi H, Itoh M, Islam MZ, et al. Sensitive, specific, and rapid detection of Leishmania donovani DNA by loop-mediated isothermal amplification. Am J Trop Med Hyg. 2009;81(4):578–82.
6. Antinori S, Calattini S, Longhi E, et al. Clinical use of polymerase chain reaction performed on peripheral blood and bone marrow samples for the diagnosis and monitoring of visceral leishmaniasis in HIV-infected and HIV-uninfected patients: a single-center, 8-year experience in Italy and review of the literature. Clin Infect Dis. 2007;44(12):1602–10.
7. Aviles H, Belli A, Armijos R, Monroy FP, Harris E. PCR detection and identification of Leishmania parasites in clinical specimens in Ecuador: a comparison with classical diagnostic methods. J Parasitol. 1999;85(2):181–7.
8. Srivastava P, Dayama A, Mehrotra S, Sundar S. Diagnosis of visceral leishmaniasis. Trans R Soc Trop Med Hyg. 2011;105(1):1–6.
9. Pereira MR, Rocha-Silva F, Graciele-Melo C, Lafuente CR, Magalhaes T, Caligiorne RB. Comparison between conventional and real-time PCR assays for diagnosis of visceral leishmaniasis. Biomed Res Int. 2014;2014:639310.
10. Dymond JS. Explanatory chapter: quantitative PCR. Methods Enzymol. 2013;529:279–89.
11. Mohammadiha A, Mohebali M, Haghighi A, et al. Comparison of real-time PCR and conventional PCR with two DNA targets for detection of Leishmania (Leishmania) infantum infection in human and dog blood samples. Exp Parasitol. 2013;133(1):89–94.
12. Hide M, Banuls AL. Species-specific PCR assay for *L. infantum*/*L. donovani* discrimination. Acta Trop. 2006;100(3):241–5.
13. Oshaghi MA, Ravasan NM, Hide M, et al. Development of species-specific PCR and PCR-restriction fragment length polymorphism assays for *L. infantum*/*L. donovani* discrimination. Exp Parasitol. 2009;122(1):61–5.
14. Goncalves-de-Albuquerque Sda C, Pessoa ESR, de Morais RC, et al. Tracking false-negative results in molecular diagnosis: proposal of a triplex-PCR based method for leishmaniasis diagnosis. J Venomous Anim Toxins Including Trop Dis. 2014;20:16.
15. de Pita-Pereira D, Cardoso MA, Alves CR, Brazil RP, Britto C. Detection of natural infection in *Lutzomyia cruzi* and *Lutzomyia forattinii* (Diptera: Psychodidae: Phlebotominae) by *Leishmania infantum* chagasi in an endemic area of visceral leishmaniasis in Brazil using a PCR multiplex assay. Acta Trop. 2008;107(1):66–9.
16. Lachaud L, Marchergui-Hammami S, Chabbert E, Dereure J, Dedet JP, Bastien P. Comparison of six PCR methods using peripheral blood for detection of canine visceral leishmaniasis. J Clin Microbiol. 2002;40(1):210–5.
17. El-Beshbishy HA, Al-Ali KH, El-Badry AA. Molecular characterization of cutaneous leishmaniasis in Al-Madinah Al-Munawarah province, western Saudi Arabia. Int J Infect Dis. 2013;17(5):e334–8.
18. de Paiva-Cavalcanti M, de Morais RC, Pessoa ESR, et al. Leishmaniases diagnosis: an update on the use of immunological and molecular tools. Cell Biosci. 2015;5:31.
19. de Ruiter CM, van der Veer C, Leeflang MM, Deborggraeve S, Lucas C, Adams ER. Molecular tools for diagnosis of visceral leishmaniasis: systematic review and meta-analysis of diagnostic test accuracy. J Clin Microbiol. 2014;52(9):3147–55.
20. Galai Y, Chabchoub N, Ben-Abid M, et al. Diagnosis of mediterranean visceral leishmaniasis by detection of leishmania antibodies and leishmania DNA in oral fluid samples collected using an Oracol device. J Clin Microbiol. 2011;49(9):3150–3.
21. Bossolasco S, Gaiera G, Olchini D, et al. Real-time PCR assay for clinical management of human immunodeficiency virus-infected patients with visceral leishmaniasis. J Clin Microbiol. 2003;41(11):5080–4.
22. Wortmann G, Hochberg L, Houng HH, et al. Rapid identification of Leishmania complexes by a real-time PCR assay. Am J Trop Med Hyg. 2005;73(6):999–1004.

23. Haralambous C, Antoniou M, Pratlong F, Dedet JP, Soteriadou K. Development of a molecular assay specific for the *Leishmania donovani* complex that discriminates *L. donovani/Leishmania infantum* zymodemes: a useful tool for typing MON-1. Diagn Microbiol Infect Dis. 2008;60(1):33–42.
24. Wong SS, Fung KS, Chau S, Poon RW, Wong SC, Yuen KY. Molecular diagnosis in clinical parasitology: when and why? Exp Biol Med (Maywood). 2014;239(11):1443–60.
25. Hide M, Banuls AL. Polymorphisms of cpb multicopy genes in the *Leishmania (Leishmania) donovani* complex. Trans R Soc Trop Med Hyg. 2008;102(2):105–6.

Part IV
Pathogenesis of Kala-Azar and PKDL

Chapter 16
Immunoglobulins in the Pathophysiology of Visceral Leishmaniasis

Satoko Omachi, Yoshitsugu Matsumoto and Yasuyuki Goto

Abstract Hypergammaglobulinemia is a common feature associated with visceral leishmaniasis (VL). It has been well accepted that cell-mediated immunity supported by Th1 responses provides protection against *Leishmania* infection. On the other hand, the role of humoral immunity in control or exacerbation of VL has been less clear, despite that the presence of strong humoral responses is the hallmark of patients with VL. Such the high level of antibody seen in VL patients is often utilized for serological diagnostic tests, including commercial enzyme-linked immunosorbent assay (ELISA) tests and rK39 dipstick tests. Because many studies have demonstrated that antibody titers in those serodiagnostic tests are higher in active VL patients compared with the asymptomatic, humoral responses may be involved in the development of symptoms during VL. However, the physiological and pathological roles of the humoral responses during VL remain elusive. In this chapter, we discuss the components of antibodies, activation mechanisms of humoral responses, and the pathological mechanisms of immunoglobulin activity during VL.

Keywords Immunoglobulin · Humoral response · Visceral leishmaniasis

16.1 High Levels of Immunoglobulins as a Hallmark of Visceral Leishmaniasis (VL)

Hypergammaglobulinemia is a clinical manifestation where increased levels of gamma globulins, mostly immunoglobulins, are present in the blood. This manifestation can often be found in autoimmune diseases, including systemic lupus erythematosus (SLE), Sjögren's syndrome, rheumatoid arthritis (RA), and polymyositis, as well as tumors of immunoglobulin-producing cells such as mul-

---

Y. Goto (✉)
Laboratory of Molecular Immunology, Graduate School of Agricultural and Life Sciences, The University of Tokyo, 1-1-1 Yayoi, Bunkyo-ku, Tokyo 113-8657, Japan
e-mail: aygoto@mail.ecc.u-tokyo.ac.jp

© Springer International Publishing 2016
E. Noiri and T.K. Jha (eds.), *Kala Azar in South Asia*,
DOI 10.1007/978-3-319-47101-3_16

tiple myeloma, Waldenström's macroglobulinemia, chronic lymphocytic leukemia, and other non-Hodgkin's lymphomas [1, 2]. In general, the main immunoglobulin classes are IgM, IgG, and IgA, but the pattern of the component ratio of these immunoglobulin classes is different in various diseases. For example, patients with Waldenström's macroglobulinemia are characterized by elevation of serum monoclonal IgM [3]. Conversely, serum IgA levels are elevated in patients with IgA nephropathy, which results from IgA-specific B-cell hyperactivity [4]. Tumors of immunoglobulin-producing cells also contribute to the elevation of certain classes of immunoglobulins, and sometimes malignant monoclonal proliferations are associated with a deficient production of all other immunoglobulin classes [5].

Hypergammaglobulinemia has also been reported as one of the characteristics of VL [6–8]. However, the majority of the previous reports are case reports, and information on the frequency of hypergammaglobulinemia during VL is limited. In France, hypergammaglobulinemia was found in 72 % of adult patients and 65 % of afflicted children [7]. The prevalence in the adults was similar to that of other major clinical symptoms such as fever (81 %), splenomegaly (81 %), and hepatomegaly (65 %) [7]. A similar prevalence of hypergammaglobulinemia was also found in Brazil; 68 % of VL patients had γ-globulin levels equal to or higher than 2.0 g/dL [8]. Another report from Brazil also demonstrated a significant increase of gamma globulin in VL patients (3.7 g/dL) compared to that found in controls (1.0 g/dL). Therefore, hypergammaglobulinemia is a common feature associated with VL.

To know what classes of immunoglobulins are responsible for the hypergammaglobulinemia is important to understanding the mechanisms of B-cell hyperactivation during VL. However, systematic analyses on immunoglobulin classes are even more limited. There is a report from India that analyzed serum IgG, IgA, and IgM levels in VL patients [9]. In that report, Ghose et al. found that IgG is the dominant class in VL patients with hypergammaglobulinemia, and the mean serum level of IgG was 4.4 times higher that of control subjects [9]. The mean IgM level was also higher in VL patients, whereas the degree of increase (2.2 times) was less than that for IgG, and there was no significant increase of IgA [9]. Another report from Brazil also demonstrated hypergammaglobulinemia due to increased IgG levels [10]. In those Brazilian VL patients, mean plasma levels of IgG, IgM, and IgA were 2.9, 0.2, and 0.2 g/dL, respectively, in contrast to those seen in healthy controls of 1.1, 0.2, and 0.2 g/dL, respectively [10]. We also reported a significant increase of serum IgG levels in Brazilian VL patients [11]. The mean serum IgG level in VL patients was 5.3 times higher than that seen in endemic controls, and 75 % of the patients had IgG levels higher than the cutoff calculated as the mean +3 SD of the controls [11]. In contrast, the increases of serum IgM or IgA levels in those patients were not significant. We also found the elevation of IgG levels in Bangladeshi patients with VL. IgG was the most abundant immunoglobulin in 19 patients with VL diagnosed at the Surya Kanta Kala-azar Research Center hospital, Mymensingh, with a mean serum level of 3.0 g/dL, in contrast to 0.27 g/dL for IgM and 0.065 g/dL for IgA (unpublished data). Together with other reports [12, 13], IgG-based hypergammaglobulinemia is a hallmark of VL that is not restricted by the geographical backgrounds or the infecting *Leishmania* species.

16.2 Mechanisms Underlying Elevation of IgG Levels in VL

What causes B-cell activation leading to hypergammaglobulinemia during VL? The activation can be T-cell dependent or independent, and also antigen specific or nonspecific. In this section, T-cell involvement, antigen specificity, and relevant molecules for the B-cell activation during VL are discussed.

Protective immunity against VL is associated with antigen-specific cell-mediated responses represented by lymphoproliferation, delayed type hypersensitivity (DTH), and the production of Th1 cytokines like IFN-γ and IL-2 upon antigen recall [14, 15]. Patients with active VL present with strong humoral responses as described earlier, whereas they are hyporesponsive in cell-mediated immunity as represented by DTH and lymphoproliferation [14, 16–18]. After treatment of VL, those individuals acquire antigen-specific, cell-mediated responses [14, 17–19]. In contrast, IL-10 that is associated with T-cell hyporesponsiveness is the predominant cytokine during active VL [20]. IL-4 is the representative Th2 cytokine that is involved in induction of humoral immunity, and there is a study that demonstrated serum levels of IL-4, as well as IL-10, are high in Indian patients with VL [21]. However, it should be noted that the role of IL-4 in experimental VL is controversial [22–24]. Nonetheless, patients with VL have elevated levels of total and antigen-specific IgE [25, 26], indicating the involvement of IL-4 in humoral responses during the disease. Therefore, the Th1/Th2 balance may be a key factor for the outcome of VL, and the Th2-dominant status may be associated with hypergammaglobulinemia during the disease.

Hypergammaglobulinemia can be classified into monoclonal and polyclonal types. Monoclonal hypergammaglobulinemia, which is characterized by the presence of M protein, can be found in patients with B-cell malignancy, including multiple myeloma and Waldenström's macroglobulinemia, as well as in people without any apparent symptoms, a condition called monoclonal gammopathy of undetermined significance (MGUS). In contrast, hypergammaglobulinemia during VL is considered to be polyclonal hypergammaglobulinemia, because serum electrophoresis of patients with VL demonstrates not a single, sharp peak but rather one that is broad [6]. However, 'polyclonal B-cell activation' is often used not only to explain the polyclonal feature of hypergammaglobulinemia but also antigen-nonspecific activation of the humoral response during VL. Polyclonal B-cell activation has been thought to occur in VL patients because they often have autoantibodies such as rheumatoid factor (RF: anti-IgG antibodies), anti-CCP antibodies, anti-nuclear antibodies, anti-smooth muscle antibodies, and anti-erythrocyte antibodies [27–33]. These reports indicate mechanisms for antigen-nonspecific activation of B cells during VL.

There are also attempts to identify *Leishmania* components that directly stimulate B cells like B-cell mitogens. *Leishmania infantum* amastigotes activate human

B cells to secrete IL-10 [34]. Rico et al. have reported that the heat shock proteins Hsp70 and Hsp83 behave as T cell-independent mitogens of B cells [35]. *Trypanosoma cruzi*, a related trypanosomatid parasite, is also known to have B-cell mitogenic molecules such as proline racemase and trans-sialidase [36, 37]. These reports suggest the capability of *Leishmania* parasites to directly activate B cells in a T cell-independent manner.

However, it is still not clear if such polyclonal B-cell activation contributes to IgG-based hypergammaglobulinemia. First, most of the autoantibodies are not IgG but rather IgM. Secondly, patients with VL have selective IgG responses to *Leishmania*-specific antigens. For example, VL and Chagas' disease are caused by *Leishmania* spp. and *T. cruzi*, respectively, and often have overlapping distributions in Latin America. When a crude lysate of *T. cruzi* is used for serodiagnosis of Chagas' disease, it also detects antibodies in patients with VL [38], and vice versa. In contrast, when species-specific antigens are used, they are capable of detecting IgG antibodies only in patients with the corresponding disease [39–41]. Because there is no IgG autoantibody to be detected as a high titer in patients with VL, unlike with those to rK39 or to other major antigens, the majority of serum IgG in patients with VL seems to be antigen specific rather than from nonspecific B-cell activation. Further studies are necessary to elucidate the contribution of polyclonal B-cell activation to hypergammaglobulinemia in VL.

B-cell activating factor (BAFF), also known as B lymphocyte stimulator (BLyS), tumor necrosis factor- and ApoL-related leukocyte expressed ligand 1 (TALL-1), and tumor necrosis factor ligand superfamily 13B (TNFSF13B), is a critical regulator of B-cell development and differentiation [42]. Although the molecule is indispensable in maintaining B-cell functions, aberrant expression of BAFF is associated with some diseases showing hypergammaglobulinemia. Those diseases include SLE, RA, and Sjögren's syndrome where inhibitors of BAFF signaling, such as belimumab and blisibimod, are approved and/or being evaluated for treatment. We have reported that serum BAFF levels are highly elevated in Brazilian patients with VL [11]. The magnitude of elevation in patients with VL was equivalent to or higher than that previously reported for other diseases [43–46]. Elevation of serum BAFF is also found in Bangladesh patients with VL (Omachi S et al., unpublished). However, a clear positive correlation between IgG and BAFF within patients with VL was not found in either region. In contrast, we have observed a statistically significant positive correlation between serum IgG and BAFF levels during *L. donovani* infection in mice as well as a lack of hypergammaglobulinemia in *L. donovani*-infected BAFF-knockout mice (Omachi S et al., unpublished). A previous report has also shown the elevation of serum immunoglobulins, with an emphasis on IgG, in BAFF transgenic mice [47]. Taken together, BAFF may be involved in IgG-based hypergammaglobulinemia during VL, but is not the sole factor in clinical cases. Because BAFF synergizes with other cytokines for B-cell activation [48], simultaneous analyses on multiple B-cell activators may be necessary to understand the mechanism for hypergammaglobulinemia.

16.3 Possible Roles of Immunoglobulins in the Pathogenesis of VL

VL and other autoimmune diseases share clinical symptoms such as fever, pancytopenia, and splenomegaly, as well as autoantibodies such as RF, anti-nuclear antibodies, and anti-erythrocyte antibodies. Therefore, some symptoms of VL have been speculated to be the result of autoantibodies. For example, one of the mechanisms for anemia during VL may be autoimmune hemolytic anemia. It has been reported that patients with VL are often Coombs test positive and IgG molecules can be detected on erythrocytes from the patients [32, 49].

Another immunoglobulin-related symptom found in VL is glomerulonephritis, and kidney injury is often found in patients with VL [10, 50–53]. In those patients with kidney injury, glomerulonephritis coupled with circulating immune complex (IC), as well as IC deposited in glomeruli, is commonly found [53]. IC has been postulated as a cause of kidney injury during experimental VL [54]. A recent study in dogs also demonstrated the deposition of *Leishmania* antigen, IgG, and C3b in glomeruli with VL and possible involvement of inflammasomes in glomerular damage [55].

Recent studies using experimental models have suggested that immunoglobulin can not only induce some symptoms directly but also serve to indirectly affect disease outcome through increased parasite burden. In mouse models of cutaneous leishmaniasis (CL), it was demonstrated that continual administration of anti-IgM antibodies, which causes B-cell depletion, enhances resistance to *L. tropica* and *L. mexicana* in BALB/c mice [56]. During *L. major* infection, IgG not only fails to provide protection, but actually contributes to disease progression through IC-mediated production of IL-10 from monocytes [16]. There are also studies demonstrating the involvement of B cells, IgG, and Fc receptors in the exacerbation of experimental CL [57–63], whereas there are conflicting reports on the roles of humoral immunity in disease exacerbation [64, 65]. There are also some studies on the role of humoral immunity in the pathogenesis of experimental VL. B cell-deficient μMT mice are more resistant to *L. donovani* infection than are wild type C57BL/6 mice [66]. Another study showed that IgM and polyclonal B cell activation are associated with disease progression in experimental VL [67]. This study was restricted to the early stage of infection; roles of immunoglobulins in late stage infection remain unclear.

16.4 Conclusion

From our study and previous studies from other groups, it has been revealed that IgG-based hypergammaglobulinemia is a hallmark of VL. However, clear relationships between IgG elevation and the clinical manifestations/pathology found in the disease remain uncertain. While serum IgG levels of VL patients are significantly higher than those of healthy controls in general, variation in the IgG levels of patients with VL is large. Therefore, factors related to IgG elevation should be investigated and clarified. As described in the previous section, BAFF is one of the possible factors related to hypergammaglobulinemia in VL. We will be able to elucidate the role of BAFF in IgG elevation through a study using BAFF knockout mice. Furthermore, the roles of immunoglobulins in CL have become more clarified through extensive research using animal models, whereas studies in VL are lagging behind. Not only the direct effects of IgG to autoimmune-like manifestations or glomerulonephritis but also indirect involvement in the progression of the disease should be clarified for further understanding of the pathogenesis of VL. Recently it has become clear that different IgG subclasses vary highly in their capacity to induce different effector pathways [68]. Concerning *Leishmania* research, it was reported that FcγRIII, to which IgG1 preferentially binds, has a specific role in suppressing protective immunity in *L. mexicana* infection, likely through macrophage IL-10 production in skin lesions [62]. Because the elevation of each IgG subclass is not uniform in human cases [13, 69], it is necessary to examining the roles of each IgG subclass in VL pathology in future studies.

References

1. Benbassat J, Fluman N, Zlotnick A. Monoclonal immunoglobulin disorders: a report of 154 cases (Translated from eng). Am J Med Sci. 1976;271(3):325–34 (in eng).
2. Ehrenstein MR, Isenberg DA. Hypergammaglobulinaemia and autoimmune rheumatic diseases (Translated from eng). Ann Rheum Dis. 1992;51(11):1185–7 (in eng).
3. Kyrtsonis MC, et al. Waldenstrom's macroglobulinemia: clinical course and prognostic factors in 60 patients. Experience from a single hematology unit (Translated from eng). Ann Hematol. 2001;80(12):722–7 (in eng).
4. Hale GM, McIntosh SL, Hiki Y, Clarkson AR, Woodroffe AJ. Evidence for IgA-specific B cell hyperactivity in patients with IgA nephropathy (Translated from eng). Kidney Int. 1986;29(3):718–24 (in eng).
5. Heremans JF, Masson PL. Specific analysis of immunoglobulins. Techniques and clinical value (Translated from eng). Clin Chem. 1973;19(3):294–300 (in eng).
6. Cooper GR, Rein CR, Beard JW. Electrophoretic analysis of kala-azar human serum; hypergammaglobulinemia associated with seronegative reactions for syphilis (Translated from eng). Proc Soc Exp Biol Med. 1946;61:179–83 (in eng).
7. Jeannel D, Tuppin P, Brucker G, Danis M, & Gentilini M. Imported and autochthonous kala-azar in France (Translated from eng). Bmj 1991;303(6798):336–338 (in eng).
8. Evans T, et al. American visceral leishmaniasis (kala-azar) (Translated from eng). West J Med. 1985;142(6):777–81 (in eng).
9. Ghose AC, Haldar JP, Pal SC, Mishra BP, Mishra KK. Serological investigations on Indian kala-azar (Translated from eng). Clin Exp Immunol. 1980;40(2):318–26 (in eng).
10. Agenor Araujo Lima Verde F, et al. Renal tubular dysfunction in human visceral leishmaniasis (Kala-azar) (Translated from eng). Clin Nephrol 2009;71(5):492–500 (in eng).
11. Goto Y, Omachi S, Sanjoba C, Matsumoto Y. Elevation of serum B-cell activating factor levels during visceral leishmaniasis (Translated from eng). Am J Trop Med Hyg. 2014;91 (5):912–4 (in eng).

12. Rezai HR, Ardehali SM, Amirhakimi G, Kharazmi A. Immunological features of kala-azar (Translated from eng). Am J Trop Med Hyg. 1978;27(6):1079–83 (in eng).
13. el Amin EM, Wright EP, Vlug A. Characterization of the humoral immune response in Sudanese leishmaniasis: specific antibody detected by class- and subclass-specific reagents (Translated from eng). Clin Exp Immunol. 1986;64(1):14–9 (in eng).
14. Haldar JP, Ghose S, Saha KC, Ghose AC. Cell-mediated immune response in Indian kala-azar and post-kala-azar dermal leishmaniasis (Translated from eng). Infect Immun. 1983;42(2):702–7 (in eng).
15. Carvalho EM, Badaro R, Reed SG, Jones TC, Johnson WD Jr. Absence of gamma interferon and interleukin 2 production during active visceral leishmaniasis. J Clin Invest. 1985;76(6):2066–9.
16. Miles SA, Conrad SM, Alves RG, Jeronimo SM, Mosser DM. A role for IgG immune complexes during infection with the intracellular pathogen *Leishmania*. J Exp Med. 2005;201(5):747–54.
17. Ho M, Koech DK, Iha DW, Bryceson AD. Immunosuppression in Kenyan visceral leishmaniasis (Translated from eng). Clin Exp Immunol. 1983;51(2):207–14 (in eng).
18. Sacks DL, Lal SL, Shrivastava SN, Blackwell J, Neva FA. An analysis of T cell responsiveness in Indian kala-azar (Translated from eng). J Immunol. 1987;138(3):908–13 (in eng).
19. Carvalho EM, Teixeira RS, Johnson WD Jr. Cell-mediated immunity in American visceral leishmaniasis: reversible immunosuppression during acute infection (Translated from eng). Infect Immun. 1981;33(2):498–500 (in eng).
20. Ghalib HW, et al. Interleukin 10 production correlates with pathology in human *Leishmania donovani* infections. J Clin Invest. 1993;92(1):324–9.
21. Sundar S, Reed SG, Sharma S, Mehrotra A, Murray HW. Circulating T helper 1 (Th1) cell- and Th2 cell-associated cytokines in Indian patients with visceral leishmaniasis (Translated from eng). Am J Trop Med Hyg. 1997;56(5):522–5 (in eng).
22. Kaye PM, Curry AJ, Blackwell JM. Differential production of Th1- and Th2-derived cytokines does not determine the genetically controlled or vaccine-induced rate of cure in murine visceral leishmaniasis (Translated from eng). J Immunol. 1991;146(8):2763–70 (in eng).
23. Melby PC, Chandrasekar B, Zhao W, Coe JE. The hamster as a model of human visceral leishmaniasis: progressive disease and impaired generation of nitric oxide in the face of a prominent Th1-like cytokine response (Translated from eng). J Immunol. 2001;166(3):1912–20 (in eng).
24. Melby PC, Tryon VV, Chandrasekar B, Freeman GL. Cloning of Syrian hamster (*Mesocricetus auratus*) cytokine cDNAs and analysis of cytokine mRNA expression in experimental visceral leishmaniasis (Translated from eng). Infect Immun. 1998;66(5):2135–42 (in eng).
25. Atta AM, et al. Anti-leishmanial IgE antibodies: a marker of active disease in visceral leishmaniasis (Translated from eng). Am J Trop Med Hyg. 1998;59(3):426–30 (in eng).
26. Saha S, et al. *Leishmania* promastigote membrane antigen-based enzyme-linked immunosorbent assay and immunoblotting for differential diagnosis of Indian post-kala-azar dermal leishmaniasis (Translated from eng). J Clin Microbiol. 2005;43(3):1269–77 (in eng).
27. Santana IU, et al. Visceral leishmaniasis mimicking systemic lupus erythematosus: Case series and a systematic literature review (Translated from eng). Semin Arthritis Rheum. 2015;44(6):658–65 (in eng).
28. Galvao-Castro B, et al. Polyclonal B cell activation, circulating immune complexes and autoimmunity in human american visceral leishmaniasis (Translated from eng). Clin Exp Immunol. 1984;56(1):58–66 (in eng).
29. Ahlin E, Elshafie AI, Nur MA, Ronnelid J. Anti-citrullinated peptide antibodies in Sudanese patients with *Leishmania donovani* infection exhibit reactivity not dependent on citrullination (Translated from eng). Scand J Immunol. 2015;81(3):201–8 (in eng).
30. Liberopoulos E, Kei A, Apostolou F, Elisaf M. Autoimmune manifestations in patients with visceral leishmaniasis (Translated from eng). J Microbiol Immunol Infect. 2013;46(4):302–5 (in eng).

31. Atta AM, Carvalho EM, Jeronimo SM, & Sousa Atta ML. Serum markers of rheumatoid arthritis in visceral leishmaniasis: rheumatoid factor and anti-cyclic citrullinated peptide antibody (Translated from eng) J Autoimmun 2007;28(1):55–58 (in eng).
32. Pontes De Carvalho LC, et al. Nature and incidence of erythrocyte-bound IgG and some aspects of the physiopathogenesis of anaemia in American visceral leishmaniasis (Translated from eng) Clin Exp Immunol 1986;64(3):495–502 (in eng).
33. Carvalho EM, Andrews BS, Martinelli R, Dutra M, Rocha H. Circulating immune complexes and rheumatoid factor in schistosomiasis and visceral leishmaniasis (Translated from eng). Am J Trop Med Hyg. 1983;32(1):61–8 (in eng).
34. Andreani G, et al. *Leishmania infantum* amastigotes trigger a subpopulation of human B cells with an immunoregulatory phenotype (Translated from eng). PLoS Negl Trop Dis. 2015;9(2): e0003543 (in eng).
35. Rico AI, Girones N, Fresno M, Alonso C, Requena JM. The heat shock proteins, Hsp70 and Hsp83, of *Leishmania infantum* are mitogens for mouse B cells (Translated from eng). Cell Stress Chaperones. 2002;7(4):339–46 (in eng).
36. Reina-San-Martin B, et al. A B-cell mitogen from a pathogenic trypanosome is a eukaryotic proline racemase (Translated from eng). Nat Med. 2000;6(8):890–7 (in eng).
37. Gao W, Wortis HH, Pereira MA. The *Trypanosoma cruzi* trans-sialidase is a T cell-independent B cell mitogen and an inducer of non-specific Ig secretion (Translated from eng). Int Immunol. 2002;14(3):299–308 (in eng).
38. Anonymous (ORTHO® *T. cruzi* ELISA Test System. (Ortho-Clinical Diagnostics, Inc.).
39. Goto Y, Carter D, Reed SG. Immunological dominance of *Trypanosoma cruzi* tandem repeat proteins (Translated from eng). Infect Immun. 2008;76(9):3967–74 (in eng).
40. Braz RF, et al. The sensitivity and specificity of *Leishmania chagasi* recombinant K39 antigen in the diagnosis of American visceral leishmaniasis and in differentiating active from subclinical infection (Translated from eng). Am J Trop Med Hyg. 2002;67(4):344–8 (in eng).
41. Houghton RL, et al. A multi-epitope synthetic peptide and recombinant protein for the detection of antibodies to *Trypanosoma cruzi* in radioimmunoprecipitation-confirmed and consensus-positive sera. J Infect Dis. 1999;179(5):1226–34.
42. Tangye SG, Bryant VL, Cuss AK, Good KL. BAFF, APRIL and human B cell disorders (Translated from eng). Semin Immunol. 2006;18(5):305–17 (in eng).
43. Briones J, Timmerman JM, Hilbert DM, Levy R. BLyS and BLyS receptor expression in non-Hodgkin's lymphoma (Translated from eng). Exp Hematol. 2002;30(2):135–41 (in eng).
44. Elsawa SF, et al. B-lymphocyte stimulator (BLyS) stimulates immunoglobulin production and malignant B-cell growth in Waldenstrom macroglobulinemia (Translated from eng). Blood. 2006;107(7):2882–8 (in eng).
45. Nduati E, et al. The plasma concentration of the B cell activating factor is increased in children with acute malaria (Translated from eng). J Infect Dis. 2011;204(6):962–70 (in eng).
46. Matsushita T, et al. Elevated serum BAFF levels in patients with systemic sclerosis: enhanced BAFF signaling in systemic sclerosis B lymphocytes (Translated from eng). Arthritis Rheum. 2006;54(1):192–201 (in eng).
47. Mackay F, et al. Mice transgenic for BAFF develop lymphocytic disorders along with autoimmune manifestations (Translated from eng). J Exp Med. 1999;190(11):1697–710 (in eng).
48. Doreau A, et al. Interleukin 17 acts in synergy with B cell-activating factor to influence B cell biology and the pathophysiology of systemic lupus erythematosus (Translated from eng). Nat Immunol. 2009;10(7):778–85 (in eng).
49. Vilela RB, Bordin JO, Chiba AK, Castelo A, Barbosa MC. RBC-associated IgG in patients with visceral leishmaniasis (kala-azar): a prospective analysis (Translated from eng). Transfusion. 2002;42(11):1442–7 (in eng).
50. Dutra M, et al. Renal involvement in visceral leishmaniasis (Translated from eng). Am J Kidney Dis. 1985;6(1):22–7 (in eng).
51. Vassallo M, et al. Visceral leishmaniasis due to *Leishmania infantum* with renal involvement in HIV-infected patients (Translated from eng). BMC Infect Dis. 2014;14:561 (in eng).

52. Lima Verde FA, Lima Verde IA, Silva Junior GB, Daher EF, Lima Verde EM. Evaluation of renal function in human visceral leishmaniasis (kala-azar): a prospective study on 50 patients from Brazil (Translated from eng) J Nephrol 2007;20(4):430–436 (in eng).
53. Clementi A et al. Renal involvement in leishmaniasis: a review of the literature (Translated from eng) NDT Plus 2011;4(3):147–152 (in eng).
54. Sartori A, De Oliveira AV, Roque-Barreira MC, Rossi MA, Campos-Neto A. Immune complex glomerulonephritis in experimental kala-azar (Translated from eng). Parasite Immunol. 1987;9(1):93–103 (in eng).
55. Esch KJ, et al. Activation of autophagy and nucleotide-binding domain leucine-rich repeat-containing-like receptor family, pyrin domain-containing 3 inflammasome during *Leishmania infantum*-associated glomerulonephritis (Translated from eng). Am J Pathol. 2015;185(8):2105–17 (in eng).
56. Sacks DL, Scott PA, Asofsky R, Sher FA. Cutaneous leishmaniasis in anti-IgM-treated mice: enhanced resistance due to functional depletion of a B cell-dependent T cell involved in the suppressor pathway. J Immunol. 1984;132(4):2072–7.
57. Kima PE, et al. Internalization of *Leishmania mexicana* complex amastigotes via the Fc receptor is required to sustain infection in murine cutaneous leishmaniasis (Translated from eng). J Exp Med. 2000;191(6):1063–8 (in eng).
58. Kane MM, Mosser DM. The role of IL-10 in promoting disease progression in leishmaniasis (Translated from eng). J Immunol. 2001;166(2):1141–7 (in eng).
59. Hoerauf A, Rollinghoff M, Solbach W. Co-transfer of B cells converts resistance into susceptibility in T cell-reconstituted, *Leishmania major*-resistant C.B-17 scid mice by a non-cognate mechanism. Int Immunol. 1996;8(10):1569–75.
60. Hoerauf A, Solbach W, Lohoff M, Rollinghoff M. The Xid defect determines an improved clinical course of murine leishmaniasis in susceptible mice (Translated from eng). Int Immunol. 1994;6(8):1117–24 (in eng).
61. Chu N, Thomas BN, Patel SR, Buxbaum LU. IgG1 is pathogenic in *Leishmania mexicana* infection (Translated from eng). J Immunol. 2010;185(11):6939–46 (in eng).
62. Thomas BN, Buxbaum LU. FcgammaRIII mediates immunoglobulin G-induced interleukin-10 and is required for chronic *Leishmania mexicana* lesions (Translated from eng). Infect Immun. 2008;76(2):623–31 (in eng).
63. Buxbaum LU, Scott P. Interleukin 10- and Fcgamma receptor-deficient mice resolve *Leishmania mexicana* lesions Infect Immun. 2005;73(4):2101–8.
64. Brown DR, Reiner SL. Polarized helper-T-cell responses against *Leishmania major* in the absence of B cells (Translated from eng). Infect Immun. 1999;67(1):266–70 (in eng).
65. Woelbing F, et al. Uptake of *Leishmania major* by dendritic cells is mediated by Fcgamma receptors and facilitates acquisition of protective immunity (Translated from eng). J Exp Med. 2006;203(1):177–88 (in eng).
66. Smelt SC, Cotterell SE, Engwerda CR, Kaye PM. B cell-deficient mice are highly resistant to *Leishmania donovani* infection, but develop neutrophil-mediated tissue pathology. J Immunol. 2000;164(7):3681–8.
67. Deak E, et al. Murine visceral leishmaniasis: IgM and polyclonal B-cell activation lead to disease exacerbation (Translated from eng). Eur J Immunol. 2010;40(5):1355–68 (in eng).
68. Nimmerjahn F, Ravetch JV. Fcgamma receptors as regulators of immune responses (Translated from eng). Nat Rev Immunol. 2008;8(1):34–47 (in eng).
69. Elassad AM, et al. The significance of blood levels of IgM, IgA, IgG and IgG subclasses in Sudanese visceral leishmaniasis patients (Translated from eng). Clin Exp Immunol. 1994;95(2):294–9 (in eng).

Part V
Knowledge and Practice for Vector Control in Kala-Azar

Chapter 17
Geographical Distribution and Ecological Aspect of Sand Fly Species in Bangladesh

Yusuf Özbel, Chizu Sanjoba and Yoshitsugu Matsumoto

Abstract Phlebotomine sand flies, which are biological vectors of *Leishmania* spp., are represented by around 400 species in the Old World and more than 600 species in the Americas. The vector sand fly species generally belong to the *Phlebotomus* genus in the Old World and the *Lutzomyia* genus in the New World. They are yellowish, long legged hairy insects and active after sunset until sunrise. Sand flies can transmit *Leishmania* parasites as well as some group of viruses called Phleboviruses and a bacterium, *Bartonella bacilliformis*. Visceral leishmaniasis (VL, Kala-azar) caused by *Leishmania donovani* is an important health problem in the Indian subcontinent including Bangladesh and *Phlebotomus argentipes* is a proven vector species of *Leishmania donovani*. In Bangladesh, a total of 13 sand fly species (3 *Phlebotomus*, 10 *Sergentomyia* spp.) were recorded so far. All studies showed the dominancy of *P. argentipes* especially in endemic areas for VL. In this chapter, besides *P. argentipes* and its biological and ecological features, other species constituting sand fly fauna and their geographical distribution in Bangladesh are discussed.

Keywords Sand fly · *Phlebotomus argentipes* · Ecology · Bangladesh

17.1 General Information on Sand Flies

Sand flies are classified under the class Insecta, ordo Diptera, subordo Nematocera, family Psychodidae and sub-family Phlebotominae. The Phlebotominae sub-family consists of 6 genera: *Phlebotomus*, *Sergentomyia*, and *Chinius* in the Old World and *Lutzomyia*, *Brumptomyia*, and *Warileya* in the New World. They have medical and veterinary importance because they are biological vectors of the various *Leishmania* species that cause pathogenicity with different clinical presentations in humans and

---

Y. Özbel (✉)
Faculty of Medicine, Department of Parasitology,
Ege University, 35100 Bornova, İzmir, Turkey
e-mail: yusuf.ozbel@ege.edu.tr

mammalian species. Vector Phlebotomine sand flies are represented by more than 40 species in the Old World and 30 species in the Americas. Besides *Leishmania*, sand flies can also transmit viruses (Phleboviruses) and bacterium (*Bartonella bacilliformis*). The vector sand fly species generally belong to the *Phlebotomus* genus in the Old World and the *Lutzomyia* genus in the New World [1].

Sand flies, which are smaller than other *Diptera*, are a light brown, yellowish color, long legged, and their entire body is hairy including the wings; the wings are in a "V" shape during rest, and because their mesonotum is longer, they appear hunchbacked. Sand flies are holometabol *Diptera* and complete their life cycle in humid soil with organic material. Four different phases occur in their life cycle: egg, larva (4 stages), pupa, and adult.

Sand flies rest in dark hidden locations, such as houses, barns, basements, tree holes, wall cracks, and rodent nests during the daytime and become active during the night. In addition, during this period, they may perform certain activities depending on the light, temperature, humidity, and wind density. Feeding in the form of biting starts immediately before dark and continues the whole night, increasing shortly before sunrise and continuing until the sun rises completely. Females of many species are predominantly exophagic (biting outdoors) and exophilic (resting outdoors during maturation of eggs) and cannot be controlled by spraying internal walls of habitations with insecticide. In contrast, species that are endophilic (resting indoors during maturation of eggs) can be attacked in this way.

They may be especially active in hot nights during the summer season until the sun rises and as the temperature decreases they will have decreased activity during the night. Though the humidity level that they need to survive varies depending on the species, the ideal humidity level is higher than 50 %. Adult females live for 3–4 weeks and males for 2 weeks; despite this, the ratio of males in the population is usually higher, presumably to increase mating possibilities. The ideal temperature, during which the adult sand flies are most active, is between 25 and 28 °C, though in some species, it may be lower. Although the adults are very sensitive to the cold, 4th stage larvae are more resistant against it and can survive the winter season under 5–10 cm of soil.

Characteristically, they fly in a "zigzag" pattern for even short distances and their maximum flying distance is 1 km. They travel by frequently landing and they can climb higher each time to reach the top floors of buildings. Females can perceive a human from under 10 m in distance.

Males of *P. argentipes* wait on the host for females coming to feed and then court the female for mating. Males of other species like *P. ariasi* may never be seen on the host but mate after the female has engorged [1]. Pairing generally happens in the air or on hosts and the female sand fly dies 8–10 days after pairing and feeding on blood; a female can deposit about 50–100 eggs. Larval development very much depends on the temperature/humidity and larvae hatch from the egg depending on the environmental temperature after 5–20 days. The fourth stage of larvae transform to pupae and become adults within 6–15 days. The cycle is completed in total within 40–50 days. In general, in the Mediterranean region the first generation is observed in May, the second in July, and the third in September.

It is known that the time from a blood meal to oviposition varies with the species and the ambient temperature. The number of eggs deposited depends on the size of the blood meal taken and may be as many as 200. There are 2 groups of females: (i) gonotrophically concordant (the ovaries develop as a single blood meal is digested, and the female does not feed a second time during the oviposition cycle, e.g. *P. perniciosus, P. ariasi*); (ii) gonotrophically discordant (no relation between digestion of a blood meal and the development of eggs, and more than 1 blood-meal may be taken during a single oviposition cycle, e.g. *P. argentipes, P. papatasi*). Two blood meals in different stages of digestion are commonly seen in the second group of females [1].

Male sand flies feed on plant juice only due to shapes of their mouthparts, but the females may feed on blood from cold- and hot-blooded hosts or on plant juice. Sugar meals serve as energy sources for sand flies and are important in the development of parasites within the gut. Some species suck blood only from the animals (zoophilic), some from both animals and humans (zoo-anthropophilic), and some only from humans (anthropophilic).

Sand flies may be caught with various methods in nature either alive (with CDC light trap, mouth aspirator, etc.) or dead (such as with oil paper), and the species may be determined after proper dissection/preparation. A stomach dissection may be performed to determine the presence of parasites or may be used for many other purposes after DNA/RNA isolation. Catch effectiveness is always influenced by environmental conditions, including wind, temperature, rain, relative humidity, atmospheric pressure, and moonlight. Capture with a mouth aspirator proved to be most reliable method for obtaining blood-fed and gravid *P. argentipes* specimens in Bangladesh.

Identification of the species caught can be done morphologically using available written keys, drawings, and photos. For the species belonging to the *Phlebotomus* genus, many, but not all, males can be identified by their morphology alone, but it is often more difficult to identify females, some of which are proven vectors. In last decade, DNA analysis techniques are helpful for identifying morphologically indistinguishable specimens of related species. The biology of each species of sand fly is unique and complex, covering all aspects of reproduction, feeding, dispersal, and other activities that have a direct bearing on the epidemiology of leishmaniasis and vector control [1].

17.1.1 Determining Sand Fly Species as Certain Disease Vectors

Sand flies have been classified as specific (restrictive) or permissive vectors. The sand fly species that support the development of multiple *Leishmania* species are called "permissive vectors" (e.g. *P. argentipes, P. pernicious*) while the sand fly species that do not or poorly support the development of different *Leishmania* species are called "specific vectors" (e.g. *P. sergenti, P. papatasi*) [1, 2]. The criteria for determining a sand fly species as a vector were first described by Dr. R. Killick-Kendrick and later were modified [3–5]. The last 2 criteria, shown next, are important for the vectors in a specific focus, like endemic areas of Bangladesh.

The criteria for incrimination [5]

- Promastigotes should be isolated from female sand fly specimens more than once
- Infective metacyclic forms should be seen in the anterior midgut and on the stomodeal valve of naturally infected female flies
- The species should be attracted to and bite humans and any reservoir hosts
- A strong ecological association needs to be shown between fly, humans, and any reservoir host
- Experimental transmission needs to be shown in the laboratory model
- Mathematical modeling demonstrates that the vector is essential for maintaining transmission with or without the involvement of other vectors
- Mathematical modeling based on a planned control program demonstrates that disease incidence significantly decreases following a significant decrease in the biting density of the specific vector

17.2 Sand Flies in Bangladesh

Visceral leishmaniasis (VL) caused by *Leishmania donovani* is an important health problem in the Indian subcontinent including Bangladesh. VL is known as an anthroponotic disease and patients with post kala-azar dermal leishmaniasis (PKDL) are a significant reservoir for anthroponotic transmission of VL [1]. The sustainable application of the integrated kala-azar control program is crucial for decreasing the morbidity and mortality of the disease. For increasing the efficacy of the vector component of the control program sand fly fauna should be studied in the endemic areas.

Sand fly studies in the Indian subcontinent started with Sinton [6] and Lewis who published a detailed identification key for *Phlebotomus* and *Sergentomyia* species, as well as information on their distribution in the oriental region [7]. This key is still useful and can be used for morphological identification.

17 Geographical Distribution and Ecological Aspect ...

Table 17.1 Sand fly species recorded in Bangladesh

Genus	Subgenus	Species	References
Phlebotomus	Euphlebotomus	P. argentipes	[7–10]
	Phlebotomus	P. papatasi	[7, 8]
	Paraphlebotomus	P. sergenti	[9]
Sergentomyia	Parrotomyia	S. africana magna	[9, 10]
		S. babu babu	[8–10]
		S. baghdadis	[8, 9]
		S. barraudi	[7–10]
		S. shorttii	[7–10]
		S. himalayensis	[9, 10]
		S. montana	[9]
	Neophlebotomus	S. malabarica	[9, 10]
		S. perturbans	[7, 10]
	Grassomyia	S. indica	[10]

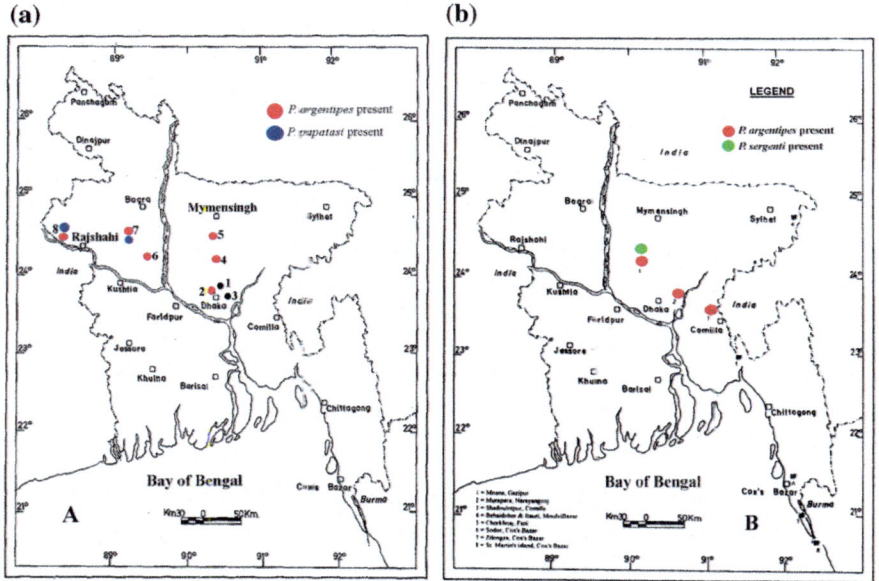

Fig. 17.1 The previous study areas and presence/absence of *Phlebotomus* species in Bangladesh (modified from [8] (**a**) and [9] (**b**))

After the publication of Lewis's key, several studies were carried out in Bangladesh in different areas of the country that recorded a total of 13 sand fly species (3 *Phlebotomus*, 10 *Sergentomyia* spp.), although there is 1 questionable, *S. theodori*, and 1 unidentified *Sergentomyia* species in Bangladesh [8–11]

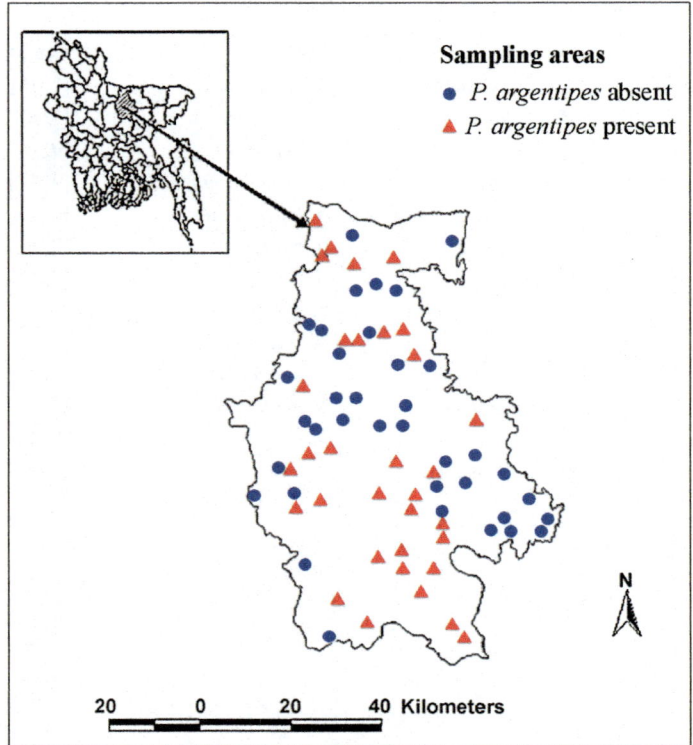

Fig. 17.2 Map of Mymensingh district showing sampling areas and absence/presence of *P. argentipes* (modified from [10])

(Table 17.1). In all studies related to sand fly fauna in Bangladesh, *P. argentipes* was found to be the dominant species among the members of the *Phlebotomus* genus. *P. papatasi* and *P. sergenti* are reported rarely from VL endemic and non-endemic areas (Fig. 17.1).

In Bangladesh, the Mymensingh district is the most highly endemic area for VL and therefore recent VL control programs, including sand fly studies, are focused on this district. In a recent study in Mymensingh, of 726 sand flies, 562 were caught from endemic areas, of which 413 were *P. argentipes*. It is emphasized that sand fly density as a whole differed significantly between endemic and non-endemic areas ($P = 0.004$). Similarly *P. argentipes* numbers were significantly higher ($P = 0.008$) in endemic areas compared with that seen in non-endemic areas [10] (Fig. 17.2). The micro-environmental and micro-climatic factors need to be investigated to better explain the reasons behind these observations. Mapping of endemic and non-endemic locations in parallel in Mymensingh and other highly endemic areas is also suggested.

In an ongoing study, similar results were obtained in VL endemic areas of Mymensingh; *P. argentipes* was the dominant species (>94 %). In 2012, a difference in the population size of sand flies between endemic and non-endemic areas was also observed. The population size was bigger in the endemic areas than that seen in the non-endemic areas. There might be some micro or nano environmental factors contributing to both sand fly abundance and kala-azar endemicity in the Mymensingh district (unpublished data).

The endophilic and peridomestic species *Phlebotomus argentipes* is the only known vector for *L. donovani* in the Indian subcontinent. This species is dominant in all endemic districts in Bangladesh and is accepted as a vector species but no infected fly has been detected so far. A unique published study was carried out in the Mymensingh highly endemic district for Kala-azar, and 376 *P. argentipes* were individually analyzed for *Leishmania* by PCR targeting *Leishmania* minicircle DNA. In addition to this, 679 *Sergentomyia* specimens were also included in the study, but none of the 1055 sand fly specimens were found positive in this study. It can be speculated that the infection rate is less than 1/376 among *P. argentipes* females in the district [11]. In 2015, in the Pabna district, around 50 *P. argentipes* females were dissected by direct microscopical method but no positive fly was found (unpublished data). Therefore, it will be needed to (i) detect *Leishmania* parasites by microscopical dissection and then PCR methods to incriminate the species as vector, (ii) discover the infection rate in sand flies, (iii) identify and compare *Leishmania* parasites isolated from humans and vectors, and (iv) perform blood-meal analysis also from *Sergentomyia* specimens for detecting their host preference.

Visceral leishmaniasis caused by *L. donovani* in the Indian subcontinent is exclusively anthroponotic in contrast to zoonotic VL caused by *L. infantum* in the New World and Mediterranean Basin where the animal reservoir, dog, is involved [1]. The vector sand fly species, *P. argentipes*, normally zoophilic on cattle, is also strongly attracted to man, making it anthropophilic and zoophilic [12]. It is reported that *P. argentipes* prefers animal bait 7 times more than human bait in a similar situation [13]. In a recent study performed in Trishal, Mymensingh, anti-*Leishmania* antibodies were detected in cattle sera by ELISA and direct agglutination test (DAT) but no *Leishmania* DNA was detected by nested PCR (Ln PCR) and a recently developed molecular technique, loop-mediated isothermal amplification (LAMP). The authors emphasized that the anti-*Leishmania* antibodies detected by serological techniques might be the result of exposure to the *Leishmania* parasite by the bite of infected sand fly despite no infection developing [14]. For better knowledge about VL epidemiology in this area, studies related to the host preference of *P. argentipes* should be performed in a larger number of cattle or other domestic animals.

Information about the daily and seasonal activity of the endophilic vector species, *P. argentipes*, is also important for control programs, mainly for the timing of indoor residual spraying (IRS) application. In the Indian subcontinent, the transmission of *L. donovani* occurs during 2 periods: a pre-winter peak in September-November and a post-winter peak in March-May. During the monsoon season from June to September, the numbers of sand flies are low. It is also reported that *P. argentipes* lives in and around houses and biting occurs at night, mainly between 21:00 and 01:00, peaking at 23:00–24:00 [13]. The infection risk elevates during the peak of the season along with daily sand fly activity. But, no study thus far has been carried out in Bangladesh to study sand fly activities. Regular, preferably monthly, sand fly collection can be done in areas where patients with VL are reported.

17.2.1 Phlebotomus Argentipes, *Annandale and Brunetti: The Vector of Visceral Leishmaniasis in Bangladesh*

Phlebotomus argentipes is taxonomically classified under the subgenus *Euphlebotomus* Theodor and its morphological features were described together with other Oriental phlebotomine sand flies by Lewis in 1978 [7] (Table 17.2) (Figs. 17.3 and 17.4).

The studies carried out so far, support the anthropophagic behavior and vectorial competence of *Phlebotomus argentipes* [15, 16]. This species can be found in Asia from Iran to Western Malaysia and Indonesia and shows some geographical variations in its morphology and feeding behavior [12, 17, 18]. The ascoids on the fourth antennal segment of females in Northeastern India are shorter than those seen on the specimens in Sri Lanka [12, 19, 20]. In the studies performed in India, the presence of 2 different local *P. argentipes* populations with 2 different morphological features was attributed to the different vectorial capacities of these 2 morphospecies [16, 21].

Table 17.2 Morphological features of *P. argentipes* used for identification

Male	Female
• Cibarium with spicules or unarmed. Pharynx usually armed. Palp extending further than antenna 3	• Pharynx of female with a small group of teeth in the middle and behind it some concentric lines
• Coxite of male without sub-basal hairy process and genital filaments 3–11 times as long as pump	• Spermatheca with differentiated end-segment (subgenus EUPHLEBOTOMUS)
• Style with five long spines	• Spermatheca distinctly segmented
• Paramere with three lobes	• Spermathecal common duct with rather thin walls. Antenna 5 without papilla
• Middle lobe of paramere thinner than main lobe	
• Main lobe of paramere much more than twice length of middle lobe, lower lobe narrow, depth of paramere about 0.29 of its length (measured to junction with coxite)	

Fig. 17.3 Male genitalia of *P. argentipes*. (*Photo* Özbel Y) Magnification: (from *left* to *right*: X200; X100; X400)

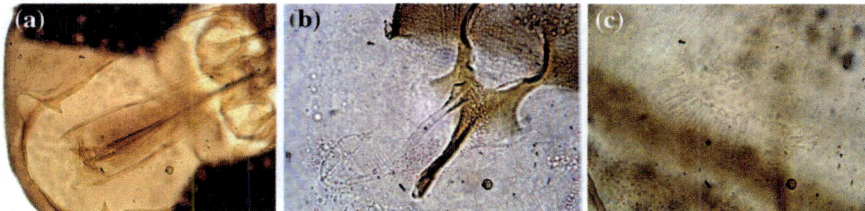

Fig. 17.4 *P. argentipes* female. **a.** Pharynx; **b** and **c**. Spermatheca (*Photo* Özbel Y) magnification (X400)

The ecological parameters affecting *P. argentipes* population size and *Leishmania* transmission were studied in 2 villages of West Bengal, India [22]. In this study, they found that the abundance of the species increased during and after monsoon, and decreased during winter and summer, probably because of the prolonged duration of the gonotrophic cycle. The researchers also assessed that 8 ecological parameters (soil temperature and moisture, rainfall, air temperature, relative humidity, soil pH, soil organic carbon, and windspeed) can affect the sand fly abundance in a particular area.

P. argentipes breeds in moist organic soil at the junction of the floor and walls of cattle sheds and earthen houses [23, 24] Soil temperature and moisture may contribute to the survival and development of the immature stages of this species. Low moisture and temperature causes drying of organic material and prolongation of the life cycle, respectively. These ecological variables are highly effective on sand fly abundance and should be taken into account while implementing vector control measures.

The ecological situation is also important for the application of vector control managements, especially for indoor and outdoor residual spraying. The success of the application very much depends on the application season and time in endemic areas. Ecological data are also useful in predicting *P. argentipes* abundance and assessing the risk of increased *Leishmania* transmission.

In Bangladesh, the following information needs to be collected:

- The composition and population density of *P. argentipes* and other sand fly species, in addition to mapping their distribution and resting habits in endemic areas,
- Infection rate of *P. argentipes*: search for *Leishmania* parasite by microscopical dissection and/or PCR methods,
- Identification of *Leishmania* parasites: comparison of *Leishmania donovani* isolates from humans and vectors
- Other possible vectors: blood-meal analysis for *Sergentomyia* specimens

The seasonal and nocturnal activities of *P. argentipes* were studied throughout a year by monthly collections in India where the climatic and environmental conditions are similar to those in Bangladesh. In this study, it was found that the numbers of *P. argentipes* collected varied according to season, being significantly higher during the summer than that seen during the rainy season or winter, as expected. Related to seasonal activity, the authors concluded that the numbers of *P. argentipes* on humans and bovine bait in Bahapur village, Bihar, India showed 2 peaks before and after the rainy season [13].

The nocturnal activity findings showed that their night activity starts around 19:00 and peaks between 22:00 and 24:00 h in the summer season (between February and May) when most of the parasite transmission occurs [13]. The other studies on nocturnal activity of this species showed that landing on human bait peaked between 21:00 and 03:00 in West Bengal [25] while it was observed between 21:00 and 01:00 h in Southern India [26].

Both issues, seasonality and nocturnal activity, have not yet been studied in the endemic areas of Bangladesh. Because micro/nano environmental/ecological/climate factors can affect the daily and nocturnal behavior of the vector species, a detailed field study needs to be performed in at least the 2 distinct endemic areas of Bangladesh. In the Indian subcontinent, *P. argentipes* is known to be endophilic, but this behavior of *P. argentipes* should also be confirmed in Bangladesh to help the IRS part of the national control program.

References

1. WHO. "Control of the leishmaniasis," Report of a meeting of the WHO expert committee on the control of Leishmaniases, World Health Organization, Geneva, Switzerland. WHO Technical Report Series No: 949; 2010.
2. Volf P, Myskova J. Sand flies and Leishmania: specific versus permissive vectors. Trends Parasitol. 2007;23:91–2.
3. Killick-Kendrick R. Phlebotomine vectors of the leishmaniases: a review. Med Vet Entomol. 1990;4:1–24.
4. Ready P. Should sandfly taxonomy predict vectorial and ecological traits? J Vector Ecol. 2011;36:S17–32.

5. Ready P. Biology of phlebotomine sand flies as vectors of disease agents. Ann Rev Entomol. 2013;58:227–50.
6. Sinton JA. Notes on some Indian species of the genus *Phlebotomus*. Part XXX. Diagnostic table for the females of the species recorded from India. Indian J Med Res. 1932;20:55–75.
7. Lewis DJ. The Phlebotomine sandflies (Diptera: Psychodidae) of the oriental region. Bull Brit Mus (Nat Hist). 1978;37:217–343.
8. Hossain MI, Khan SA, Ameen M. Phlebotomine sandflies of Bangladesh: recent surveys (short communication). Med Vet Entomol. 1993;7:99–101.
9. Salahuddin M, Ameen M, Hossain M. Faunistic survey of sandflies in six districts of Bangladesh. Bangladesh J Zool. 2000;28:109–18.
10. Alam MS, Wagatsuma Y, Mondal D, Khanum H, Haque R. Relationship between sand fly fauna and kala-azar endemicity in Bangladesh. Acta Trop. 2009;112:23–5.
11. Alam MS, Kato H, Fukushige M, Wagatsuma Y, Itoh M. Application of RFLP-PCR-based identification for sand fly surveillance in an area endemic for Kala-Azar in Mymensingh, Bangladesh. J Parasitol Res. 2012;2012:467821. doi:10.1155/2012/467821.
12. Lewis DJ. Phlebotomine sandflies (Diptera: Pscychodidae) from the oriental region. Syst Entomol. 1987;12:163–80.
13. Dinesh DS, Ranjan A, Palit A, Kishore K, Kar SK. Seasonal and nocturnal landing/biting behaviour of *Phlebotomus argentipes* (Diptera: Psychodidae). Ann Trop Med Parasitol. 2001;95:197–202.
14. Alam MS, Ghoch D, Khan MG, Islam MF, Mondal D, Itoh M, et al. Survey of domestic cattle for anti-*Leishmania* antibodies and *Leishmania* DNA in a visceral leishmaniasis endemic area of Bangladesh. BMC Vet Res. 2011;7:27. doi:10.1186/1746-6148-7-27.
15. Lane RP, Pile MM, Amerasinghe FP. Anthropophagy and aggregation behaviour of the sandfly *Phlebotomus argentipes* in Sri Lanka. Med Vet Entomol. 1990;4:79–88.
16. Surendran SN, Kajatheepan A, Hawkes NJ, Ramasamy R. First report on the presence of morphospecies A and B of *Phlebotomus argentipes* sensu lato (Diptera: Psychodidae) in Sri Lanka—implication for leishmaniasis transmission. J Vector Borne Dis. 2005;42:155–8.
17. Illango K. Morphological characteristics of antennal flagellum and its sensilla chaetica with character displacement in sand fly *Phlebotomus argentipes* Annandale and Brunetti sensu lato (Diptera: Psychodidae). J Biol Sci. 2000;25:163–72.
18. Dinesh DS, Kishore K, Singh VP, Bhattacharya SK. Morphological variations in *Phlebotomus argentipes* Annandale and Brunetti (Diptera: Psychodidae). J Commun Dis. 2005;37:35–8.
19. Lewis DJ, Killick-Kendrick R. Some phlebotomid sand flies and other Diptera of Malaysia and Sri Lanka. Trans R Soc Trop Med Hyg. 1973;67:4–5.
20. Ilango K. A taxonomic reassessment of the *Phlebotomus argentipes* species complex (Diptera: Psychodidae: Phlebotominae). J Med Entomol. 2010;47:1–15.
21. Lane RP, Rahman SJ. Variation in the ascoids of the sandfly *Phlebotomus argentipes* in a population from Patna, northern India. J Com Dis. 1980;124:216–8.
22. Ghosh K, Mukhopadhyay J, Desai MM, Senroy S, Bhattacharya A. Population ecology of *Phlebotomus argentipes* (Diptera: Psychodidae) in West Bengal, India. J Med Entomol. 1999;36:588–94.
23. Singh R, Lal S, Saxena VK. Breeding ecology of visceral leishmaniasis vector sandfly in Bihar state of India. Acta Trop. 2008;107:117–20.
24. Bern C, Courtenay O, Alvar J. Of cattle, sand flies and men: a systematic review of risk factor analyses for south Asian visceral leishmaniasis and implications for elimination. PLOS Neglected Dis. 2010;4:e599. doi:10.1371/journal.pntd.0000599.
25. Hati AK, Ghosh KK, Das S, De N, Sur S. A longitudinal study on *Phlebotomus argentipes*/man contact in a village in West Bengal. Document WHO/VBC/81.801. Geneva: World Health Organization; 1981.
26. Rahman SJ, Menon PKM, Rajgopal R, Mathur KK. Behaviour of *Phlebotomus argentipes* in foothills of Nilgiris (Tamilnadu), South India. J Comm Dis. 1986;18:35–44.

Chapter 18
The Efficacy of Long Lasting Insecticidal Nets for Leishmaniasis in Asia

Chizu Sanjoba, Yusuf Özbel and Yoshitsugu Matsumoto

Abstract Vector control is an important part of controlling arthropod-transmitted diseases such as leishmaniasis, dengue, lymphatic filariasis, and malaria. Selecting the most appropriate vector control measures is essential. The current vector control strategy for visceral leishmaniasis (VL), which is the most severe form of leishmaniasis, is based on indoor residual spraying (IRS) in the Indian subcontinent. However, this technique has received critique, not only on its effectiveness and sustainability, but also for its side effects on health and the environment. Long-lasting insecticidal nets (LLINs) have been proposed as an alternative measure to IRS; however, the effectiveness of LLINs is still under the evaluation for vector control of leishmaniasis. This review aims to examine the potential of LLINs for controlling VL in the Indian subcontinent, areas that are some of the most highly endemic for VL.

Keywords Leishmaniasis · Sand fly · Vector control · Long-lasting insecticidal net · Bangladesh

18.1 Vector Control Against Leishmaniasis

Phlebotomine sand flies are the vectors of leishmaniasis, and comprise more than 40 species of *Phlebotmus* in the Old World and 30 species of *Lutzomyia* in the Americas [1]. *Leishmania* parasites are transmitted through the bites of infected female phlebotomine sand flies. The epidemiology of leishmaniasis depends on the behavior of the sand fly species, the characteristics of the parasite species, the reservoir species, ecological conditions at the transmission sites, and the symptoms and cultural practices of the human host. Therefore, the most appropriate control

---

C. Sanjoba (✉)
Faculty of Medicine, Department of Parasitology, Ege University, İzmir, Turkey
e-mail: asanjoba@mail.ecc.u-tokyo.ac.jp

measures will differ by country. Several factors to be considered when choosing control measures against sand flies are listed as follows [2]:

1. Has the vector been identified?
2. Is the transmission cycle partially or totally anthroponotic?
3. If zoonotic, is the reservoir species known?
4. Does transmission occur in or around houses, or in some extradomiciliary situation such as inside a forest?
5. Is transmission by sand flies seasonal or year-round?
6. Is there an infrastructure present that would allow organized, sustainable measures to be used?
7. Are the human communities at risk willing to participate in the control measures proposed?
8. What methods are available and are there practical, legal, environmental, or cultural constrains on their use?

Sand fly breeding sites remain unknown and are difficult to find in nature, thus source reduction targeted at pupae or larvae is not an option for practical control measures against sand flies. Most vector control efforts are primarily focused on the adult sand flies to reduce their contact with humans. Indoor residual spraying (IRS) and insecticide-treated bed nets (ITNs) are key components of vector control programs for leishmaniasis.

There is historic evidence that house spraying with DDT during the malaria eradication campaigns of the 1950s–1960s had a drastic impact on transmission of *Leishmania donovani* in the Indian subcontinent [3], but it returned to previous levels, or higher, when spraying was stopped [4, 5]. There is another experience in the early 1990s, where IRS campaigns in the VL endemic sites of Bihar and West Bengal in India reduced the number of reported cases, but their effect was limited in time as VL cases started to rise again early in the 21st century [6]. Although IRS is a reasonably effective method against endophilic sand fly species such as *Phlebotomus papatasi*, *P. sergenti*, and *P. argentipes* in the Old World [7], IRS has been criticized for being costly, not easily accepted, and not sustainable [8, 9] (Fig. 18.1). The kala-azar control program in the Indian subcontinent has applied the strategy of 2 rounds of insecticide spraying per year because the effectiveness of IRS does not last very long, mainly related to housing materials used in the area [10]. IRS also needs trained personnel for not only the insecticide application, but also for the appropriate application of safety procedures. Nevertheless, IRS was recommended as the main vector control strategy against visceral leishmaniasis (VL) in the Indian subcontinent [11].

Insecticide treated nets (ITNs) serve as a physical barrier that functions to reduce human-vector contact. ITNs is the most sustainable method of reducing intradomiciliary transmission of leishmaniasis, and it does not require the technical training and special implements required for IRS. However, there are few observational studies on ITNs conducted in the Indian subcontinent. Herein we review the current

Fig. 18.1 Insecticide sprays of IRS used in Bangladesh for vector control programs. A large number of sprayers and educated workers are needed

knowledge on the efficacy of ITN against sand fly vectors and inquire into the potential of ITN for controlling VL in the Indian subcontinent, especially in Bangladesh.

18.2 Vector Control by Long-Lasting Insecticidal Nets Against Visceral Leishmaniasis

There are 2 types of insecticide treated nets: conventional treated nets and long-lasting insecticidal nets (LLINs). A conventional treated net is a mosquito net that has been treated by dipping it in one of the insecticides recommended by the World Health Organization (WHO). At the present moment, alpha-cypermethrin, cyfluthrin, deltamethrin, etofenprox, lambda-cyhalothrin, and permethrin are WHO recommended insecticide products for the treatment of mosquito nets for malaria vector control [12]. To maintain the efficacy of the applied insecticide, the net must be treated again after it has been washed 3 times, or at least once a year even if it is not washed [13].

LLINs have been developed to overcome these problems. LLINs have the insecticide coated on, or incorporated within the fibers of the netting fabric (polyester, polyethylene, or polypropylene) during the manufacturing process. The WHO Pesticide Evaluation Scheme (WHOPES) evaluated the efficacy, wash resistance, and safety of LLINs and 15 products are recommended for the prevention and control of malaria (Table 18.1). The net must retain its effective biological activity after 20 washes under laboratory conditions and 3 years of recommended use under field conditions [14]. Seven products of 15 are deltamethrin-treated LLINs, 6 products are alpha-cypermethrin-treated LLINs, and 2 products of LLINs are treated with permethrin. The efficacy of these LLINs is

Table 18.1 WHO recommended long-lasting insecticidal nets

Product name	Product type	Status of WHO recommendation	Status of publication of WHO specification
DawaPlus 2.0	Deltamethrin coated on polyester	Interim	Published
Duranet	Alpha-cypermethrin incorporated into polyethylene	Full	Published
Interceptor	Alpha-cypermethrin coated on polyester	Full	Published
LifeNet	Deltamethrin incorporated into polypropylene	Interim	Published
MAGNet	Alpha-cypermethrin incorporated into polyethylene	Full	Published
MiraNet	Alpha-cypermethrin incorporated into polyethylene	Interim	Published
Olyset Net	Permethrin incorporated into polyethylene	Full	Published
Olyset Plus	Permethrin and PBO incorporated into polyethylene	Interim	Published
Panda Net 2.0	Deltamethrin incorporated into polyethylene	Interim	Published
PermaNet 2.0	Deltamethrin coated on polyester	Full	Published
PermaNet 3.0	Combination of deltamethrin coated on polyester with strengthened border (side panels), and deltamethrin and PBO incorporated into polyethylene (roof)	Interim	Published
Royal Sentry	Alpha-cypermethrin incorporated into polyethylene	Full	Published
SafeNet	Alpha-cypermethrin coated on polyester	Full	Published
Yahe	Deltamethrin coated on polyester	Interim	Published
Yorkool	Deltamethrin coated on polyester	Full	Published

Updated 9 November 2015. http://www.who.int/whopes/Long-lasting_insecticidal_nets_November_2015.pdf?ua=1

well studied in both laboratory and field testing for mosquito control, but not for sand fly control. The number of LLINs distributed reached more than 143 million annually by 2013 [15] with the majority of nets being distributed in sub-Saharan Africa where malaria is prevalent [16]. The most massive community intervention trial to test the effectiveness of LLINs is the KALANET project that was conducted between 2006 to 2008 in India and Nepal [17]. LLINs treated with deltamethrin were distributed in the 13 intervention clusters, and 12,691 people were tested for infection by direct agglutination testing, including the people in the 13 control clusters. There was no significant difference in the risk of seroconversion over 24 months in the intervention group (5.4 %; 347/6372) compared with the control group (5.5 %; 345/6319) [17]. The most biologically plausible explanation for this result is that a substantial fraction of *L. donovani* transmission occurs outside the house, where any nets would have less impact on preventing sand fly-human contact [17]. Human behavior is also argued to be a factor in that adults are not going to sleep before 9 p.m. [18]. *Phlebotomus argentipes,* which is the vector species in India and Nepal, live in and around houses and biting occurs at night, mainly between 9 p.m. and 1 a.m., peaking at 11–12 p.m. [19]. It should be also noted that a very high percentage of families in the control clusters used untreated nets [18], which may lead to a complex interpretation of the results. The antibody level of approximately 150 people from each cluster (intervention/control) against *P. argentipes* saliva was tested to determine the effect of LLINs on vector sand fly exposure. Although the distribution of LLINs reduced exposure to *P. argentipes* by 12 % at 12 months and 9 % at 24 months in the intervention cluster compared to that seen in the control after adjusting for baseline values, LLINs had a limited effect on sand fly exposure in VL endemic communities [20]. In an entomological survey of the KALANET project, the density/house of *P. argentipes* is significantly reduced by 24.9 % after distribution of LLINs [21]. This suggests a 25 % reduction in sand fly density is not enough to have an impact on disease outcome in India and Nepal [18].

Field trials of ITNs have been evaluated Phlebotomine sand flies besides *P. argentipes* and leishmaniasis in some countries. Two field trials in Adana and Sanliurfa, Turkey, where cutaneous leishmaniasis (CL) is endemic, showed a significant reduction in CL incidence in the intervention areas after the introduction of permethrin and PBO impregnated bed nets and K-OTAB impregnated bed nets, respectively [22, 23]. Three trials in Syria where *P. sergenti* is the main vector of anthroponotic CL, showed a significant impact of ITNs use versus that seen with either untreated nets or no intervention, including on confirmed CL incidence [24, 25]. A cluster-randomized trial was conducted over 5 years to compare the relative efficacy and cost effectiveness of IRS and LLINs relative to standard of care environmental management (SoC-EM) in Morocco for CL control. The main findings of this study indicate that both IRS with alpha-cypermethrin and the use of deltamethrin treated LLINs reduced the incidence of CL; however, the reduction due to LLINs did not reach statistical significance and the protective effect size associated with IRS was much larger [26]. These contrasting results suggest that field trials are necessary with different species of sand fly and parasite before adopting ITNs as a means of vector control.

18.3 The Efficacy of Long-Lasting Insecticidal Nets in Bangladesh

In 2005, the governments of India, Nepal, and Bangladesh, in collaboration with the WHO, developed a strategic framework to eliminate VL as a public health problem by 2015. This was defined as reducing the annual VL incidence below 1/10,000 people at the block level via a combination of case management and vector control.

The efficacy of 3 different interventions for vector control for VL: IRS, deltamethrin-treated LLIN (PermaNet®), and environment modification through plastering of walls with lime or mud was evaluated in India, Nepal, and Bangladesh [27]. Insecticides for IRS used in India, Nepal, and Bangladesh are DDT 5 % (target concentration 1 g/m^2), alpha-cypermethrin (target concentration 0.025 gm/m^2), and deltamethrin (K-Otrine 5 %, target concentration 20 mg active ingredient per square meter) respectively. A reduction of intra-domestic sand fly densities measured in the study households by overnight CDC light trap captures was the main outcome measure in this study. IRS was effective in all study areas but LLINs were only effective in Bangladesh and India [27]. In Bangladesh, IRS and LLINs were associated with a 70–80 % decrease in male and female *P. argentipes* density up to

Fig. 18.2 a, b Permethrin-treated LLINs were distributed as a pilot study in the Pabna district, Bangladesh, in August 2015 under the VL control project of SATREPS. **c, d** Housing style in VL endemic area of Bangladesh

5 months after intervention [28]. Sand fly density rebounded by 11 months post-IRS, whereas LLIN-treated households continued to show significantly lower density compared to that seen in households without intervention [28]. Mud plastering of wall and floor cracks did not reduce sand fly density in Bangladesh [27, 28]. The efficacy of LLIN (KO Tab 123, Bayer (Ply) Ltd.) was also evaluated in the VL endemic area of Bangladesh [29]. A significant reduction in sand fly densities was observed for at least 18 months in houses with LLINs compared to that seen in the houses in untreated control communities [29]. The distribution of KO Tab 123 also reduced VL incidence by 66.5 % in the intervention areas [30]. The use of conventional untreated bed nets has become widespread among people in Bangladesh [31]. Permethrin-treated LLINs were distributed as a pilot study in the Pabna district, Bangladesh, in August 2015 under the VL control project of SATREPS, and a utilization rate of distributed LLINs was 100 % up to the present (unpublished data) (Fig. 18.2). The distribution of pyrethroid-treated LLINs with/without IRS will probably be the most effective measure in reducing the vector populations and human-vector contact in Bangladesh.

18.4 Conclusion

The best vector control measure is not an easy decision to make for each leishmaniasis endemic country. There are 4 different types of interactions between humans and *Leishmania* foci that we have to consider before we apply a vector control; (a) an accidental host, (b) the principal or only host, (c) one of several hosts in a stable amphixenosis or (d) a potential host exposed to increased risk of transmission due to rapid multiplication of vectors and reservoir hosts [2]. These interactions are not clear in some endemic countries; for example, the existence of an animal host is a still matter of debate in Bangladesh.

Risk of the appearance of insecticide resistant sand flies due to IRS is also a concern. Various researchers have reported resistance to insecticides such as DDT, deltamethrin, or permethrin among *P. argentipes* in endemic areas of the Indian subcontinent [2, 7]. In a recent susceptibility test of sand flies in 6 districts of the state of Bihar, India, the *P. argentipes* population developed a resistance to DDT in 16 (38.1 %) of 42 villages surveyed and a resistance to malathion in 1 (4.5 %) of 22 villages surveyed [32]. A further fear is that populations of *P. argentipes* may have changed their behavior from being predominantly endophilic (resting indoors) to exophilic (resting outdoors) as a consequence of DDT-based IRS [33]. Although continued research on the biology and behavior of *P. argentipes* in relation to *L. donovani* transmission would be necessary to refine the intervention strategies for VL, LLINs will be a useful tool to reduce or interrupt transmission of VL as alternative or complement to IRS in the Indian subcontinent, especially in Bangladesh. The use of LLINs may also be the most sustainable method of reducing intradomiciliary transmission of *Leishmania* in communities where the diurnal resting sites of vectors or vector species are unknown.

References

1. WHO. Control of the leishmaniasis report of a meeting of the WHO expert committee on the control of leishmaniases. WHO technical report series no: 949. Geneva, Switzerland: World Health Organization; 2010.
2. Alexander B, Maroli M. Control of phlebotomine sandflies. Med Vet Entomol. 2003;17:1–18.
3. Killick-Kendrick R. The biology and control of Phlebotomine sand flies. Clin Dermatol. 1999;17:279–89.
4. Sen Gupta PC. Return of Kala-azar. J Indian Med Ass. 1975;65:89–90.
5. Mukhopadhyay Ak, Chakravarty AK, Kureel VR, Shivraj V. Resurgence of *Phlebotomus argentipes* and *P. papatasi* in parts of Bihar (India) after DDT spraying. J Indian Med Res. 1987;85:158–60.
6. Picado A, Dash AP, Bhattacharya S, Boelaert M. Vector control interventions for visceral leishmaniasis elimination initiative in South Asia, 2005–2010. Indian J Med Res. 2012;136:22–31.
7. Ostyn B, Vanlerberghe V, Picado A, Dinesh DS, Sundar S, Chappuis F, Rijal S, Dujardin J-C, Coosemans M, Boelaert M, Davies C. Vector control by insecticide-treated nets in the fight against visceral leishmaniasis in the Indian subcontinent, what is the evidence? Trop Med Int Health. 2008;13:1073–85.
8. Bhattacharya SK, Sur D, Shnha PK, Karbwang J. Elimination of leishmaniasis (kara-azar) from the Indian subcontinent is technically feasible and operationally achievable. Indian J Med Res. 2006;123:195–6.
9. Picado A, Singh SP, Rijal S, Sundar S, Ostyn B, Chappuis F, Uranw S, Gidwani K, Khanal B, Rai M, Paudel IS, Das ML, Kumar R, Srivastava P, Dujardin JC, Vanlerberghe V, Andersen EW, Davies CR, Boelaert M. Longlasting insecticidal nets for prevention of Leishmania donovani infection in India and Nepal: paired cluster randomised trial. BMJ. 2010;29(341):c6760. doi:10.1136/bmj.c6760.
10. WHO. WHO recommended insecticides for indoor residual spraying against malaria vectors. 2015. Updated: 2 March 2015. http://www.who.int/whopes/Insecticides_IRS_2.
11. WHO. Regional strategic framework for elimination of kala-azar from the South-East Asia region (2005–2015). New Delhi: Regional Office for South-East Asia SEA-VEC-85 (Rev-1); 2005.
12. WHO. WHO recommended insecticide products for treatment of mosquito nets for malaria vector control. 2014. Updated: 17 Nov 2014. http://www.who.int/whopes/Insecticides_ITN_Malaria_Nov2014.pdf.
13. WHO. Instructions for treatment and use of insecticide-treated mosquito nets. 2002. WHO/CDS/WHOPES/GCDpp/2005.11.
14. WHO. Malaria vector control commodities landscape. 2nd ed. WHO; 2012.
15. Ohashi K, Shono Y. Recent progress in the research and development of new products for malaria and dengue vector control. Sumitomo Kagaku. 2015;1–13.
16. Picado A, Singh SP, Rijal S, Sundar S, Ostyn B, Chappuis F, Uranw S, Gidwani K, Khanal B, Rai M, Paudel IS, Das ML, Kumar R, Srivastava P, Dujardin JC, Vanlerberghe V, Andersen EW, Davies CR, Boelaert M. Long insecticidal nets for prevention of *Leishmania donovani* infection India and Nepal: paired cluster randomized trial. BMJ. 2010;341:c6760. doi:10.1136/bmj.c6760.
17. WHO. Regional technical advisory group on kala-azar elimination report of the third meeting. Dhaka, Bangladesh: WHO Regional office for South-East Asia; 8–11 Dec 2009.
18. Dinesh DS, Ranjan A, Palit A, Kishore K, Kar SK. Seasonal and nocturnal landing/biting behavior of *Phlebotomus argentipes* (Diptera: Psychodidae). Annals Trop Med Parasitol. 2001;95:197–202.
19. Gidwani K, Picado A, Rijal S, Singh SP, Roy L, Volfova V, Andersen EW, Uranw S, Ostyn B, Sudarshan M, Chakravarty J, Volf P, Sundar S, Boelaert M, Rogers ME. Serological makers of sand fly exposure to evaluate insecticidal nets against visceral leishmaniasis in

India and Nepal: a cluster-randomized trial. PLoS Negl Trop Dis. Sep 2011;5(9):e1296. doi:10.1371/journal.pntd.0001296 (Epub 2011 Sep 13).
20. Gidwani K, Picado A, Rijal S, Singh SP, Roy L, Volfova V, Andersen EW, Uranw S, Ostyn B, Sudarshan M, Chakravarty J, Volf P, Sundar S, Boelaert M, Rogers ME. Serological markers of sand fly exposure to evaluate insecticidal nets against visceral leishmaniasis in India and Nepal: a cluster-randomized trial. PLoS Negl Trop Dis. 2011;5(9):e1296. doi:10.1371/journal.pntd.0001296.
21. Gunay F, Karakus M, Oguz G, Dogan M, Karakaya Y, Ergan G, Kaynas S, Kasap OE, Ozbel Y, Alten B. Evaluation of the efficacy of Olyset® Plus in a village-based cohort study in the Cukurova Plain, Turkey, in an area of hyperendemic cutaneous leishmaniasis. J Vector Ecol. 2014;39:395–405.
22. Alten B, Caglar SS, Kaynas S, Simsek FM. Evaluation of protective efficacy of K-OTAB impregnated bednets for cutaneous leishmaniasis control in Southeast Anatolia-Turkey. J Vector Ecol. 2003;28:53–54.
23. Tayeh A. A cutaneous leishmaniasis control trial using pyrethroid-impregnated bednets in villages near Aleppo, Syria. WHO/Leish/97.41. Geneva, Switzerland: WHO; 1997.
24. Jalouk L, Al Ahmed M, Gradoni L, Maroli M. Insecticide-treated bednets to prevent anthroponotic cutaneous leishmaniasis in Aleppo Governorate, Syria: results from two trials. Trans R Soc Trop Med Hyg. 2007;101:360–7.
25. Faraj C, Yukich J, Adlaoui EB, Wahabi R, Kaddaf M, El Idrissi AL, Ameur B, Kleinschmidt I. Effectiveness and cost of insecticide-treated bed nets and indoor residual spraying for the control of cutaneous leishmaniasis: a cluster-randomized control trial in Morocco. Am J Trop Med Hyg. 2016;. doi:10.4269/ajtmh.14-0510.
26. Faraj C, Yukich J, Adlaoui EB, Wahabi R, Kaddaf M, El Idrissi AL, Ameur B, Kleinschmidt I. Effectiveness and cost of insecticide-treated bed nets and indoor residual spraying for the control of cutaneous leishmaniasis: a cluster-randomized control trial in Morocco. Am J Trop Med Hyg. 2016;14–0510.
27. Joshi AB, Das ML, Akhter S, Chowdhury R, Mondal D, Kumar V, Das P, Kroeger A, Boelaert M, Petzold M. Chemical and environmental vector control as a contribution to the elimination of visceral leishmaniasis on the Indian subcontinent: cluster randomized controlled trials in Bangladesh, India and Nepal. BMC Med. Oct 5 2009;7:54. doi:10.1186/1741-7015-7-54.
28. Chowdhury R, Dotson E, Blackstock AJ, McClintock S, Maheswary NP, Faria S, Islam S, Akter T, Kroeger A, Akhter S, Bern C. Comparison of insecticide-treated nets and indoor residual spraying to control the vector of visceral leishmaniasis in Mymensingh District, Bangladesh. Am J Trop Med Hyg. 2011;84:662–7.
29. Mondal D, Chowdhury R, Huda MM, Maheswary NP, Akther S, Petzold M, Kumar V, Das ML, Gurung CK, Ghosh D, Kroeger A. Insecticide-treated bed nets in rural Bangladesh: their potential role in the visceral leishmaniasis elimination programme. Trop Med Int Health. 2010;15:1382–9.
30. Mondal D, Huda MM, Karmoker MK, Ghosh D, Matlashewski G, Nabi SG, Kroeger A. Reducing visceral leishmaniasis by insecticide impregnation of bet-nets, Bangladesh. Emerg Infect Dis. 2013;19(7):1131–4.
31. Mondal D, Alam SM, Karim A, Haque R, Kroeger BM. Present situation of vector control in Bangladesh: a wake up call. Health Policy. 2008;87:369–76.
32. Singh R, Kumar P. Susceptibility of the sandfly *Phlebotomus argentipes* Annandale and Brunetti (Diptera: Psychodidae) to insecticides in endemic areas of visceral leishmaniasis in Bihar, India. Jpn J Infect Dis. 2015;68:33–7.
33. Kumar V, Shankar L, Rama A, Kesari S, Dinesh DS, Bhunia GS, et al. Analysing host preference behavior of *Phlebotomus argentipes* (Diptera: Psychodidae) under the impact of indoor residual spray. Int J Trop Dis Health. 2015;7:69–79.

Part VI
New Challenges in Kala-Azar Elimination Programme

Chapter 19
Environmental Change and Kala-Azar with Particular Reference to Bangladesh

Ashraf Dewan, Masahiro Hashizume, Md. Masudur Rahman, Abu Yousuf Md. Abdullah, Robert J. Corner, Md. Rakibul Islam Shogib and Md. Faruk Hossain

Abstract Global environmental change is expected to have significant impacts on public health. Low income countries will be disproportionately affected by such change even though their contribution to global greenhouse gas emission is insignificant relative to developed countries. Visceral leishmaniasis, a vector-borne disease known as kala-azar on the Indian subcontinent, causes a significant burden of mortality and morbidity every year across the world. Approximately 200 million people are said to be at risk of kala-azar in the Indian subcontinent with 25,000–40,000 cases reported per year. This chapter examines past and present kala-azar incidences in Bangladesh and explores the future vulnerability of the country and its inhabitants under conditions of environmental change. Using published research and 2014 reported kala-azar cases from the Disease Control Unit of the Directorate General of Health Services, this study also attempts to present the epidemiology of kala-azar in Bangladesh. It shows that global environmental changes have the potential to cause deterioration to the social and ecological systems of the country. As a result, the country may suffer from an upsurge of all vector-borne and water-borne diseases, including kala-azar. Because climatic change in combination with land use change driven by population pressure, is expected to increase, so too will the occurrence of vector-borne diseases such as kala-azar. Additional measures will thus be required to control the vectors, their abundance, and disease transmission. More research is needed if we are to reduce the impact of this fatal disease.

Keywords Bangladesh · Climate variability · Environmental change · Epidemiology · Leishmaniasis

---

A. Dewan (✉)
Department of Spatial Sciences, Curtin University, Perth, Australia
e-mail: A.Dewan@curtin.edu.au

19.1 Introduction

Because of increasing human activities many aspects of the natural environment, such as biophysical and ecological systems, are constantly being impaired [75], causing global environmental change. There is no question that many environmental issues such as climate change, shortage of freshwater, emergence and spread of infectious diseases, etc., are directly or indirectly linked to human alteration of the biosphere. In their Fifth Assessment Report, the Intergovernmental Panel on Climate Change (IPCC) noted that the climate system of earth is significantly influenced by human activities, leading to warming of the atmosphere and ocean, and consequent shrinking of snow and ice [57]. Many have expressed the opinion that the modification of the environment by humans is resulting in a persistent pattern of both depletion and degradation [112] that has considerable implications for environmental sustainability. One of the many current concerns of the scientific community is that global environmental change has the potential to alter the distribution of diseases, transmission pathways, and seasonality leading to the emergence or re-emergence of infectious diseases [82, 118]. Although it remains uncertain as to what extent environmental change will lead to a shift in disease patterns, studies have shown that socioeconomic, environmental, and ecological drivers are closely associated with the spatial distribution of infectious diseases [61], particularly parasitic diseases [27]. Guernier et al. [45] reports that climatic factors are of importance in understanding the linkage between latitude and spatial patterns of human pathogens. Further, the combined effects of land use change and climatic variability can overwhelm natural ecosystems, which will affect human health in many ways [40]. For instance, coastal *Vibro* infections are thought to be closely connected with zooplankton booms, warmer water, and severe storms [11]; likewise, deforestation has been linked to the incidence of leishmaniasis in some parts of Latin America [82]. Using ecological niche models of vector and reservoir species, Gonzalez et al. [44] predict that climate change has the potential to worsen the ecological risk of human exposure to leishmaniasis and it is highly likely that the disease could spread in areas outside of the present range and may affect some parts of the United States and Canada. An upsurge of malaria has been associated with human-induced land use change in many parts of the world, including Southeast Asia, South America, and Africa [109]. With climate warming, malaria could be a serious public health concern in current malaria-free countries as temperature is one of the major habitat determinants of the main vector, *Anopheles* spp. [118].

Leishmaniasis, caused by protozoan parasites, is a serious public health issue across the world. There are about 20 species of *Leishmania* and many of them can cause human leishmaniasis [93]. It is endemic in 88 countries with a global burden of 300 million people, largely affecting the marginalised segments of people [8, 96, 116]. Using an overall case-fatality rate of 10 %, Alvar et al. [7] estimated that somewhere between 20,000 and 40,000 deaths per annum can be attributed to leishmaniasis globally with an annual incidence of 2 million new cases [31]. There are 4 clinical manifestations of leishmaniasis: cutaneous leishmaniasis (CL), mucocutaneous leishmaniasis (MCL), visceral leishmaniasis (VL), and post-kala-azar dermal leishmaniasis (PKDL) [104]. Visceral leishmaniasis, known locally as kala-azar or black fever, is widespread in the Indian subcontinent and is caused by *Leishmania donovani*, which is transmitted to humans by the bite of an infected female sandfly, *Phlebotomus argentipes* [69]. It can be fatal in the absence of treatment and the transmission is anthroponotic, because humans are the only known reservoir [55, 93]. More than 67 % of the total VL cases in the world are found in the Indian subcontinent with an estimated 200,000 cases per annum that disproportionately affects the poorest subset of the population [7, 19, 107]. PKDL, a subsequent condition to VL, is an important skin complication [120]. It appears that patients with PKDL may be a source of further transmission [69] as 5–10 % of patients who have recovered from VL develop PKDL within 5 years of kala-azar treatment, though a small number of PKDL cases appear without any prior history of VL (Ramesh 1995 cited in [14]). Addy and Nandy [1] suggest that unlike VL, PKDL is not fatal and does not cause systematic illness, yet patients can play an important role in inter-epidemic transmission of the infection because the duration of PKDL can be years or even decades [78].

Historical records reveal that the first VL cases were reported on the Indian Subcontinent during the British colonial regime [48]. However, the etiological agent of the disease *Leishmania donovani* was not discovered until the early 20th century in Madras, India [34]. Since then VL has become a significant threat to public health as it occurs in various places in South Asia. The Indian states of Bihar, Jharkand, West Bengal, and Uttar Pradesh, the south-eastern Terai region of Nepal, and western and central districts of Bangladesh are recognised endemic locations, and the courses of the Ganges and Brahmaputra appear to have played a considerable role in spreading VL in the Indian subcontinent [13]. A large number of VL cases are not reported for many reasons, such as the lack of a proper surveillance system [55, 69, 104]. In 2005, India, Bangladesh, and Nepal signed a memorandum of understanding (MoU) to eliminate VL by reducing the incidence rate to below 1 case per 10,000 people at the district level in India by 2010 and sub-district level in Bangladesh and Nepal by the year 2015 [107], and this was endorsed by the World Health Organisation (WHO) [25]. However, the targets have not been achieved and the program seems to have been extended until 2020 [76].

The aim of this chapter is to examine the incidence of VL in Bangladesh from both historical and current perspectives. In addition, it attempts to explore the linkage between environmental change and the incidence of disease, particularly commenting on how global environmental change may place the country and its inhabitants at greater risk of VL. Based, primarily, on existing research and 2014 VL incidence data from the Disease Control Unit of the Directorate General of Health Services, Government of Bangladesh, this study also attempts to present the epidemiology of VL in Bangladesh. Potential research gaps are also identified.

19.2 VL Epidemiology in Bangladesh

Located in South Asia, Bangladesh is bordered by India to the north, east, and west, the Bay of Bengal to the south, and to the southeast by Myanmar (Fig. 19.1). With a total population of 159.1 million (1,222 per km^2), it is one of the most densely populated countries in the world [114]. Apart from experiencing high population growth, the country is also prone to floods of large magnitude, tropical cyclones, river bank erosion, and occasional local severe storms that severely disrupt its economic progress and bring immense suffering to the people, including episodic

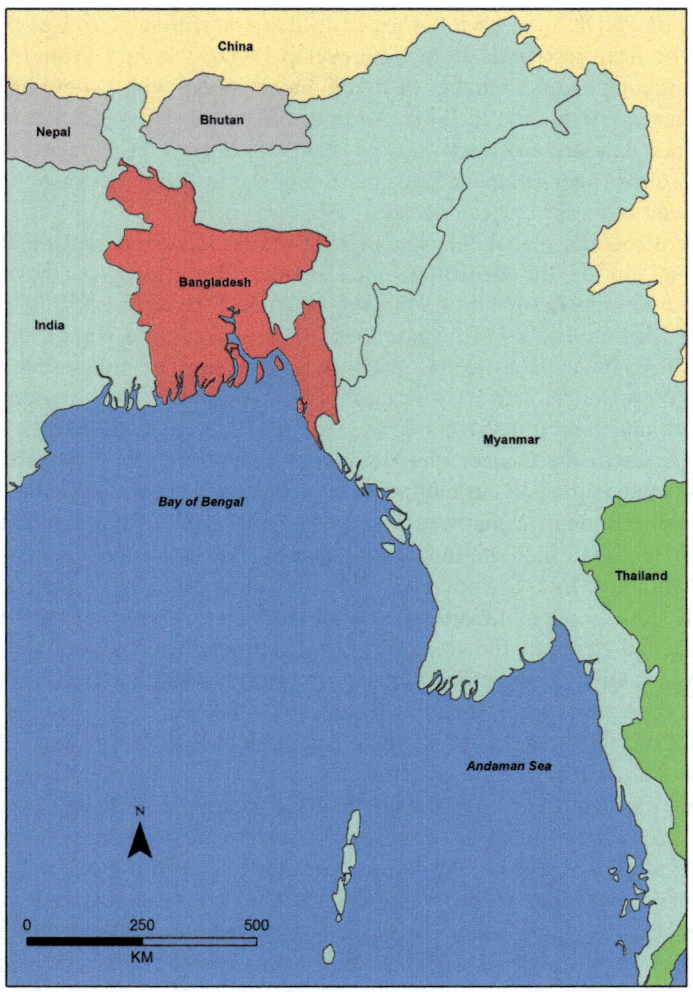

Fig. 19.1 Location of Bangladesh

surges of various vector-borne and water-borne diseases. For instance, Hashizume et al. [51] suggest that hospitalisation in relation to dengue fever increased between 2005 and 2009, which may be attributable to increased water levels in the rivers surrounding Dhaka. Likewise, diarrheal epidemics have been found to be associated with 3 major flood events of 1988, 1998, and 2004 [97].

Historically, outbreaks of VL in the alluvial parts of the Ganges-Brahmaputra plain have been reported since 1820 with an epidemic cycle of at least 15–20 years [8]. The earliest documented VL outbreak was in the Jessore district of Bangladesh (the then colonial Bengal Presidency) in 1824 that killed at least 75,000 people between 1824 and 1825 [13, 48]. A similar epidemic is thought to have occurred in the Burdwan district between 1850 and 1875, resulting in severe mortality that led to depopulation [102]. The disease appeared to have migrated eastward with high mortality documented in the Rangpur district in the northeast of Bengal. It then shifted to Assam, and gradually continued to spread eastwards along the valley of the Brahmaputra River, resulting in a number of epidemics in the sub-divisions of Garo Hills, Nowgong, Goalghat, and Shibsagar between 1882 and 1917 [102, 117]. By 1873, the epidemic began to decrease in Bengal and the next VL epidemic did not occur until 1943 and lasted till 1949 (Desowitz 1991 cited in [37]). However Shortt [102] notes that a treatment, called tartar emetic, was introduced in 1915 that helped saved a very large number of lives.

A total of 130,952 cases were recorded in Bengal between 1923 and 1924 of which nearly 11,000 cases were located in the Rajshahi district [110]. Again, at the end of 1944 and the beginning of 1945, a marked increase of the disease occurred in the districts of Chittagong, Dacca, and Faridpur of Bengal [47]. Between 1959 and 1963, more than 15,000 cases were thought to have occurred in Bangladesh [35]. The near absence of VL cases during the 1970s was because of wide-scale DDT (dichloro-diphenyl-trichloroethane) house spraying under a malaria eradication program in the 1950s and 1960s that also reduced the sand-fly population [13, 35, 36, 90]. With the cessation of malaria control activities in the late 1970s, a resurgence of the disease (including PKDL) occurred [90], sporadically affecting various parts of the country, and estimates indicate that a total of 148 VL cases were reported between 1968 and 1980 [35, 90] (Table 19.1). At least 224 confirmed VL cases were found in the Shahjadpur *thana* of Sirajganj district during 1980–1981 [72]; however, ISPAN [59] reported that 583 cases were recorded nationwide from 1980 to 1985. Elias et al. [35] show that there were 351 VL cases between 1981 and 1985, and an increase of VL cases has continued since the early 1990s. According to Emch [37], the total number of cases was found to be 991, 1123, and 1809 in 1989, 1991, and 1992, respectively (Table 19.1).

A major resurgence of VL seems to have occurred since the 1990s [25]. Table 19.1 shows the number of VL cases in Bangladesh from 1968 to 2014, compiled from various sources, including literature and data from secondary sources. It shows that a total of 114,070 VL cases were found in Bangladesh, and that 2006 had the highest occurrence (9,379 cases) followed by 1997 (9,000 cases). In 2014, only 379 cases were detected, which is the lowest number since 1994 in Bangladesh. At least 329 deaths from VL had occurred between 1994 and 2014

Table 19.1 The total number of visceral leishmaniasis (VL) or kala-azar cases and deaths from 1968 to 2014 in Bangladesh

Year	Reported cases	Number of VL deaths
1968–1980	148	–
1981–1985	351	–
1989	991	–
1991	1123	–
1992	1809	–
1994	3900	10
1995	4200	19
1996	6800	30
1997	9000	16
1998	7000	24
1999	5799	23
2000	7640	24
2001	4283	6
2002	8110	36
2003	6113	27
2004	5920	23
2005	6892	16
2006	9379	23
2007	4932	17
2008	4824	17
2009	4293	14
2010	3806	0
2011	3376	2
2012	1902	0
2013	1100	2
2014	379	0
Total	114,070	329

Compiled from various sources

(Table 19.1) and the highest fatalities were in 2002 when 36 people were reportedly killed due to the VL infection. As found in Elias et al. [35], only 2 VL cases were reported from Mymensingh district during 1968–1980 which increased to about 82 cases during 1981–1985, and more than 50,000 cases between 1994 and 2014 in the same district. Since 2010, the number of deaths appears to have declined significantly, which may be attributed to the introduction of the VL elimination program from the Indian Subcontinent. One needs to interpret these data very carefully because there remains a huge under reporting of VL cases in Bangladesh [39]. It is important to note that Rahman et al. [91] estimated around 50 % of VL cases are not detected by the existing passive surveillance system of the government.

A variety of risk factors are found to be closely linked with the incidence of VL in Bangladesh, and those of significance are proximity to patients with recent VL history, patterns of health seeking behaviour, income, attitude towards women, poor housing condition, flood control works, use of bed nets, cattle density, damp soil, education, presence of peri-domestic vegetation, etc. (e.g. Bern et al. [14, 15]).

19.2.1 Spatial Distribution

Due to data paucity and an ineffective passive surveillance system, it is obviously a difficult task to portray the exact geographical distribution of VL cases in Bangladesh. As noted by Chowdhury et al. [25], VL data at the upazila (sub-district) level are unfortunately not available before 2007. In addition, serious under reporting of cases (e.g., Khatun et al. [64]) is a persistent issue that needs to be accounted for in understanding the accurate geographic distribution of this disease. Keeping these in mind, we have developed a district-wise distribution map, based on the available data from literature and statistics, that clearly indicates that VL cases are confined to central, western, and northwest districts of Bangladesh (Fig. 19.2). In addition to this, 9 districts, namely Mymensingh, Tangail, Gazipur, Pabna, Jamalpur, Khulna, Panchagar, Rajshahi, and Sirajganj showed heavy occurrence of VL from 2003 to 2014. The Mymensingh district has consistently been the highest contributor to the total number of occurrences since 1994 [13, 25]. For example between 2008 and 2014, 12,963 cases (more than 81 %) were from Mymensingh district, suggesting that it is a sustained epidemic focus in the country [13]. The Tangail and Jamalpur districts, with a much lower 6.04 and 3.09 % of the cases, respectively, are the second and third contributors to the total occurrence of VL in Bangladesh over the same time period. Upazila-wise data from 2008 to 2014 shows that by 2014, 8 previously uninfected new upazilas were infected (Table 19.2). These all have relatively low numbers of cases; Melandaha (4 cases), Jamalpur Sadar (1 case), and Sharishabari (3 cases) in the Jamalpur district; Gopalpur (1 case) and Ghatail (3 cases) in the Tangail district; Gazipur Sadar (2 cases) and Kaliakoir (1 case) in the Gazipur district; and Mymensingh Sadar (5 cases) in the Mymensingh district.

19.2.2 Seasonality

No marked seasonality can be identified in the occurrence of VL because the incubation period in humans is highly variable (usually 2–6 months but occasionally as long as 24 months). However, peak occurrence of the disease is often used to characterise seasonality of the disease. Bern et al. [15], for example, suggest that the incidence is higher during the pre-monsoon and post-monsoon season. Chowdhury et al. [24] reported that the seasonality of VL cases in Bangladesh is consistent with studies from Nepal and India, showing that the hot season (March–June) has the highest incidences when sand fly populations, and particularly the proportion of gravid females, are at their highest. The study further showed that

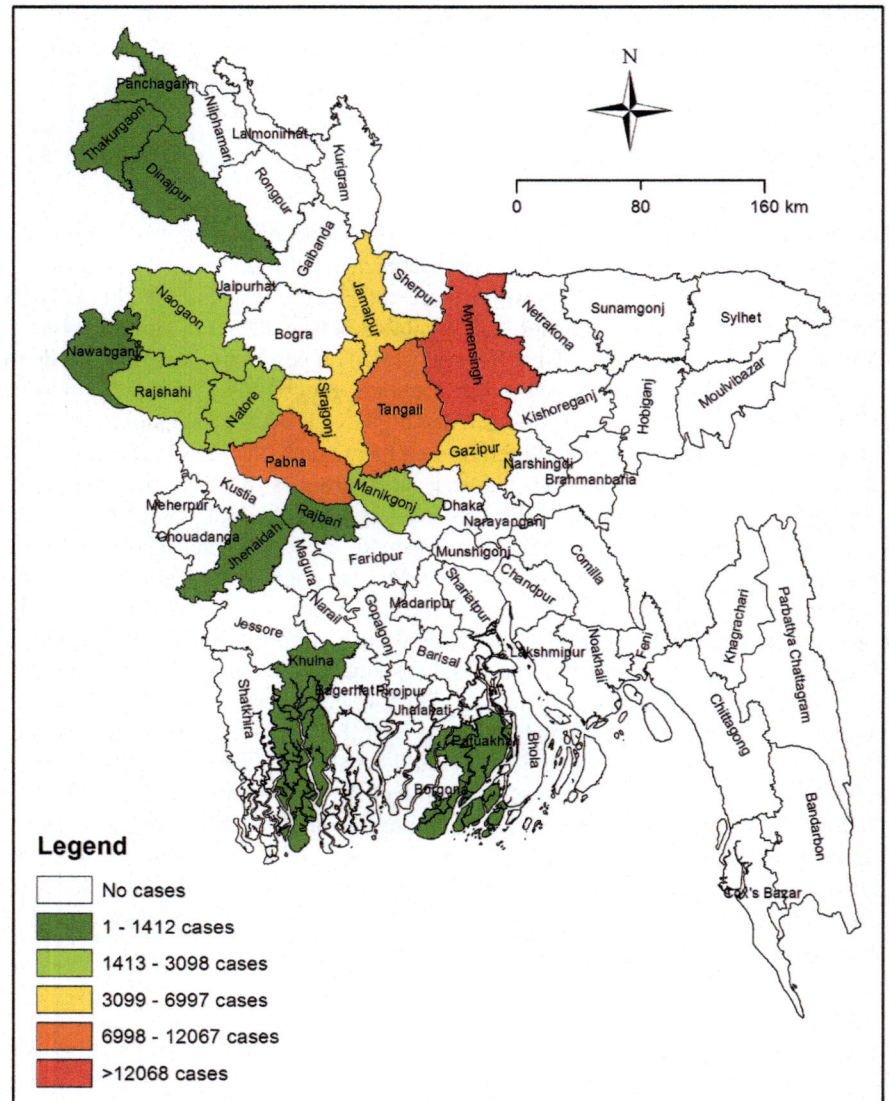

Fig. 19.2 The total number of kala-azar cases by Zila (district) in Bangladesh, 2008–2014

temperature and humidity of an area play an important role in the seasonality of sand fly vectors, while rainfall appears to be less significant. A similar observation was made by Young [117] based on meteorological and disease data from the Assam and Sylhet districts.

National surveillance data for 2014 is plotted in Fig. 19.3 to further investigate the seasonality of VL incidence in Bangladesh. It shows that there are 2 distinct

Table 19.2 Upazila-wise VL or kala-azar cases, 2008–2014

Zila	Upazila	2008	2009	2010	2011	2012	2013	2014	Total
Mymensingh	Trishal	1492	1279	735	564	252	235	113	4670
	Fulbaria	1315	781	456	1608	397	211	90	4858
	Gaffargaon	260	240	260	292	241	82	51	1426
	Muktagacha	198	343	214	99	82	127	17	1080
	Mymensingh Sadar	–	–	–	–	–	–	5	5
	Bhaluka	235	285	125	107	86	73	13	924
Tangail	Shakhipur	70	58	103	73	25	32	18	379
	Madhupur	80	45	55	50	48	49	22	349
	Nagorpur	39	47	25	39	38	37	0	225
	Gopalpur	–	–	–	–	–	–	1	1
	Ghatail	–	–	–	–	–	–	3	3
Pabna	Faridpur	23	23	19	18	20	17	0	120
	Chatmohor	27	36	36	31	30	13	0	173
Gazipur	Gazipur sadar	–	–	–	–	–	–	2	2
	Kaliakoir	–	–	–	–	–	–	1	1
	Sreepur	55	53	42	58	15	49	11	283
Jamalpur	Madarganj	115	109	47	88	64	35	24	482
	Melandaha	–	–	–	–	–	–	4	4
	Jamalpur Sadar	–	–	–	–	–	–	1	1
	Sharishabari	–	–	–	–	–	–	3	3
Khulna	Terokhada	46	43	39	17	16	21	0	182
Panchagar	Debiganj	18	49	44	98	36	16	0	261
Rajshahi	Godagari	70	102	14	47	17	24	0	274
Sirajganj	Chauhali	26	30	19	20	26	23	0	144

Source Malaria and Vector-Borne Disease Control Unit, Director General of Health Services, Government of Bangladesh

peaks, one in February and the other in September. Note that VL in Bangladesh seems to be prevalent throughout the year, though a few months (e.g. December) have a lower number of cases.

19.2.3 Patient Demographics

Based on 17,725 VL cases in 1922, Young [117] suggests that 49.34 % cases represent the age group <1–15 years, 45.4 % were within the age group 15–50 years, and only 6 % of cases were in the 50 and above age group. Compared

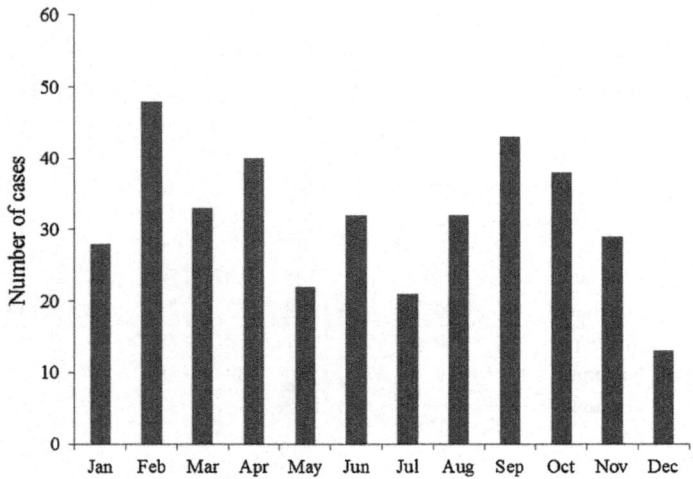

Fig. 19.3 Monthly number of kala-azar cases from the national surveillance data in 2014

with this early 20th century data, recent case-control studies [4] have demonstrated that the infection rate is higher for persons aged ≥ 15. In an another study, Ahluwalia et al. [5] show the rate was higher in the adult category (≥ 45 of age) with the <1–14 age group ranked second, and the 15–44 age group ranked third for both sexes. Khatun et al. [64] also report a somewhat similar result. The study by Bern et al. [15] showed that the incidence was higher among children and young adults than in the youngest and oldest age group (>45). The 2014 data indicates that 55.1 % (209 cases) of patients with VL were from the age group of 15–49 years old, 33.2 % (126 cases) were between 0–14, and 11.6 % (44 cases) were from the adult age group (50+), which is in agreement with another recent case study of Bangladesh [39].

The sex distribution of patients with VL from 2014 showed that 56.5 % (214 cases) were male whereas 43.5 % (165 cases) were female across Bangladesh. Many studies [4, 5, 15, 35, 39, 91, 117] have also demonstrated that the incidence is higher among males than females, which is believed to be related to sex inequity due to the tendency of female patients with VL not to seek help with heath issues [64], which is also related to societal structures and attitudes towards women [4, 5]. This tends to lead to a high case-fatality rate among adult women. However, Ferdousi et al. [39] did not find any association between VL incidence and sex. A plausible explanation of the higher incidence of VL in males is that, according to this study, men keep the upper part of their body exposed and wear fewer clothes than women during summer in the countryside, which could make them more vulnerable to *Leishmania donovani* bites.

19.3 Future Vulnerability of Bangladesh to VL

Despite the fact that Bangladesh contributes only 0.4 tonnes per capita of CO_2 to global greenhouse emissions [114], far below the regional figure, it would nevertheless be one of the worst affected countries in the world by climate change [81]. Among other sectors, public health is regarded as one of those most affected by global environmental change [89], and the risk to human health will be greater than thought previously [29, 98].

Although climate is a key factor, it is not the only determinant of the incidence of disease in an area. Other factors such as ecological change, socioeconomic development, and public health infrastructure influence the public health effects of climate change [45, 49, 71]. Disease agents, human hosts, and disease vectors are all linked in a multifactorial relationship to the environment, therefore, any changes in the coupled human and natural systems may result in the outbreak of disease [26] in which climate modification by human activities plays a significant role [58, 84]. For example, deaths from heat stress during the 2003 heatwave in Europe were the result of human-induced climate change [105]. Similarly, development of the natural environment for irrigation purposes may drive an increase in vector population and disease [26]. Although a growing body of literature has clearly described how human activities can drive rapid ecological change [6, 67, 88] and the emergence of disease, there is still a paucity of evidence linking environmental change and the emergence of diseases (such as VL) [33]. However, alternative explanations between environmental change (e.g., land use change, shift in demographic profile, habitat fragmentation, etc.) and the incidence of vector and parasitic borne diseases are plausible. Environmental change may, therefore, affect the incidence of VL both directly and indirectly, and the consequences could have serious implications since Bangladesh is a low income country. Direct impacts include changes to climate variables (e.g., rainfall), whilst an indirect impact can result from crop failure due to climate change leading to malnutrition and an increased disease vulnerability of the population. The following sections explain some of the direct and indirect factors that may affect geographical distribution and incidence of VL in Bangladesh under conditions of environmental change.

19.3.1 Climate Variability

One of the obvious mechanisms by which the incidence of a disease, either vector borne or water borne, and its distribution, could intensify would be an increase in global average temperature. The Fifth Assessment Report by the IPCC [58] states that since 19th century the global average temperature has increased, with combined land and ocean temperature data showing a warming of 0.85 °C over the period of 1880–2012. It is estimated that, by the end of this century, the global average temperature will have increased by between 1.4–5.8 °C [56]. As diseases, particularly vector-borne infections, are sensitive to weather conditions, a small change in temperature can shift seasonality, transmission patterns, and geographic distribution. For instance, the pattern of cholera in Bangladesh is found to vary with climatic fluctuations and sea surface temperatures, El Nino/Southern Oscillation (ENSO) phenomenon, and the extent of flooding [65]. However, the relationship between climate change and disease may be scale dependent as Sultan et al. [106] have shown that at a small scale the correlation is weak, because of other local factors, but at a large spatial scale long term climate change affects seasonality as well as transmission pattern. The nature of the vector populations in an area is also dependent on the aggregation of temperature, rainfall, humidity, and solar radiation [111]. Hence disease dynamics are likely to be affected by a climate-induced shift in phenology and species movement [23]. Because the transmission of many parasitic diseases is confined to the rainy season, changes to the length of rainy and dry seasons, together with changes to the interval between seasons, would affect larvae and adult vector development and abundance [82].

Shahid [99] showed that the average, mean maximum and mean minimum temperatures of Bangladesh are increasing at a rate of 0.103, 0.091, and 0.097 °C, respectively, per decade. Another study indicates that seasonal variations in temperature will be higher in winter than in summer, suggesting a 1.3 °C increase in winter and 0.7 °C in summer for 2030 [3]. As far as rainfall is concerned, it is estimated that annual and pre-monsoon rainfall are increasing at the rate of 5.53 and 2.47 mm per year, respectively [99]. Likewise, the number of heavy precipitation days is expected to increase [100]. The sand fly vector appears to be climate sensitive and thrives in a temperature range of between 7–37 °C with a relative humidity of above 70 % [80]. A recent study examined the relationship between VL cases and temperature and humidity in 2 endemic areas of Bangladesh, which showed a strong correlation between VL incidence and both mean maximum temperature and annual mean relative humidity in the Fulbaria upazila in Mymensingh district (the most endemic site in Bangladesh), whereas in the Santhia upazila in Pabna district, annual average humidity was found to have a strong relationship with the incidence of the disease [9]. Though it is not conclusive, the recent western migration of VL cases in India is believed to be driven by climatic warming [16]. If temperature increases and rainfall regime changes as a result of climate warming then there would be a significant effect on the vector populations, as rainfall helps to increase in relative humidity and alterations of temperature could support longevity of the vector and transmission of the disease (Molineaux and Gramiccia 1980 cited in [33]). Although the ENSO is a naturally occurring event and has wide range consequences for weather around the world [66, 95], an increase in VL incidence was observed in Brazil [42] to be related to ENSO's influences. This sort of study is scarce in the Indian subcontinent and could enhance understanding of the relationship between VL and climatic warming.

19.3.2 Extreme Events

Changes in climate parameters may also modify the frequency and intensity of climate extremes [17, 57] and studies have shown that extreme weather events have profound impact on infectious diseases such as VL [84]. Because more frequent extreme weather events are expected in coming years in the region [98], changes to the potential incidence, seasonality, and geographic range of parasitic borne diseases might cause those diseases to migrate to areas that were previously free from infection [73]. For example, increased rainfall under climate warming could lead to more devastating floods and an increase in waterlogged areas in Bangladesh, and therefore affect vector distribution. For example, Mirza [77] predicts that the maximum change in mean flooded areas would occur with an increase in mean annual temperature of between >0–2 °C in relation to the increase in peak discharges of the Ganges, Brahmaputra, and Meghna rivers. He further states that for each degree of warming, the flooded area would increase by about 0.44–0.55×10^6 ha which may have direct and indirect effect on all insect-borne diseases like leishmaniasis. Mukhopadhyay et al. [79] examined the effect of flooding on sand flies in a flood-prone VL endemic village in Bihar, a highly endemic area of the Indian Subcontinent, and concluded that during floods larvae were found in earth scrapings as high as 90 cm up house walls and after floods they returned to ground level. This means flooding has an adverse effect on breeding sites; sand flies are compelled to escape upwards to find a suitable place for the survival. Although some of the health outcomes associated with extreme weather events to a larger extent are now clear (e.g., effect of heat waves and the impact of floods on water borne diseases), little exploration of the extent to which extreme events might affect the activity of *Phlebotomus argentipes* has been done. This question requires further investigation.

19.3.3 Sea-Level Rise

Owing to low lying land on a relatively unstable deltaic plain, Bangladesh's coast is under serious threat of sea level rise induced by global environmental change. Rises in sea level may have 3 potential adverse effects on Bangladesh, as follows: (i) increase in soil and water salinity through inundation; (ii) loss of land; and (iii) displacement of people, all of which will have devastating consequences to public health. Although precise estimation is not available, a study shows that the rate of sea level rise is significantly higher along the western part of the coast [86]. In another study, Hossain et al. [52] report that the mean tidal level is increasing between 6.0 and 7.8 mm per year at Char Changa and Cox's Bazar stations, respectively. They further indicate that if sea level rises by 100 mm and 300 mm then 2 % (2,500 km^2) and 5 % (8,000 km^2) of the land, respectively, will be lost, although this estimate is contentious [21]. Apart from land loss, sporadic coastal flooding will be another concern since a study indicates that brackish water induced by coastal flooding is conducive to the abundance of vectors breeding [74].

Bhuiyan and Dutta [18] used a hydrodynamic model to show that coastal inundation will increase many fold with a 59 cm rise in sea level along the western coastal zone of Bangladesh. An increase in groundwater salinity may lead to a shortage of freshwater supply and subsequent loss of crop production, which would have significant localised public health effects. Further coastal inundation may convert freshwater habitats into salt marsh areas that may support increased vector activities. Population displacement caused by sea level change also often leads to nutritional, physical, infectious disease, and mental health risks [73].

19.3.4 Human Movement

Increased human movement is likely to introduce "pathogen pollution" [28] in new areas, which in turn drives the emergence of infectious diseases. Because of massive migration from southern Sudan to northern Sudan as a result of the civil war in the late 1990s, an outbreak of VL occurred that was possibly attributed to a combination of the introduction of the parasite from an endemic area to a non-immune population and ecological changes favourable to the sand fly vector [119].

Climate induced coastal inundation could displace as many as 10–30 million people in Bangladesh [54], which may have a number of negative consequences on public health in terms of disease distribution, immune status, and fatalities. Moreover, internal migration could also lead to the occurrence of VL. If seasonal migration, for instance, from an endemic area occurs during the transmission season to another area which is free from infection, it implies that both areas are likely to be affected by the disease. Maroli et al. [70] demonstrate that sand fly diffusion and density could increase due to human migration. Because neighbouring districts of India have high rates of VL incidence, population movement across borders for business, travel, and cultural purposes are likely to increase the risk of VL in Bangladesh in the future. Desjuex [30] showed that cross border migration is an important risk factor in transmitting VL to the Nepalese Terai region from India. In 1862, the disease was introduced into (the then) Dacca district of Bengal by the crew of a country boat who travelled from the upstream inland port of Jageer on the Dhaleshwari River [48]. In addition, bordering regions with high endemicity of climate sensitive diseases may be at greater risk under a warmer climate [84]. Hence, there remains huge potential to examine how human migration, particularly internal movement in response to environmental change, could affect future distribution of VL in Bangladesh.

19.3.5 Land Use Change, Population Pressure and Habitat Fragmentation

Land use change appears to have cascading effects both on climatic change and on the spread of various infectious diseases [40]. Rapid changes in land use due to an ever increasing population are a major factor causing habitat fragmentation, urban sprawl, and loss of biodiversity in Bangladesh. While land use/cover changes might affect the risk of parasitic disease, there is strong evidence that urban sprawl substantially alters local microclimates and leads to a decline in biodiversity through habitat fragmentation, degradation of soil and water, etc. [87]. Habitat fragmentation results in small patches that often do not have adequate prey for predators, and consequently prey species increase at the cost of the extinction of local predator species [83]. In addition, agricultural activities can also lead to habitat fragmentation through the "edge effect" that eventually promotes interaction between pathogens, vectors, and hosts. This edge effect is found to be associated with Lyme disease in the US [43].

Clearing of forests and the expansion of human settlements into cleared forest patches can magnify the activity of vector borne disease and may alter ecological niches that provide a favourable environment for vectors and parasites. For example, the increase of leishmaniasis in some parts of Latin America has been linked to deforestation [82], while an increase of forest edge areas associated with deforestation led to the introduction of leishmaniasis into the Brazilian Amazon [46]. By the same token, expansion of human settlements to newly cleared areas may lead to disorganised occupation of new areas with new settlers having reduced immunity [85]. In the longer term, land use change or forest clearing for resource extraction can result in drought leading to malnutrition and increased vulnerability to a wide range of fatal diseases [38]. Because South Asian countries, particularly India and Bangladesh are experiencing rapid population growth, environmental changes due to increased human activities such as reduction of fallow lands, intensification of agricultural activities, and encroachment onto vegetative surfaces, etc., may lead to greater environmental change in the region, which obviously could contribute to changes in the vector community, their abundance, and their behaviour.

As mentioned earlier, Bangladesh is one of the most densely populated countries in the world. Because proximity to an existing VL case has been found to be a significant risk factor in endemic areas, in addition to other risk factors [15], larger household size as well as denser human settlements may have an impact on the incidence of disease. However, Barnett et al. [12] suggest the opposite, which warrants further investigation.

19.3.6 Water Control Works

Various types of seasonal and perennial water bodies exist on the floodplains of Bangladesh, which have supported the livelihoods of millions since time immemorial. As flooding is a serious problem in Bangladesh, to date 12,850 km of embankments (including 7,500 km of embankments along the major rivers), 25,580 km of drainage channels, and 4,190 sluices/regulators have been constructed across the country [32, 63], constituting around 30 % of the country's total land area [50]. These water control works have created conditions that are conducive to VL as shown by the work of Emch [37] and ISPAN [59, 60]. Although the mechanisms are not fully clear, a reasonable explanation is that although floods have the ability to wash away organic matter and larvae sites, this is impeded by water control works that evidently facilitate larval development and breeding of sand flies [59]. As a result, the probability of VL infection among people living within or close to an embankment is significantly higher than people living outside of the embankment [30] due to differential dispersion of *Phlebotomus argentipes* larvae breeding sites by floodwaters [37]. Land use changes coupled with climate warming may induce more flooding through intensified precipitation and more numerous rainfall events—all of which will create the humid environment required for sand fly larval development. Flood control works may act as a catalyst to VL incidences in many parts of Bangladesh, particularly in areas prone to seasonal flooding. Further study is therefore required to investigate the relationship between flood control schemes and adverse health outcomes because flood control structures are believed to alter the ecological environment of an area [37]. In addition, the construction of roads, dams, riverine pools, or artificial logging areas can also have detrimental effects on the ecological environment.

19.3.7 Poverty and Malnutrition

Several studies from South Asia [2, 5, 15, 19, 92, 94] indicate that VL disproportionately affects those marginalised segments of the society whose income is less than \$1/day [8]. This is believed to be one of the major risk factors among many, causing high morbidity and mortality [107], and an important barrier for eliminating the disease from hyperendemic areas such as the Indian Subcontinent [108]. For example Sharma et al. [101] show that VL treatment causes a major economic burden in affected families, and probably because of this most female patients are left untreated [5]. Badaro et al. [10] remarked that the classic infection gradually progresses to severe disease because the host's ability to control the infection is reduced by malnutrition and poverty.

Poverty and malnutrition may directly or indirectly affect the disease. For example, malnutrition is responsible for immune dysfunction before infection occurs and therefore acts as a predisposing factor to leishmaniasis in a human population [22]. Although Bangladesh has made significant progress in reducing the number of poor people from 63 million people in 2000 to 47 million in 2010 [113], rates of malnutrition are still among the highest in the world [41]. Due to differences in the prevalence of poverty, undernutrition, and exposure to improvised environments, the rate of VL is higher in Fulbaria upazila, which is the most endemic area in Bangladesh. The relationship between poverty and VL incidence has been observed by many case control studies [20] that unequivocally indicate that a single VL case in a family puts a considerable economic burden, with the largest cost driver being the loss of income due to illness. Another issue that has been brought to light by a recent study [55] is that socioeconomically poorer tribal communities are at greater risk of VL compared to other local people. This study also identifies co-infection of VL and tuberculosis—obviously a new phenomenon for Bangladesh. Because global warming is expected to bring changes in agricultural production due to changes in climatic variables [99], proper measures need to be taken to minimise the uncertainty of climate change on public health. Alvar et al. [7] have shown that the incidence of VL has reduced where living standards have improved and proper nutrition is believed to mitigate the progression to fatal VL [22].

19.4 Public Health Strategies

As described earlier, global environmental change has the potential to overwhelm the ecological and societal systems that in turn are likely to affect the incidence, seasonal transmission, and geographic coverage of VL in Bangladesh. As Smith et al. [103] pointed out, official statistics in some countries, including Bangladesh, are so inconsistent in quality and coverage that it is difficult to come to a definite conclusion about health trends. Although some improvement in Bangladesh in terms of recording, surveillance, and diagnosis of other diseases, such as cholera, has been shown, VL detection and recording is still very inconsistent. As a result, it is difficult to predict the conditions under which climatic shift will shape host and pathogen evolution [6] in the coming years. It can however be said that ongoing modification of the environment, with increasing demand of resources such as food and energy will definitely cause a 'domino effect' that is likely to considerably increase the health burden above the current level.

19.4.1 Vector Control and Management

Vector control is an effective strategy for reducing the incidence of VL and, if applied correctly, could potentially eliminate VL from the Indian Subcontinent because the disease has unique epidemiological features [62]. Although integrated vector management is one of the 5 pillars of the South Asian elimination initiative [115], there remains some challenges [24, 93]. Indoor residual spraying (IRS), insecticide-treated nets (ITNs), and stoppage of wall and floor cracks are implemented in Bangladesh, Nepal, and India as part of the environmental vector management (EVM) of the VL elimination program [62]. A study conducted to examine their efficacy in the Mymensingh district of Bangladesh showed that mud plastering of wall and floor cracks did not have any impact, while IRS and ITNs were effective in reducing vector density by 70–80 % up to 5 months after intervention [24]. However, vector density rebounded 11 months after IRS treatment, whereas households with ITNs continued to show lower vector density. Another study [68] suggests that IRS may be suitable for low and medium endemic settings but additional intervention is required in highly endemic settings to achieve the target of elimination. These measures could lead to an immediate reduction in vector populations, but effective strategies and operational research is needed to adequately control the vector [62]. Another issue can be the identification of the exact terrestrial sites of sand fly habitat, which are largely unknown; DDT spraying under the malaria control program supressed the incidence of VL in the Indian subcontinent during the 1960s–1980s, but this could have been coincidental [93]. There have been various experimental trials to control vectors but none have had enough success to be able to inform public health groups on suitable approaches to effectively control sand fly vectors [93]. Hence, a microenvironmental study involving satellite data, climatic records, and other geographic information could be useful in providing support for the effective management of sand fly vectors in Bangladesh and elsewhere.

19.4.2 Surveillance System and Social Factors

One of the first and crucial steps in reducing the VL burden is early diagnosis and treatment of cases [76, 104], which currently appears to be unsatisfactory, particularly in Bangladesh [64]. Almost all case-control studies, conducted in the Indian Subcontinent, clearly emphasise the needs of early case detection to keep the burden of the disease at an acceptable level. Although the rK39 test is used for early case detection in India and Nepal as part of the elimination initiative, for instance, this is largely absent in Bangladesh [62]. Under the Bangladeshi government's existing passive surveillance system around 50 % of VL cases are not detectable due to considerable complexity in the system and factor such as lack of resources and variation in staff experience [91]. In addition, tribal ethnic population seems to be at higher risk of death as many VL patients never visit to hospital for proper healthcare [55]. Therefore, the national VL elimination program needs to be strengthened to detect cases early and refer them for early treatment, perhaps the accelerated active case detection (AACD) method [64] could be useful in this regard. In addition, geotagging of cases would help identify microenvironmental factors influencing the incidence of VL. This type of study has yet to be carried out in Bangladesh.

Apart from the problem with the existing surveillance system, the healthcare-seeking behaviour of VL patients in the Indian Subcontinent is highly spatially variable [7] because it is determined by sex and socioeconomic status. This is again influenced by the remoteness of case locations [52], accessibility to health facilities [5, 55] and cost of treatment [4, 101]. Out of these, high out-of-pocket expense of treatment may lead to further impoverishment as suggested by Sharma et al. [101]. In addition, there appears to be a lack of uniform diagnostic facilities in upazilas that are considered endemic. Although considerable recent progress has been made towards reducing VL burden, improving the healthcare system together with political commitment is essential to generate credible information on VL cases, vector distributions, and habitats.

19.4.3 Public Awareness

Local action such as increasing public awareness could be taken to reduce the vulnerability of poor communities, located in endemic areas. Although this may vary from locality to locality, taking prevailing socioeconomic conditions and available resources into consideration this will surely help in eliminating the disease. Medley et al. [76] emphasise the importance of maintaining population and healthcare system awareness to reduce the burden of the disease. Because communities are willing to take collective action to confront the problems it causes, involving communities to supplement effective treatment and early case detection may be an important strategy for the country [4]. An example of this is the Government of Bangladesh initiative that involves communities to raise awareness about the signs and symptoms of dengue, and made the community participate in controlling mosquito breeding sites during the 2000 dengue outbreak in Dhaka [53].

19.5 Conclusion

The objectives of this chapter were to review the past and present epidemiology of VL in Bangladesh. The potential effect of global environmental change on VL was also reviewed based, primarily, on existing research. Despite the fact that the incidence of VL progressively declined between the 1960s and 1980s due to household spraying of DDT, a resurgence occurred in the 1990s. It is evident from this work that global environmental change could shift the seasonality, geographic range and the transmission dynamics of VL in Bangladesh, and extreme weather events may introduce additional outbreaks. Vector-borne disease is highly influenced by both meteorological parameters and local factors. Because environmental degradation caused by both depletion and degradation of environmental resources will exacerbate the situation, a correlation between high incidence and global warming is assumed.

Currently, the disease is receiving more attention in the light of the regional elimination program. However, in the absence of modern surveillance system that involve both early detection and early treatment the disease may continue to re-emerge. To lower the incidence, we believe that social factors are also important. Hence increasing public awareness could be an important strategy that can reinforce the regional elimination program, otherwise demographic changes, introduction of roads, dams and flood control works may overwhelm the present initiative. Multidisciplinary research involving people from social, ecological, health and physical sciences is therefore essential to aid in determining vector ecology, and the microenvironmental factors that account for disease outbreaks. In short, to transform knowledge to actual mitigation actions in endemic areas, socioeconomic, disease occurrences and environmental information should be integrated through spatial analysis that could enable landscape-based modelling and risk mapping. Further, prevention strategies may include micro, macro, and meso scale planning.

References

1. Addy M, Nandy A. Ten years of kala-azar in west Bengal, Part I. Did post-kala-azar dermal leishmaniasis initiate the outbreak in 24-Parganas? Bull World Health Org. 1992;70(3):341–6.
2. Adhikari SR, Supakankunti S, Khan MM. Kala azar in Nepal: estimating the effects of socioeconomic factors on disease incidence. Kathmandu Univ Med J. 2010;8(1):73–9.
3. Agrawala S, Ota T, Ahmed AU, Smith J, Van Aalst M. Development and climate change in Bangladesh: focus on coastal flooding and the Sundarbans. Organization for Economic Co-operation and Development Report (OECOD), Paris, France. http://s3.documentcloud.org/documents/15705/climate-change-in-bangladesh-oecd-2003.pdf (2003). Accessed 11 March 2016.
4. Ahluwalia IB, Bern C, Costa C, Akter T, Chowdhury R, Ali M, Alam D, Kenah E, Amann J, Islam M, Wagatsuma Y. Visceral leishmaniasis: consequences of a neglected disease in a Bangladeshi community. Am J Tropic Med Hygiene. 2003;69(6):624–8.
5. Ahluwalia IB, Bern C, Wagatsuma Y, Costa C, Chowdhury R, Ali M, Amann J, Haque R, Breiman R, Maguire JH. Visceral leishmaniasis: consequences to women in a Bangladeshi community. J Women's Health. 2004;13(4):360–4.
6. Altizer S, Ostfeld RS, Johnson PT, Kutz S, Harvell CD. Climate change and infectious diseases: from evidence to a predictive framework. Science. 2013;341(6145):514–9.
7. Alvar J, Velez ID, Bern C, Herrero M, Desjeux P, Cano J, Jannin J, den Boer M, Leishmaniasis Control Team WHO. Leishmaniasis worldwide and global estimates of its incidence. PLoS ONE. 2012;7(5):e35671.
8. Alvar J, Yactayo S, Bern C. Leishmaniasis and poverty. Trends Parasitol. 2006;22(12):552–7.
9. Amin MR, Tareq SM, Rahman SH, Uddin MR. Effects of temperature, rainfall and relative humidity on Visceral Leishmaniasis prevalence at two highly affected upazilas in Bangladesh. Life Sci J. 2013;10(4):1440–6.
10. Badaró R, Rocha H, Carvalho E, Queiroz A, Jones T. Leishmania donovani: an opportunistic microbe associated with progressive disease in three immunocompromised patients. Lancet. 1986;327(8482):647–9.
11. Baker-Austin C, Trinanes JA, Taylor NG, Hartnell R, Siitonen A, Martinez-Urtaza J. Emerging Vibrio risk at high latitudes in response to ocean warming. Nat Clim Change. 2013;3(1):73–7.
12. Barnett PG, Singh SP, Bern C, Hightower AW, Sundar S. Virgin soil: the spread of visceral leishmaniasis into Uttar Pradesh, India. Am J Tropic Med Hyg. 2005;73(4):7 20–725.
13. Bern C, Chowdhury R. The epidemiology of visceral leishmaniasis in Bangladesh: prospects for improved control. Indian J Med Res. 2006;123(3):275.
14. Bern C, Courtenay O, Alvar J. Of cattle, sand flies and men: a systematic review of risk factor analyses for South Asian visceral leishmaniasis and implications for elimination. PLoS Negl Trop Dis. 2010;4(2):e599.
15. Bern C, Hightower AW, Chowdhury R, Ali M, Amann J, Wagatsuma Y, Haque R, Kurkjian K, Vaz LE, Begum M, Akter T. Risk factors for kala-azar in Bangladesh. Emerg Infect Dis. 2005;11(1):655–62.
16. Bhat K, Pandita K, Khajuria A, Wani S. Visceral leishmaniasis (kalazar) migrating West: a new autochthonous case from sub-Himalayas. Indian J Med Microbiol. 2014;32(1):94–5.
17. Bhatt D, Mall RK, Banerjee T. Climate change, climate extremes and disaster risk reduction. Nat Hazards. 2015;78(1):775–8.

18. Bhuiyan MJAN, Dutta D. Analysis of flood vulnerability and assessment of the impacts in coastal zones of Bangladesh due to potential sea-level rise. Nat Hazards. 2012;61(2):729–43.
20. Boelaert M, Meheus F, Sanchez A, Singh SP, Vanlerberghe V, Picado A, Meessen B, Sundar S. The poorest of the poor: a poverty appraisal of households affected by visceral leishmaniasis in Bihar, India. Trop Med Int Health. 2009;14(6):639–44.
19. Boelaert M, Meheus F, Robays J, Lutumba P. Socio-economic aspects of neglected diseases: sleeping sickness and visceral leishmaniasis. Ann Tropic Med Parasitol. 2010;104(7): 535–42.
21. Brammer H. Bangladesh's dynamic coastal regions and sea-level rise. Clim Risk Manag. 2014;1:51–62.
22. Cerf BJ, Jones TC, Badaro R, Sampaio D, Teixeira R, Johnson WD. Malnutrition as a risk factor for severe visceral leishmaniasis. J Infect Dis. 1987 156(6):1030–3.
23. Chen IC, Hill JK, Ohlemüller R, Roy DB, Thomas CD. Rapid range shifts of species associated with high levels of climate warming. Science. 2011;333(6045):1024–6.
24. Chowdhury R, Dotson E, Blackstock AJ, McClintock S, Maheswary NP, Faria S, Islam S, Akter T, Kroeger A, Akhter S, Bern C. Comparison of insecticide-treated nets and indoor residual spraying to control the vector of visceral leishmaniasis in Mymensingh District, Bangladesh. Am J Tropic Med Hyg. 2011;84(5):662–7.
25. Chowdhury R, Mondal D, Chowdhury V, Faria S, Avar J, Nabi SG, Boelaert M, Dash AP. How far are we from visceral leishmaniasis elimination in Bangladesh? an assessment of epidemiological surveillance data. PLoS Negl Trop Dis. 2014;8(8):e3020.
26. Comrie A. Climate change and human health. Geogr Compass. 2007;1(3):325–39.
27. Confalonieri UE, Margonari C, Quintão AF. Environmental change and the dynamics of parasitic diseases in the Amazon. Acta Trop. 2014;129:33–41.
28. Daszak P, Cunningham AA, Hyatt AD. Emerging infectious diseases of wildlife-threats to biodiversity and human health. Science. 2000;287(5452):443–9.
29. Department of Environment (DoE). Climate change and health impacts in Bangladesh. DoE, Dhaka: Climate Change Cell; 2009 72 pp.
30. Desjeux P. The increase in risk factors for leishmaniasis worldwide. Trans R Soc Tropic Med Hyg. 2001;95(3):239–43.
31. Desjeux P. Leishmaniasis: current situation and new perspectives. Comp Immunol Microbiol Infect Dis. 2004;27(5):305–18.
32. Dewan AM. Floods in a megacity: geospatial techniques in assessing hazards, risk and vulnerability. Dordrecht: Springer; 2013 187 pp.
33. Dhiman RC, Pahwa S, Dhillon GPS, Dash AP. Climate change and threat of vector-borne diseases in India: are we prepared? Parasitol Res. 2010;106(4):763–73.
34. Donovan C. The aetiology of one the heterogeneous fevers of India. Br Med J. 1903;2 (2239):1401.
35. Elias M, Rahman AJ, Khan NI. Visceral leishmaniasis and its control in Bangladesh. Bull World Health Org. 1989;67(1):43–9.
36. El-Masum MA, Evans DA, Minter DM, El Harith A. Visceral leishmaniasis in Bangladesh: the value of DAT as a diagnostic tool. Trans R Soc Tropic Med Hyg. 1995;89(2):185–6.
37. Emch M. Relationships between flood control, kala-azar, and diarrheal disease in Bangladesh. Env Plann A. 2000;32(6):1051–63.
38. Esrey SA, Potash JB, Roberts L, Shiff C. Effects of improved water supply and sanitation on ascariasis, diarrhoea, dracunculiasis, hookworm infection, schistosomiasis, and trachoma. Bull World Health Org. 1991;69(5):609–21.
39. Ferdousi F, Alam MS, Hossain MS, Ma E, Itoh M, Mondal D, Haque R, Wagatsuma Y. Visceral leishmaniasis eradication is a reality: data from a community-based active surveillance in Bangladesh. Tropic Med Health. 2012;40(4) 133–9.
40. Foley JA, DeFries R, Asner GP, Barford C, Bonan G, Carpenter SR, Chapin FS, Coe MT, Daily GC, Gibbs HK, Helkowski JH. Global consequences of land use. Science. 2005;309 (5734):570–4.

41. Food and Agricultural Organisation (FAO). ftp://fao.org/es/esn/nutrition/ncp/BGDmap.pdf (1999). Accessed 12 March 2016.
42. Franke CR, Ziller M, Staubach C, Latif M. Impact of El Niño/Southern Oscillation on Visceral Leishmaniasis, Brazil. Emerg Infect Dis. 2002;8(9):914–7.
43. Glass GE, Schwartz BS, Morgan JM III, Johnson DT, Noy PM, Israel E. Environmental risk factors for Lyme disease identified with geographic information systems. Am J Public Health. 1995;85(7):944–8.
44. González C, Wang O, Strutz SE, González-Salazar C, Sánchez-Cordero V, Sarkar S. Climate change and risk of leishmaniasis in North America: predictions from ecological niche models of vector and reservoir species. PLoS Negl Trop Dis. 2010;4(1):e585.
45. Guernier V, Hochberg ME, Guégan JF. Ecology drives the worldwide distribution of human diseases. PLoS Biol. 2004;2(6):e141.
46. Gunkel G, Lange U, Walde D, Rosa JW. The environmental and operational impacts of Curuá-Una, a reservoir in the Amazon region of Pará, Brazil. Lakes Reservoirs Res Manage. 2003;8(3–4):201–16.
47. Gupta PCS. Kala-azar in Bengal: its incidence and trends. Indian Med Gazette. 1944;79(11):547–52.
48. Gupta PCS. History of kala-azar in India. Indian J Med Res. 1947;123(3):281–6.
49. Haines A, Kovats RS, Campbell-Lendrum D, Corvalán C. Climate change and human health: impacts, vulnerability and public health. Public Health. 2006;120(7):585–96.
50. Halls AS, Payne AI, Alam SS, Barman SK. Impacts of flood control schemes on inland fisheries in Bangladesh: guidelines for mitigation. Hydrobiologia. 2008;609(1):45–58.
51. Hashizume M, Dewan AM, Sunahara T, Rahman MZ, Yamamoto T. Hydroclimatological variability and dengue transmission in Dhaka, Bangladesh: a time-series study. BMC Infect Dis. 2012;12(1):1.
52. Hossain M, Noiri E, Moji K. Climate change and kala-azar. Kala Azar in South Asia. The Netherlands: Springer; 2011. p. 127–37.
53. Hossain MI, Wagatsuma Y, Chowdhury MA, Ahmed TU, Uddin MA, Sohel SN, Kittayapong P. Analysis of some socio-demographic factors related to DF/DHF outbreak in Dhaka city. Dengue Bull. 2000;24:34–41.
54. Houghton J. Global warming. Cambridge, UK: CUP; 2009.
55. Huda MM, Chowdhury R, Ghosh D, Dash AP, Bhattacharya SK, Mondal D. Visceral leishmaniasis-associated mortality in Bangladesh: a retrospective cross-sectional study. BMJ Open. 2014;4(7):e005408.
56. Intergovernmental Panel on Climate Change (IPCC). Climate change 2001: the scientific basis: contributions of working group i to the third assessment report. Cambridge, UK: Cambridge University Press; 2001 944 pp.
57. Intergovernmental Panel on Climate Change (IPCC). Climate change: synthesis report. http://ar5-syr.ipcc.ch/ (2014). Accessed 6 March 2016.
58. Intergovernmental Panel on Climate Change (IPCC). Climate change: synthesis report. contribution of working groups I, II and III to the fifth assessment report of the intergovernmental panel on climate change. Geneva, Switzerland; 2014. 151 pp.
59. Irrigation Support Project for Asia and the Near East (ISPAN). The kala-azar epidemic in Bangladesh and its relationship to flood control embankments. Bangladesh Flood Action Plan (FAP-16). Ministry of Water Resources. Flood Plan Coordination Organisation (FPCO), Dhaka; 1995. 24 pp.
60. Irrigation Support Project for Asia and the Near East (ISPAN). Impacts of flood control and drainage on vector-borne disease incidence in Bangladesh. Special Studies Series, Flood Action Plan (FAP-16). Ministry of Water Resources, Flood Plan Coordination Organisation (FPCO), Dhaka; 1992. 35 pp.
61. Jones KE, Patel NG, Levy MA, Storeygard A, Balk D, Gittleman JL, Daszak P. Global trends in emerging infectious diseases. Nature. 2008;451(7181):990–3.
62. Joshi A, Narain JP, Prasittisuk C, Bhatia R, Hashim G, Jorge A, Banjara M, Kroeger A. Can visceral leishmaniasis be eliminated from Asia? J Vector Borne Dis. 2008;45(2):105–11.

63. Khan MSA. Disaster preparedness for sustainable development in Bangladesh. Disaster Prev Manage Int J. 2008;17(5):662–71.
64. Khatun J, Huda MM, Hossain MS, Presber W, Ghosh D, Kroeger A, Matlashewski G, Mondal D. Accelerated active case detection of visceral leishmaniasis patients in endemic villages of Bangladesh. PLoS ONE. 2014;9(8):e103678.
65. Koelle K, Rodó X, Pascual M, Yunus M, Mostafa G. Refractory periods and climate forcing in cholera dynamics. Nature. 2005;436(7051):696–700.
66. Kovats RS, Bouma MJ, Hajat S Worrall E, Haines A. El Niño and health. Lancet. 2003;362 (9394):1481–9.
67. Lafferty KD. The ecology of climate change and infectious diseases. Ecology. 2009;90 (4):888–900.
68. Le Rutte EA, Coffeng LE, Bontje DM, Hasker EC, Postigo JAR, Argaw D, Boelaert MC, De Vlas SJ. Feasibility of eliminating visceral leishmaniasis from the Indian subcontinent: explorations with a set of deterministic age-structured transmission models. Parasites Vectors. 2016;9(1):1–14.
69. Lobo DA, Velayudhan R, Chatterjee P, Kohli H, Hotez PJ. The neglected tropical diseases of India and South Asia: review of their prevalence, distribution, and control or elimination. PLoS Negl Trop Dis. 2011 5(10):e1222.
70. Maroli M, Feliciangeli MD, Bichaud L, Charrel RN, Gradoni L. Phlebotomine sandflies and the spreading of leishmaniases and other diseases of public health concern. Med Vet Entomol. 2013;27(2):123–47.
71. Martens WJM, Jetten TH, Rotmans J, Niessen LW. Climate change and vector-borne diseases: a global modelling perspective. Glob Env Change. 1995;5(3):195–209.
72. Masum MA, Alam B, Ahmed R. An epidemiological investigation of a kala-azar outbreak in Kalahati upazila of Tangail district, Bangladesh. J Prev Soc Med. 1990;4–9(1):13–4.
74. McMichael AJ, Woodruff RE. Climate change and human health: present and future risks. Lancet. 2006;367:859–69.
75. McMichael AJ, Githeko A, Akhtar R, Carcavallo R, Gübler D, Leary N, Dokken D, White K, editors. Climate change 2001: impacts, adaptation and vulnerability. New York: Cambridge University Press; 2001.
73. McMichael AJ, Friel S, Nyong A, Corvalan C. Global environmental change and health: impacts, inequalities, and the health sector. BMJ. 2008;336:191–4.
76. Medley GF, Hollingsworth TD, Olliaro PL, Adams ER. Health-seeking behaviour, diagnostics and transmission dynamics in the control of visceral leishmaniasis in the Indian subcontinent. Nature. 2015;528(7580):S102–8.
77. Mirza MMQ. Climate change, flooding in South Asia and implications. Reg Env Change. 2011;11(1):95–107.
78. Mondal D, Hamano S, Hasnain MG, Satoskar AR. Challenges for management of post kala-azar dermal leishmaniasis and future directions. Res Rep Trop Med. 2014;5:105–11.
79. Mukhopadhyay AK, Rahaman SJ, Chakravarty AK. Effect of flood on immature stages of sandflies in a flood-prone kala-azar endemic village of North Bihar, India. World Health Organisation, WHO/VBC/90.986, Geneva; 1990. pp. 1–5.
80. Napier LE, Gupta CRD. An epidemiological investigation of kala-azar in a rural area in Bengal. Indian J Med Res, 1931 XIX(1), 295–341.
81. Parry M, Canziani O, Palutikof J, van der Linden P, Hanson C (eds). Climate change 2007: impacts, adaptation and vulnerability. In: Contribution of working group II to the fourth assessment report of the intergovernmental panel on climate change. Cambridge, UK: Cambridge University Press; 2007. 982 pp.
84. Patz JA, Graczyk TK, Geller N, Vittor AY. Effects of environmental change on emerging parasitic diseases. Int J Parasitol. 2000;30(12):1395–405.
83. Patz JA, Daszak P, Tabor GM, Aguirre AA, Pearl M, Epstein J, Wolfe ND, Kilpatrick AM, Foufopoulos J, Molyneux D, Bradley DJ. Unhealthy landscapes: policy recommendations on land use change and infectious disease emergence. Env Health Perspect. 2004;112 (10):1092–8.

82. Patz JA, Campbell-Lendrum D, Holloway T, Foley JA. Impact of regional climate change on human health. Nature. 2005;438(7066):310–7.
85. Penna G, Pinto LF, Soranz D, Glatt R. High incidence of diseases endemic to the Amazon region of Brazil, 2001–2006. Emerg Infect Dis. 2009;15(4):626–32.
86. Pethick J, Orford JD. Rapid rise in effective sea-level in southwest Bangladesh: its causes and contemporary rates. Glob Planetary Change. 2013;111:237–45.
87. Pimm SL, Raven P. Biodiversity: extinction by numbers. Nature. 2000;403(6772):843–5.
88. Plowright RK, Sokolow SH, Gorman ME, Daszak P, Foley JE. Causal inference in disease ecology: investigating ecological drivers of disease emergence. Frontiers Ecol Env. 2008;6(8):420–9.
89. Rahman A. Climate change and its impact on health in Bangladesh. Reg Health Forum. 2008;12(1):16–26.
90. Rahman KM, Islam N. Resurgence of visceral leishmaniasis in Bangladesh. Bull World Health Org. 1983;61(1):113–6.
91. Rahman KM, Samarawickrema IV, Harley D, Olsen A, Butler CD, Sumon SA, Biswas SK, Luby SP, Sleigh AC. Performance of kala-azar surveillance in Gaffargaon subdistrict of Mymensingh, Bangladesh. PLoS Negl Trop Dis. 2015;9(4):e0003531.
92. Ranjan A, Sur D, Singh VP, Siddique NA, Manna B, Lal CS, Sinha PK, Kishore K, Bhattacharya SK. Risk factors for Indian kala-azar. Am J Tropic Med Hyg. 2005;73(1):74–8.
93. Ready PD. Epidemiology of visceral leishmaniasis. Clin Epidemiol. 2014;6(6):147–54.
94. Rijal S, Koirala S, Van der Stuyft P, Boelaert M. The economic burden of visceral leishmaniasis for households in Nepal. Trans R Soc Tropic Med Hyg. 2006;100(9):838–41.
95. Ropelewski CF, Halpert MS. Global and regional scale precipitation patterns associated with the El Niño/Southern Oscillation. Mon Weather Rev. 1987;115(8):1606–26.
96. Savioli L, Daumerie D. First WHO report on neglected tropical diseases: working to overcome the global impact of neglected tropical diseases. Geneva, Switzerland: World Health Organisation; 2010 172 pp.
97. Schwartz BS, Harris JB, Khan AI, Larocque RC, Sack DA, Malek MA, Faruque AS, Qadri F, Calderwood SB, Luby SP, Ryan ET. Diarrheal epidemics in Dhaka, Bangladesh, during three consecutive floods: 1988, 1998, and 2004. Am J Tropic Med Hyg. 2006;74(6):1067–73.
98. Shahid S. Probable impacts of climate change on public health in Bangladesh. Asia-Pacific J Public Health. 2009;23:37S–45S.
99. Shahid S. Recent trends in the climate of Bangladesh. Clim Res. 2010;42(3):185–93.
100. Shahid S. Trends in extreme rainfall events of Bangladesh. Theoret Appl Climatol. 2011;104(3–4):489–99.
101. Sharma DA, Bern C, Varghese B, Chowdhury R, Haque R, Ali M, Amann J, Ahluwalia IB, Wagatsuma Y, Breiman RF, Maguire JH. The economic impact of visceral leishmaniasis on households in Bangladesh. Tropic Med Int Health. 2006;11(5):757–64.
102. Shortt HE. Recent research on kala-azar in India. Trans R Soc Tropic Med Hyg. 1945;39(1):13–31.
103. Smith KR, Woodward A, Campbell-Lendrum D, Chadee DD, Honda Y, Liu Q, Olwoch JM, Revich B, Sauerborn R. Human health: impacts, adaptation, and co-benefits. In: Field CB, Barros VR, Dokken DJ, Mach KJ, Mastrandrea MD, Bilir TE, Chatterjee M, Ebi KL, Estrada YO, Genova RC, Girma B, Kissel ES, Levy AN, MacCracken S, Mastrandrea PR, White LL, editors. Climate change 2014: impacts, adaptation, and vulnerability, Part A: Global and Sectoral Aspects, contribution of Working Group II to the fifth assessment report of the intergovernmental panel on climate change. Cambridge, UK, and New York, USA: Cambridge University Press; 2014. pp. 709–754.
104. Stauch A, Sarkar RR, Picado A, Ostyn B, Sundar S, Rijal S, Boelaert M, Dujardin JC, Duerr HP. Visceral leishmaniasis in the Indian subcontinent: modelling epidemiology and control. PLoS Negl Trop Dis. 2011;5(11):e1405.
105. Stott PA, Stone DA, Allen MR. Human contribution to the European heatwave of 2003. Nature. 2004;432(7017):610–4.

106. Sultan B, Labadi K, Guégan JF, Janicot S. Climate drives the meningitis epidemics onset in West Africa. PLoS Med. 2005;2(1):e6.
107. Sundar S, Mondal D, Rijal S, Bhattacharya S, Ghalib H, Kroeger A, Boelaert M, Desjeux P, Richter-Airijoki H, Harms G. Implementation research to support the initiative on the elimination of kala azar from Bangladesh, India and Nepal-the challenges for diagnosis and treatment. Tropic Med Int Health. 2008;13(1):2–5.
108. Thornton SJ, Wasan KM, Piecuch A, Lynd LLD, Wasan EK. Barriers to treatment for visceral leishmaniasis in hyperendemic areas: India, Bangladesh, Nepal, Brazil and Sudan. Drug Dev Ind Pharmacy. 2010;36(11):1312–9.
109. Vittor AY, Gilman RH, Tielsch J, Glass G, Shields TIM, Lozano WS, Pinedo-Cancino V, Patz JA. The effect of deforestation on the human-biting rate of Anopheles darlingi, the primary vector of falciparum malaria in the Peruvian Amazon. Am J Tropic Med Hyg. 2006;74(1):3–11.
110. Ward RD. Bangladesh flood action programme: sandflies and kala-azar, The Health Impact Program (HIP/92.08). UK: Liverpool School of Tropical Medicine; 1992.
111. Washino RK, Wood BL. Application of remote sensing to vector arthropod surveillance and control. Am J Tropic Med Hyg. 1993;50:134–44.
112. Wilson G, Furniss P, Kimbowa R. Environment, development and sustainability. Oxford, The UK: Oxford University Press; 2010.
113. World Bank. Bangladesh poverty assessment: a decade of progress in reducing poverty, 2000–2010. http://www.worldbank.org/en/news/feature/2013/06/20/bangladesh-poverty-assessment-a-decade-of-progress-in-reducing-poverty-2000-2010 (2013). Accessed 12 March 2016.
114. World Bank. Bangladesh country at a glance. http://www.worldbank.org/en/country/bangladesh (2016). Accessed 9 March 2016.
115. World Health Organisation (WHO). Regional strategies framework for elimination of kala-azar from South East Asia Region (2005–2015). New Delhi: WHO; 2005.
116. Yangzom T, Cruz I, Bern C, Argaw D, den Boer M, Vélez ID, Bhattacharya SK, Molina R, Alvar J. Endemic transmission of visceral leishmaniasis in Bhutan. Am J Tropic Med Hyg. 2012;87(6):1028–37.
117. Young TM. Fourteen years' experience with kala-azar work in Assam. Trans R Soc Tropic Med Hyg. 1924;18(3):81–97.
118. Zhang Y, Bi P, Hiller JE. Climate change and the transmission of vector-borne diseases: a review. Asia-Pacific J Public Health. 2008;20(1):64–76.
119. Zijlstra EE, Ali MS, El-Hassan AM, El-Toum IA, Satti M, Ghalib HW, Sondorp E, Winkler A. kala-azar in displaced people from southern Sudan: epidemiological, clinical and therapeutic findings. Trans R Soc Tropic Med Hyg. 1991;85(3):365–9.
120. Zijlstra EE, Musa AM, Khalil EAG, El Hassan IM, El-Hassan AM. Post-kala-azar dermal leishmaniasis. Lancet Infect Dis. 2003;3(2):87–98.

Chapter 20
Drug Safety Monitoring for Liposomal Amphotericin B

Eisei Noiri and Kosuke Minami

Abstract Liposomal amphotericin B is now the key drug for disease control of visceral leishmaniasis, also known as kala-azar, in the sub-Indian continent. However, a number of generic products are currently available. Compared that for with other chemical compounds, liposomal formation is a crucial process for appropriate drug delivery in this particular medicine. We recently received isolated clinical findings after treatment using one generic product of liposomal amphotericin B from an endemic area. Herein, we report the different characteristics of generic products of liposomal amphotericin B to the genuine brand product. The size of the liposomal particles of the generic product is not uniform and far from the expected drug delivery and toxicity values. This review introduces a potential inspection approach for when policy makers adopt generic liposomal amphotericin B products into their own country.

Keywords Particle dispersion · Potassium · Electron microscope · Drug toxicity

20.1 Introduction

Amphotericin B (AMPH-B) belongs to the polyene macrolide group of antibiotics and is often prescribed to treat fungal and parasitic diseases. AMPH-B binds to ergosterol, a component of fungal and parasitic cell membranes, and increases permeability of the membrane to cause fatal injury. Ergosterol is ubiquitously expressed in the cell membrane of fungus and parasites, and therefore provides

---

E. Noiri (✉)
Hemodialysis and Apheresis, Nephrology 107 Lab,
The University of Tokyo Hospital, Tokyo, Japan
e-mail: noiri-tky@umin.ac.jp

© Springer International Publishing 2016
E. Noiri and T.K. Jha (eds.), *Kala Azar in South Asia*,
DOI 10.1007/978-3-319-47101-3_20

broad anti-spectrum treatment against fungus and parasites. But in higher concentrations, AMPH-B can bind to cholesterol, a key component of mammalian cell membranes, which can often cause injuries, and prevents long term use because of side effects. To reduce these side effects, a formulation of AMPH-B with a lipid complex has been examined and it was found that phospholipid bilayers are able to hold AMPH-B for a longer time. Liposomal AMPH-B enables suitable drug delivery to fungal ergosterol without causing cellular damage of mammalians.

The physicochemical properties of liposome are diverse and based on different phospholipids and methods of adjustment. Liposomal AMPH-B (L-AMB, AmBisome® [Gilead Sciences]) features the combination of a molar ratio at 2 (hydrogen added soybean phospholipid): 1 (cholesterol): 0.8 (distearoyl phosphatidylglycerol [DSPG]): 0.4 (AMPH-B) referring the content rate of AMPH-B, physicochemical stabilities, and mammalian toxicity. Table 20.1 demonstrates the suitable ratio findings. The conditions to reduce toxicity are (i) the occupancies of cholesterol should be more than 25 % of the lipid contents, (ii) DSPG is a preferable long-chain fatty acid to dilauryl phosphatidylglycerol, and (iii) a molar ratio between DSPG and AMPH-B of 2:1 for better holding. L-AMB is more preferentially distributed to the infectious site. The diameter of L-AMB is ca. 100 nm, which is not filterable from endothelial fenestration, including glomerular endothelial cells. But L-AMB could be easily liberated from the infectious vessel because of the wider fenestration induced by inflammatory cytokines; thus it is site selective compared with AMPH-B.

AMPH-B and L-AMB are preferred visceral leishmaniasis (VL) treatments on the Indian subcontinent, and L-AMB is now more often used after the positive report of a single dose L-AMB strategy in India [1]. We report on different patient conditions after L-AMB treatment from 2 different companies, one from Gilead Sciences and the other a generic product from an Indian pharmaceutical company. Both L-AMB products were prepared by strictly following the manufacture's protocol. All the drugs for VL are legally controlled by the Centers for Disease Control. The potential side effects by AMPH-B and L-AMB are summarized in

Table 20.1 Molar ratio of L-AMB and in vivo toxicity. The molar ratio of AmBisome® is superior to other combinations

	Molar ratio	LD_{50} in mice (mg/kg)
HEPC:DSPG:AMPH-B	2:0.8:0.4	<20
HEPC:Chol:AMPH-B	2:1:0.4	<20
HEPC:Chol:DSPG:AMPH-B	2:1:0.4:0.4	<25
HEPC:Chol:DSPG:AMPH-B	2:1:0.8:0.4	>30
HEPC:Chol:DLPG:AMPH-B	2:1:0.8:0.4	<20
AmBisome® HSPC:Chol:DSPG:AMPH-B	2:1:0.8:0.4	>50

Chol cholesterol, *HEPC* hydrogenated egg-phospholipid, *HSPC* hydrogenated soy-phospholipid, *DSPG* distearoyl phosphatidylglycerol, *DLPG* dilauryl phosphatidylglycerol
This table was slightly modified from the manuscript described by Adler-Moore JP, et al. (*J. Liposome Research* 3:429–450, 1993)

Table 20.2 Occurrence of adverse events with AMPH-B or L-AMB

Adverse events	L-AMB (n = 343)	AMPH-B (n = 344)	P value
Infusion-related reactions			
Fever following infusion (increase of ≥ 1.0 °C)	58 (16.9 %)	150 (43.6 %)	≤ 0.001
Chills or rigors	63 (18.4 %)	187 (54.4 %)	≤ 0.001
Nausea	42 (12.2 %)	35 (10.2 %)	NS
Vomiting	21 (6.1 %)	28 (8.1 %)	NS
Dyspnea, hypotension, hypertension, tachycardia, diaphoresis, and flushing	57 (16.6 %)	82 (23.8 %)	≤ 0.05
Serum creatinine during therapy			
>1.5 times baseline	101 (29.4 %)	170 (49.4 %)	<0.001
>2.0 times baseline	64 (18.7 %)	116 (37.7 %)	<0.001
Hypokalemia ≤ 2.5 mmol/L	23 (6.7 %)	40 (11.6 %)	≤ 0.05
Hepatotoxicity	61 (17.8 %)	70 (20.3 %)	NS

Table 20.2 referring to the report from Walsh et al. [2]. The side effects listed are mostly resolved soon after finishing the therapy. After starting one particular generic L-AMB treatment in the endemic area of Bangladesh, the incidence of dyspnea was reportedly increased and temporal convulsion was occasionally seen during infusion. Even worse, relapse cases became more common when this generic was prescribed. Herein, we describe the comparative study of 2 L-AMB drugs; AmBisome® [Gilead Sciences] and the other a generic product from an Indian pharmaceutical company (herein anonymously called FunBisome). The name FunBisome has not been a proprietary product name until today.

20.2 Particle Size Analysis

The particle size was examined using transmission electron microscope (TEM) to compare AmBisome and FunBisome. Figure 20.1 exhibits the TEM images and found that the liposomal particle size was considerably more uniform in AmBisome® and the average diameter of liposomal particles was 125 nm determined by dynamic light scattering (DLS) analyses. In Fig. 20.2, the poly dispersity index (PdI) of AmBisome® was 0.226. But, the liposomal particles of FunBisome were often large and not uniform. The average diameter of FunBisome liposomal particles was 206 nm and the PdI was 0.434. As seen at the arrow of Fig. 20.2, there are many larger diameter liposomal particles, out of range of the particle dispersion analysis, even after the manufacture's preparation method of FunBisome was followed. This generic product showed a huge size dispersion. It is well known that the diameter of a pulmonary capillary is around 5.5 μm. A small numbers of the liposomal fragments might interact with the pulmonary capillary endothelium

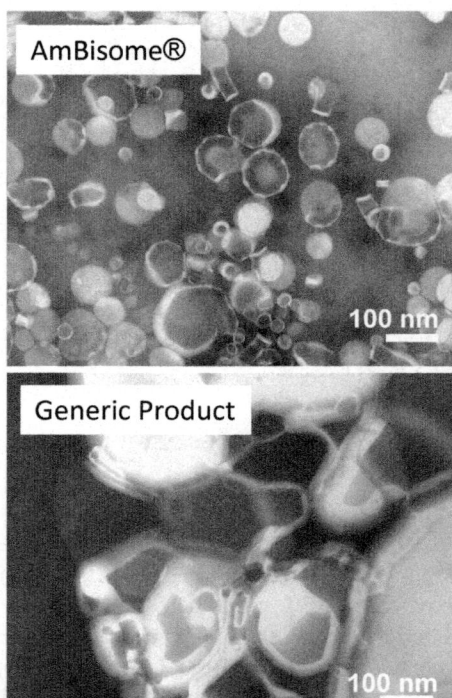

Fig. 20.1 Comparison of L-AMB on TEM images. TEM: transmission electron microscope. Generic product was discussed as FunBisome in the text

and interfere with alveolar gas exchange, though the average diameter of FunBisome particles is far below that size. In addition, FumBisome consists of a small number of liposomes and does not have high enough numbers of the correct sized liposomal particles (approx. 100 nm). Therefore, it is unlikely FunBisome can properly be delivered to the infection site because its size is much larger than the fenestration of the endothelium. TEM clearly demonstrates the characteristics of the poly-layer liposomal particle wall of the generic product, FunBisome. Conversely, AmBisome® clearly demonstrates its bilayer particles with 100 nm diameter in TEM.

20.3 Potassium Release Assay

AMPH-B has the potential to bind to cellular membrane cholesterol in higher concentrations. The convenient way to examine this potential toxicity is to utilize red blood cell (RBC) membranes. If the toxicity of AMPH-B increases, potassium will be released from RBC due to cell membrane damage by AMPH-B. L-AMB is superior to AMPH-B and presumably can be applied to RBC in higher concentration without causing potassium release. Five milliliters of human blood was

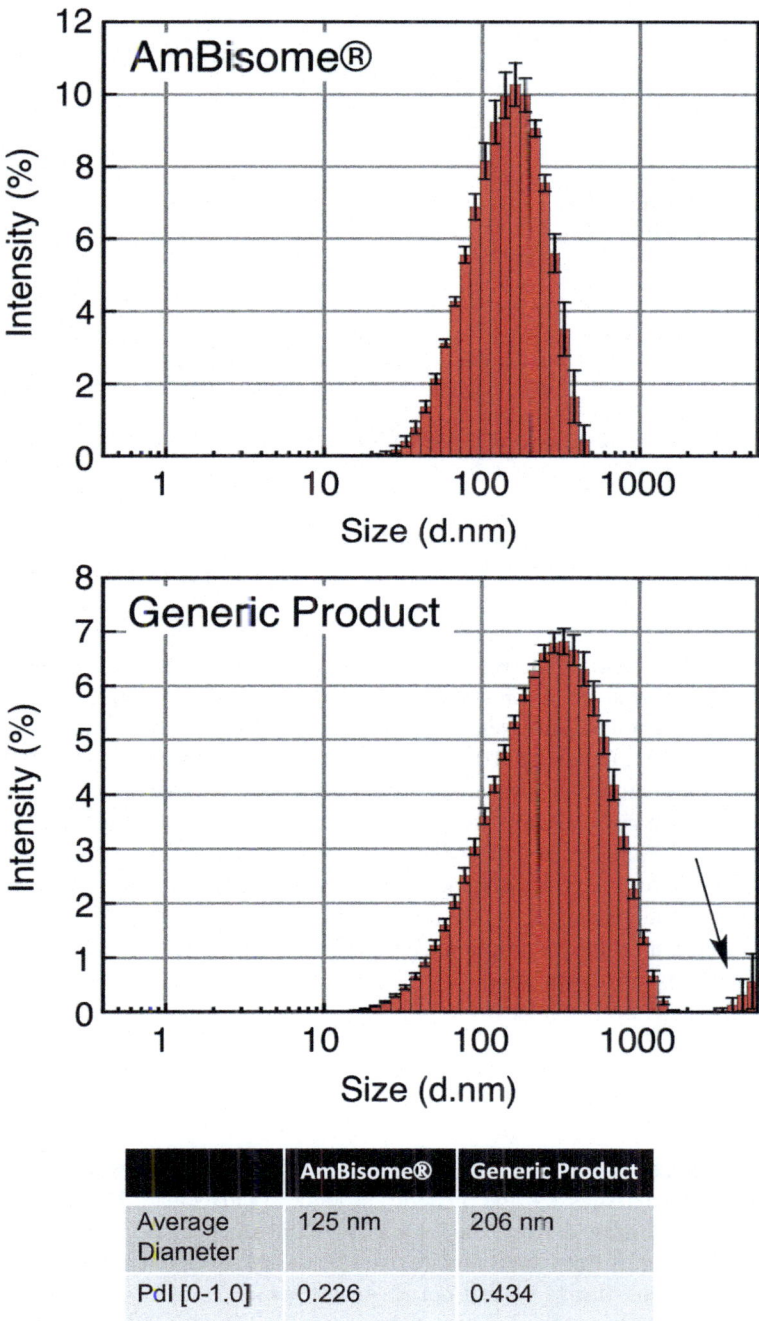

Fig. 20.2 Comparison of L-AMB on particle size dispersion by DLS analyses. PdI: poly dispersion index. Average diameter is Z-average. Generic product was discussed as FunBisome in the text. Of note x-axis is logarithmic scale

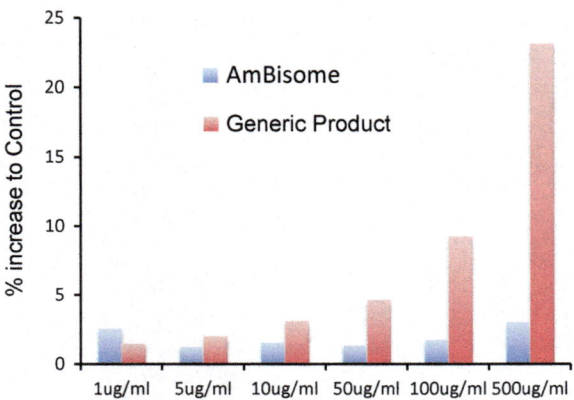

Fig. 20.3 Potassium release assay of L-AMB using human RBC. Drawing blood once transferred to an EDTA-2Na tube was centrifuged and the plasma was removed. Then sufficient PBS was applied to the RBCs for washing purposes and removed after centrifugation. The RBCs were then dissolved in 5 mL of MEM. The same numbers of RBCs were seeded to each well of a 96-well plate and kept in a CO_2 incubator overnight after adding the different concentrations of L-AMB. The control received vehicle only. Potassium level was demonstrated as the percent increase of the control level

drawn by syringe and transferred to an EDTA-2Na tube. The blood was centrifuged and the plasma removed. Then sufficient PBS was applied to the RBCs for washing purposes and removed after centrifugation. The RBC were dissolved in 5 mL of MEM. The same numbers of RBC were seeded to each well of a 96-well plate and kept in a CO_2 incubator overnight after adding the different concentrations of L-AMB. Figure 20.3 shows the result. When the concentrations of L-AMB increased, the increase of potassium was seen mostly in the generic product, FunBisome. Conversely, AmBisome® showed virtually the same level of potassium within the dosage range of 1–500 μg/mL. Patients with severe VL often exhibit severe anemia and the prescription of FunBisome will exacerbate this condition, which is clearly life threatening.

20.4 Conclusion

L-AMB is becoming the key drug to treat VL. Our data show the significant negative characteristics of the generic product, FunBisome. FumBisome will exhibit different drug delivery and toxicity from AmBisome®, and this is also supported by reports from the Bangladesh endemic area of different clinical findings after the treatment. It may be that not all generic products of L-AMB demonstrate these significant problems but we recommend policymakers and stakeholders investigate these drugs thoroughly using the methods demonstrated previously. This is because liposomal particle formation is a delicate pharmaceutical process and there is a need to carefully adhere to the established protocols. Of note, drug delivery of generic products is out of 505(b) for a US analogy. It is a concern that prescription of faulty generic products instead of AmBisome® is occurring because the earlier mentioned differences can cause potential injury to patients and furthermore insufficient drug delivery will often induce treatment failure and relapse.

References

1. Sundar S, Chakravarty J, Agarwal D, Rai M, Murray HW. Single-dose liposomal amphotericin B for visceral leishmaniasis in India. N Engl J Med. 2010;362(6):504–12.
2. Walsh TJ, Finberg RW, Arndt C, Hiemenz J, Schwartz C, Bodensteiner D, Pappas P, Seibel N, Greenberg RN, Dummer S, Schuster M, Holcenberg JS. Liposomal amphotericin B for empirical therapy in patients with persistent fever and neutropenia. National Institute of Allergy and Infectious Diseases Mycoses Study Group. N Engl J Med. 1999;340(10):764–71.

Chapter 21
High Resolution Mapping of Kala-Azar Hot Spots Using GPS Logger and Urinary Antibody Measurements

Bumpei Tojo, Makoto Itoh, Mohammad Sohel Samad, Emi Ogasawara, Dinesh Mondal, Rashidul Haque and Eisei Noiri

Abstract The findings of the epidemiological study about visceral leishmaniasis (VL), also known as kala-azar, in Bangladesh performed under the JICA/SATREPS project during the period of 2011–2015 is introduced in this chapter. Through the study of this project a new diagnosis technique was developed that allowed for urine to be usable for the detection of the VL infection. Specifically, we devised a method that could make a highly precise distribution map of the urine antibody value for the VL protozoan. There is a possibility that this high resolution map of the VL endemic foci (hot spot) may contribute to the active surveillance or vector control of VL.

Keywords Urine-based mass screening for VL · Visceral leishmaniasis · rKRP42 · GPS logger · GIS · Infection map · Hot spot analysis

21.1 General Background

In the report of the Centers for Disease Control (CDC), there are 16 visceral leishmaniasis (VL), also known as kala-azar, endemic upazila from 2008 to 2013 where the incidence rate is greater than 1 per 10,000 people in Bangladesh. Trishal is the second highest endemic upazila where the average incidence rate is 17.82 per 10,000 population in this period [1]. During 2011–2014, a total of 753 VL cases were recorded at the Upazila Health Complex (UHC) of Trishal. Upazila is the name of the administrative units in Bangladesh, and under each upazila there are 10–20 unions. Unions can be subdivided into more plural villages. The Trishal upazila is constructed from 12 Unions, 159 villages.

B. Tojo (✉)
Department of Hemodialysis and Apheresis, The University of Tokyo Hospital, 7-3-1, Hongo, Bunkyo-ku, Tokyo 113-8655, Japan
e-mail: bptojo-tky@umin.ac.jp

It depends on the population, but approximately 1–3 primary schools are generally established for every village, and the 200–300 students go to school from a radius of around 1 km. According to the upazila map provided by the Local Government Engineering Department (LGED), the total number of the primary schools in the Trishal upazila are 118 (1 school per 1.3 villages). However, the LGED map is incomplete, and each village actually has between 1–3 primary school.

Magurjura, Harirampur and Chikna monohor (MG, HP, CM) School located in the village in Trishal where the VL incidence was ranked high to middle were selected as research targets (Fig. 21.1). The CM school was located in the CM village of Trishal union, the HP school and MG school in the HL village and MG village, respectively, of Harirumpur Union. The average VL incidence (2011–2014) in Trishal was 4.27 (Median 1, Max 80, Min 0, Stdv 9.1). The incidence rank of each research target village in Trishal is 3rd (MG: 36 cases), 10th (HP: 16 cases), and 57–67th (CM: 2 cases).

21.2 Methodology

21.2.1 Urine ELISA

The procedure for producing the antigen and the method of ELISA were described elsewhere [2–4]. Briefly, flat-bottomed 96-well microtiter plates (MaxiSorp, Nunc, Denmark) were coated with 1 μg/mL rKRP42 antigen (100 μL/well) overnight at 4 °C. After blocking with casein buffer (1 % casein in 0.05 M Tris-HCl buffer with 0.15 M NaCl, pH 7.6) for 2 h at room temperature, the wells were loaded with 100 μL urine and incubated at 37 °C for 1 h. After washing with phosphate-buffered saline (PBS; pH 7.4) containing 0.05 % Tween 20, peroxidase-conjugated goat anti-human IgG (Tago, Camarillo, CA) diluted 1:4,000 with casein buffer was added and incubated at 37 °C for 1 h. After washing, 2,2′-azino-bis(3-ethylbenzothiazoline-6-sulphonic acid) (ABTS)substrate (KPL Inc., Gaithersburg, MD) was used for coloration. The optical density (OD) was measured at 415 nm, and an OD of 492 nm was used as a reference. Each sample was assayed in duplicate; if the absorbance values of the duplicates differed >40 % from their average, the sample was retested. Antibody levels were expressed arbitrarily as units, which was estimated from the standard curve constructed for each plate with serially diluted positive sera. The cutoff value was 57.9 U. The cut-off point for anti-rKRP42 IgG was first calculated as the mean plus 3 standard deviations (SD) of log (unit + 1) values of the non-endemic healthy controls, and [antilog of (the mean + 3 SDs) − 1] was regarded as the cut-off unit.

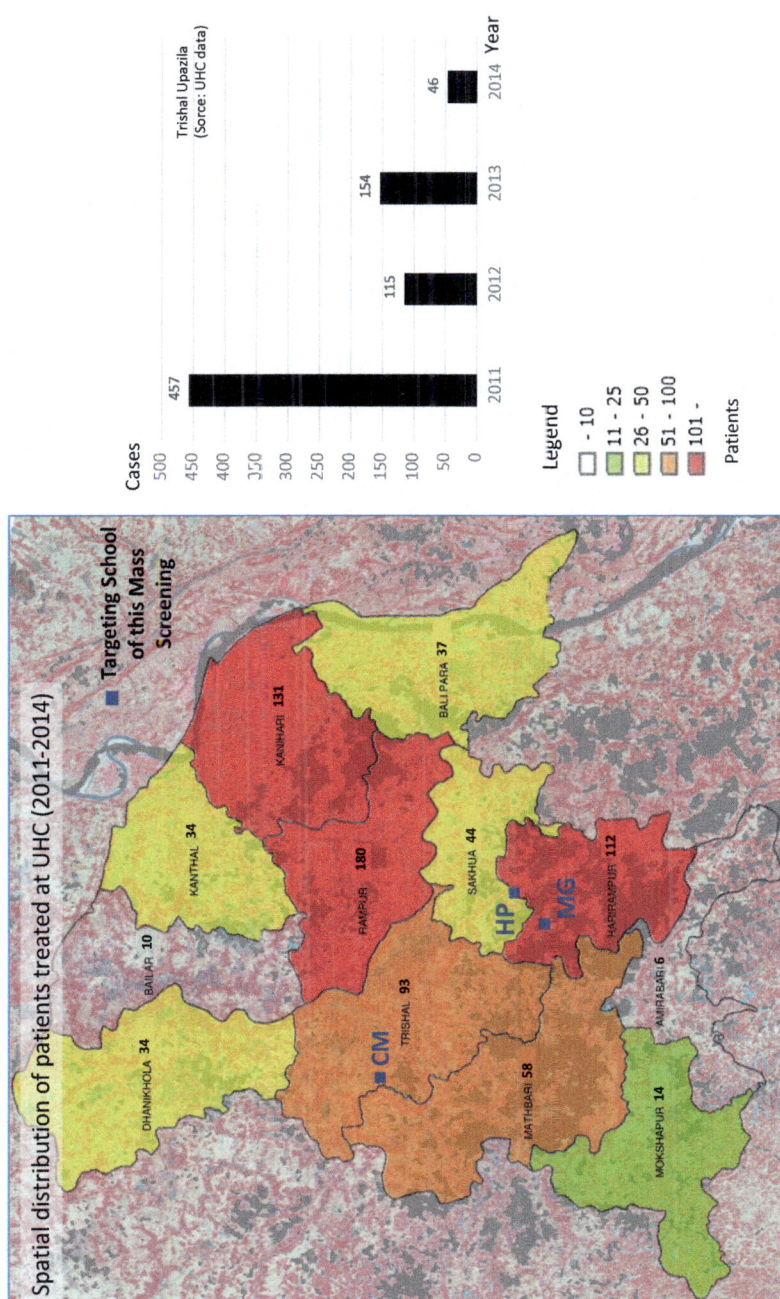

Fig. 21.1 Number of patients treated at UHC (2011–14). Magurjura, Harirampur and Chikna monohor (MG, HP, CM) School located in the village in Trishal where the VL incidence was ranked high to middle were selected as research targets

21.2.2 GPS Logger and Student Housing Location Survey

A relatively inexpensive (per unit price less than 100 dollars) and small size GPS logger model was chosen (GT-600, Mobile Action Technology). There is no Internet upload function for acquired data; collected positional data (movement course) are transferred to the PC through a USB cable connection. The logger has waterproofing and also an automatic ON/OFF function of the power supply controlled by the vibration sensor. When the setting is changed on a PC, the physical key of the logger (ON/OFF of the power supply) is locked from use. Through these functions it was possible to collect data from young schoolchildren without difficulty.

The tracking log of these GPS loggers was displayed and the last arrival point of the movement course of each student of the day was pinpointed on the geographic information system (GIS). The tracking log stops automatically when a student arrives at the home and parts with the logger. All of the positions of the houses of each student were decided on a GIS map by the result of the overlay analysis between the stop position of the tracking log and the high-resolution satellite image (Google Earth).

21.2.3 Paperless Questionnaire Using a Tablet Device and PDF

A standard function (questionnaire form making) of the Adobe Acrobat 11 standard software was used, and the necessary number of electronic questionnaire forms were prepared. These electronic questionnaire forms were transferred to a tablet device, and the question items (full name, sex, school year, parents full name, past VL medical history, and history of treatment) for each schoolchild in the screening were recorded in this form (Fig. 21.2).

The unique Serial ID (GPS ID) of each GPS logger converted into a bar code was printed on waterproof tape and affixed to the GPS logger. A bar code reader was used when inputting the ID, including the urine specimen ID, on the electronic form, to shorten input time while also removing human input error. As a result of having a computerized questionnaire, the questionnaire input data could be transferred to a PC immediately after screening.

After introducing this *paperless registration, a few staff can handle around 300-400 students for mass screening within 1 day*.

Fig. 21.2 Paperless registration and GPS logger distribution at mass screening. A standard function (questionnaire form making) of the Adobe Acrobat 11 standard software was used, and the necessary number of electronic questionnaire forms (using Tablet device) were prepared. The unique Serial ID (GPS ID) of each GPS logger converted into a bar code was printed on waterproof tape and affixed to the GPS logger. A bar code reader was used when inputting the ID, including the urine specimen ID, on the electronic form, to shorten input time while also removing human input error

21.3 Results and Discussion

21.3.1 Positive Rate of Target School

Three schools in Trishal upazila were surveyed, and the urine of 866 schoolchildren was collected on August 2013 and March 2014. The rKRP42 antibody measurement from urine using the ELISA was performed using the SKKRC laboratory facility. As a result of mass screening, 5.3 % ($n = 46$) of positive students were detected. In regards to the positive rate for each school, CM was 0.6 % ($n = 2/344$), 5.3 % for HP ($n = 15/283$), and 12.1 % for MG ($n = 29/239$) (Fig. 21.3). In CM, where the incidence ranked only 57–67 of 159 villages, the number of schoolchildren with extremely low (<10) antibody titers exceeded 90 %, and there was not enough significance to determine spatial distribution of the antibodies. Therefore, to map the urine antibody titer distribution, only 2 schools (MG and HP) were selected.

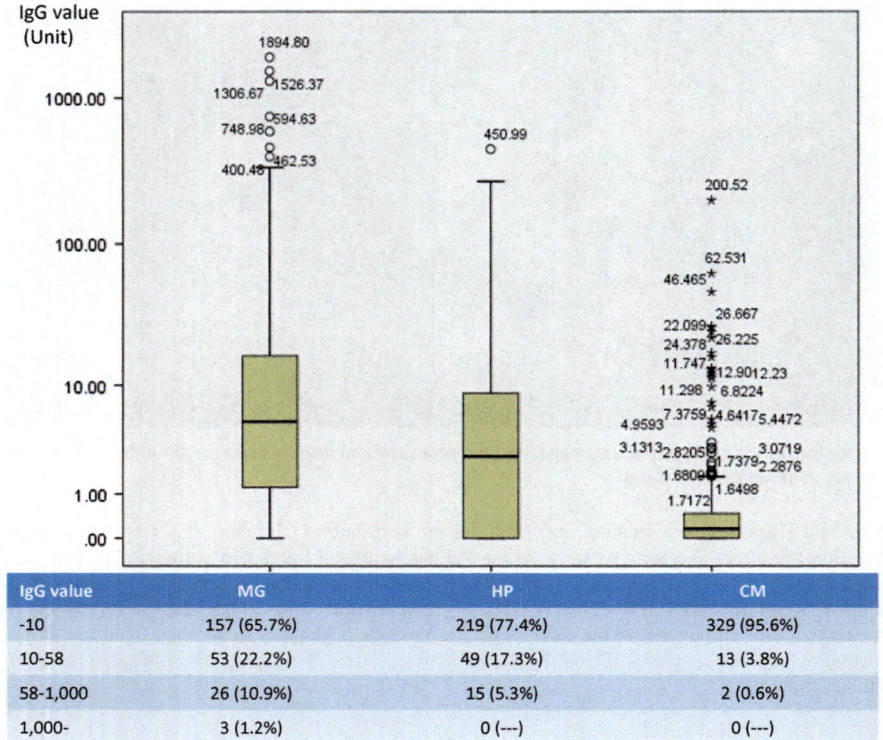

IgG value	MG	HP	CM
-10	157 (65.7%)	219 (77.4%)	329 (95.6%)
10-58	53 (22.2%)	49 (17.3%)	13 (3.8%)
58-1,000	26 (10.9%)	15 (5.3%)	2 (0.6%)
1,000-	3 (1.2%)	0 (---)	0 (---)

Fig. 21.3 Three schools in Trishal upazila were surveyed, and the urine of 866 schoolchildren was collected on August 2013 and March 2014. The rKRP42 antibody measurement from urine using the ELISA was performed using the SKKRC laboratory facility. As a result of mass screening, 5.3 % of positive students were detected. In regards to the positive rate for each school, CM was 0.6, 5.3 % for HP, and 12.1 % for MG

21.3.2 Accuracy of Student Location Data

For MG and the HP, the distribution of the GPS logger is carried out with the urine collection for all students, and 522 log data are acquired. Field testing surveyed 138 (26.4 %) of 522 student locations, and positional data with a less than 5 m positional error was acquired using a GPS receiver (eTrex20, Garmin) for calculating the error between the GPS logger based student location and their true location. Except for 4 location data (error 8,171, 705, 608, and 382 m), the mean error was 31.8 m (median 24.7 m, 95 % confidence interval lower limit 27.1 m/upper limit 36.5 m of the mean) (Fig. 21.4). It might be said from this result that the estimation of each student location using the GPS logger had the enough accuracy for geographical analysis.

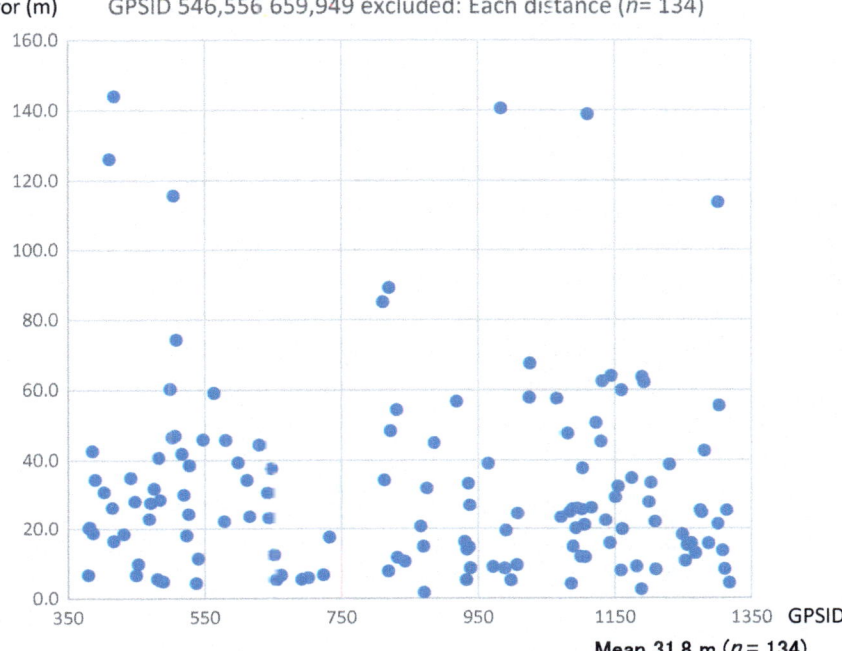

Fig. 21.4 The positional information precision from the GPS logger. Except for 4 location data (error 8,171, 705, 608, and 382 m), the mean error was 31.8 m (median 24.7 m, 95 % confidence interval lower limit 27.1 m/upper limit 36.5 m of the mean)

21.3.3 Zone and Population Mapping of the Precinct

As a result of having distributed a GPS logger for the mass screening, the house location distribution of school children (and family members) of MG or HP became clear (Fig. 21.5). The residential area in Fig. 21.5 is a natural embankment distinguished from the neighboring low altitude zones. This boundary line was decided by the visual distinction of the AVNIR-2 image (Fig. 21.6). In Bangladesh, this belt-shaped natural embankment (it becomes a slight highlands rising several meters above the neighboring area) has been developed as residential area. Neighboring low altitude zones have poor drainage covered by flood water every rainy season (Jun–Jul); the main land use of these lowlands is swamps and ponds (waterbodies), or rice fields.

The precinct of MG occupies the central part of the MG village, and the HP precinct occupies a part at the west end of the HP village. Both precincts occupy about 1/3 of the whole village area (Fig. 21.6). Figure 21.7 is the population of MG and HP village by population census [5]. The total population of both villages is

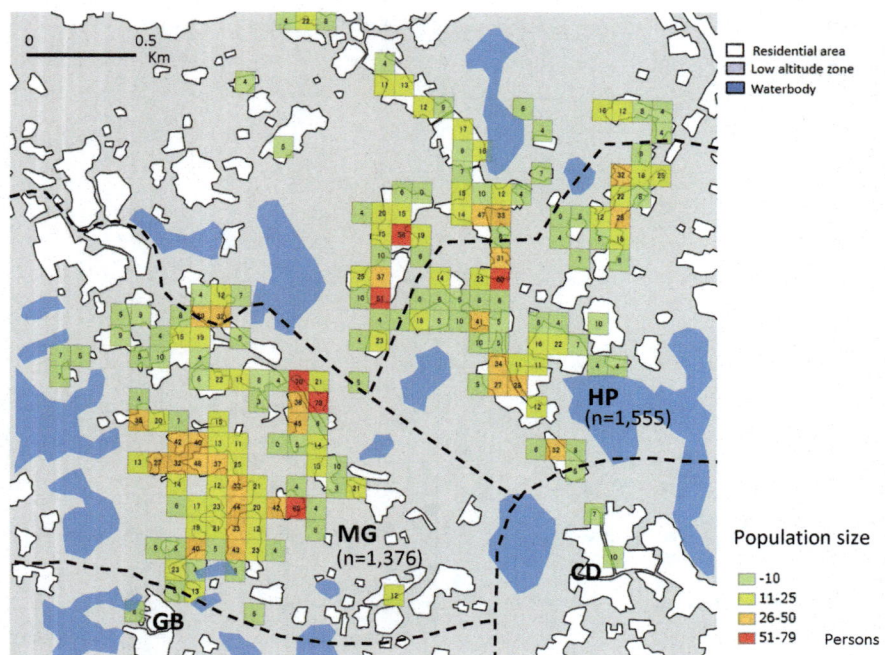

Fig. 21.5 Zone and population distribution of MG and HP precincts. As a result of having distributed a GPS logger for the mass screening, the house location distribution of schoolchildren (and family members) of MG or HP became clear. The residential area in this figure is a natural embankment distinguished from the neighboring low altitude zones. In Bangladesh, this belt-shaped natural embankment has been developed as residential area. Neighboring low altitude zones have poor drainage covered by flood water every rainy season

approximately 5,500, and the population of the primary school age (5–14 years old) is 1,600–1,700 (approximately 30 % of the whole population). Furthermore, the non-school attendance percentage of the 5–14-year-old age group in both villages is 35–37 % according to the census. It means the actual population of the student (regularly attending the school) becomes 1,000–1,100. MG and HP schools include an area of 1/3 of the village. Thus, the estimated number of the schoolchildren is approximately 330 (in MG) and 360 (in HP). From this figure, it is calculated that the schoolchild who joined this mass screening is 72.4 % (in MG) and 78.6 % (in HP). The age of the population at the time of mass screening is not really the same as the age of the population at the time of census data. However, this error is ignored in this chapter. Each precinct population can be estimated as 1,376 (MG) and 1,555 (HP) from Fig. 21.5. It was equivalent to 25.2 % (MG) and 28.4 % (HP) of the total population of the village.

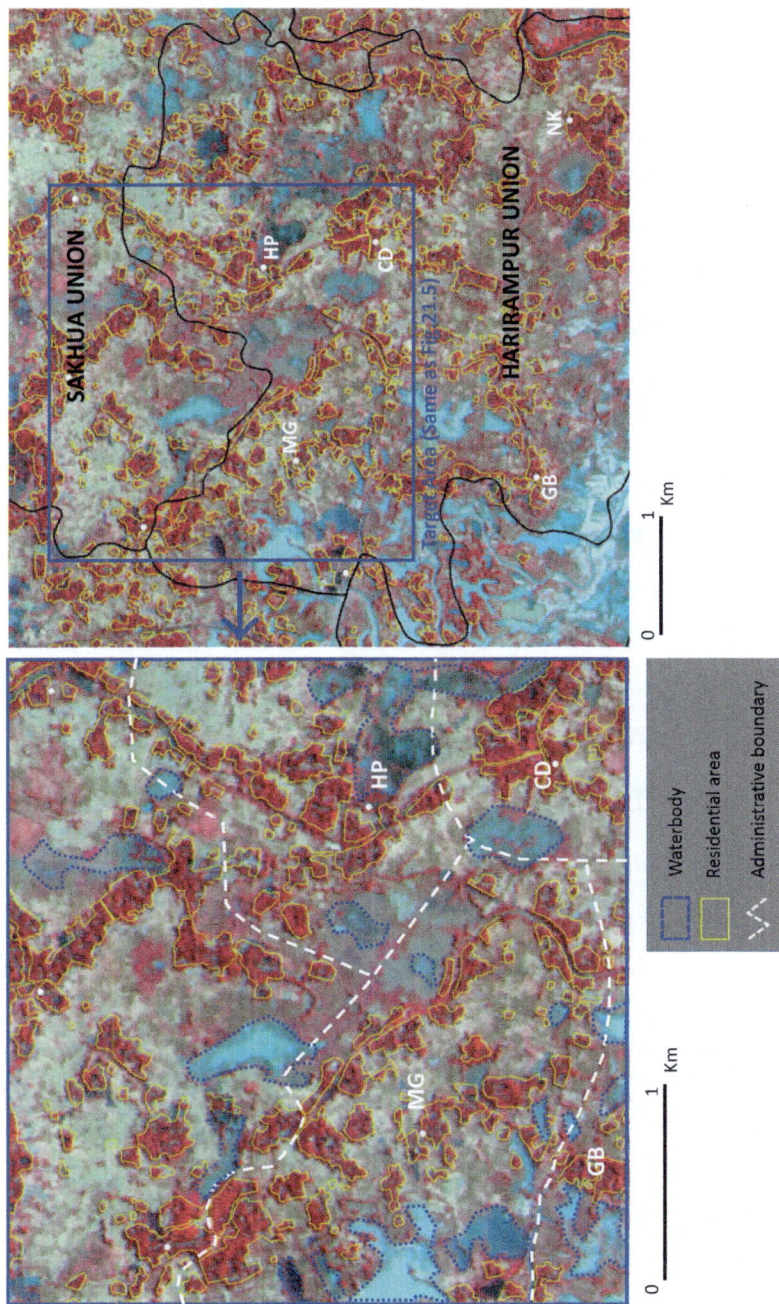

Fig. 21.6 Residential area, waterbody, union and village boundary in target area. The precinct of MG occupies the central part of the MG village, and the HP precinct occupies a part at the west end of the HP village. Both precincts occupy about 1/3 of the whole village area

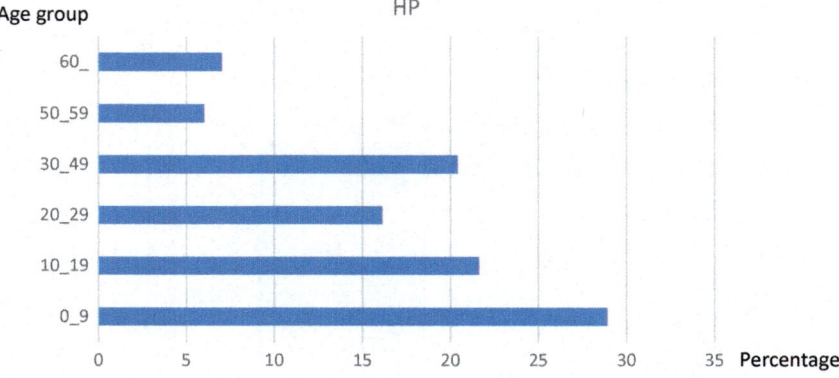

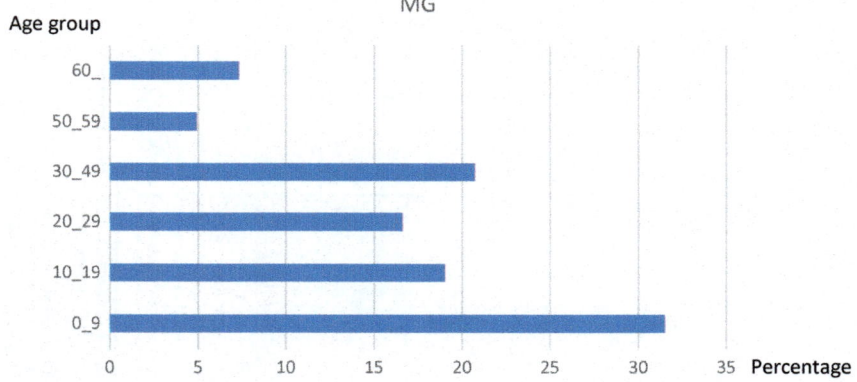

Fig. 21.7 Age population composition of MG and HP village according population census (2011) [5]. The total population of both villages is approximately 5,500, and the population of the primary school age (5–14 years old) is 1,600–1,700 (approximately 30 % of the whole population). The non-school attendance percentage of the 5–14-year-old age group in both villages is 35–37 % according to the census. It means the actual population of the student (regularly attending the school) becomes 1,000–1,100. MG and HP schools include an area of 1/3 of the village. Thus, the estimated number of the schoolchildren is approximately 330 (in MG) and 360 (in HP). From this figure, it is calculated that the schoolchild who joined this mass screening is 72.4 % (in MG) and 78.6 % (in HP)

21.3.4 Geographical Hot Spots of Urine Antibodies for L. Donovani

As well as the data shown in Fig. 21.5, the distribution of the number of patients with VL (included family members of schoolchildren) was mapped using questionnaire data (Fig. 21.8). The HP precinct is approximately equal to the MG

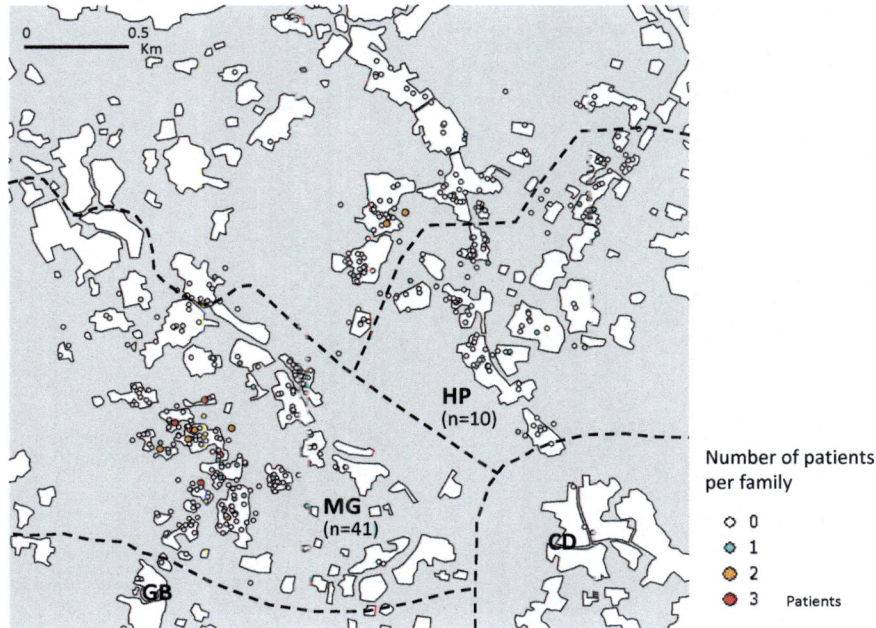

Fig. 21.8 The distribution of the number of patients with VL (included family members of schoolchildren) was mapped using questionnaire data. The HP precinct is approximately equal to the MG precinct by area and population size. However, the incidence of VL has a gap of approximately 4 times between HP and MG. In addition, though the spatial accumulation of patients does not occur in the HP precinct, patients seemed to accumulate on the western part of the precinct in MG

precinct by area and population size. However, the incidence of VL has a gap of approximately 4 times between HP and MG. In addition, though the spatial accumulation of patients does not occur in the HP precinct, patients seemed to accumulate on the western part of the precinct in MG.

Figure 21.9 mapped the result of the measurement of the rKRP42 antibody from urine. The mapping values were converted into a logarithm, measurements were spatially interpolated based on the inverse distance weighted (IDW) method. The rKRP42 antibody positive data corresponds to the class after class 5 (value = 1.5–2) on this distribution map. It is seen in Fig. 21.9 that class 5–7 is scattered from the south part to the western part of the MG precinct. However, the HP precinct had little distribution of these classes. Class 4 (antibody titer ≥ 10) was widely distributed over half of the bottom of the MG precinct. In the HP precinct, class 4 shows slight distribution only at the border with the MG precinct (the west end).

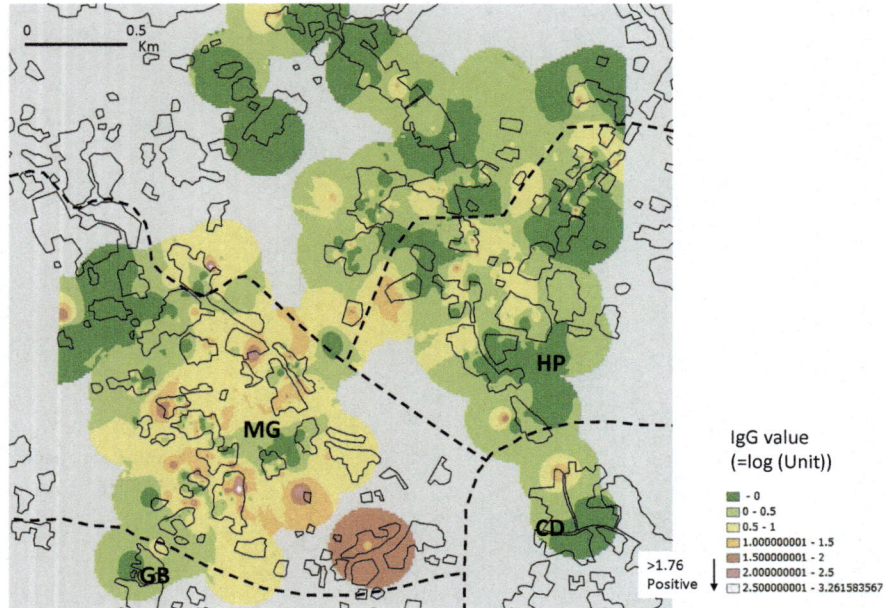

Fig. 21.9 Spatial distribution of urinary antibody for *L. donovani*. This figure mapped the result of the measurement of the rKRP42 antibody from urine. The mapping values were converted into a logarithm, measurements were spatially interpolated based on the inverse distance weighted (IDW) method. The rKRP42 antibody positive data corresponds to the class after class 5 (value = 1.5–2) on this distribution map. Class 5–7 is scattered from the south part to the western part of the MG precinct. However, the HP precinct had little distribution of these classes

Figure 21.10 shows the space cluster of a meaningful high or low value statistically using hot spot analysis (Getis-Ord Gi^*) concerning the spatial distribution of Figs. 21.8 and 21.9. In regards to the patient population distribution considered in Fig. 21.8, the intensive outbreak cluster (hotspot) of the patients with statistical meaningfulness was seen in only the western part of the MG precinct. However, related to the distribution of rKRP42 antibody considered in Fig. 21.9, it was observed that the hotspot was escalating around from the hotspot of the patient outbreak. The hotspot of the rKRP42 antibody did not exist in the HP precinct, instead it is observed that the distribution of the cluster of low antibody titer (cold spot) is around the north side of the precinct. In other words, in the HP precinct, the spatial distribution of positive schoolchildren emerged accidentally. However, it appears clearly in the MG precinct that the hotspot of the patient outbreak becomes the nucleus of the infection, the hotspot of the rKRP42 antibody positive gradually transmitted to neighboring area. The differences of both precincts are more than the difference of the simple positive rate.

21 High Resolution Mapping of Kala-Azar Hot Spots ...

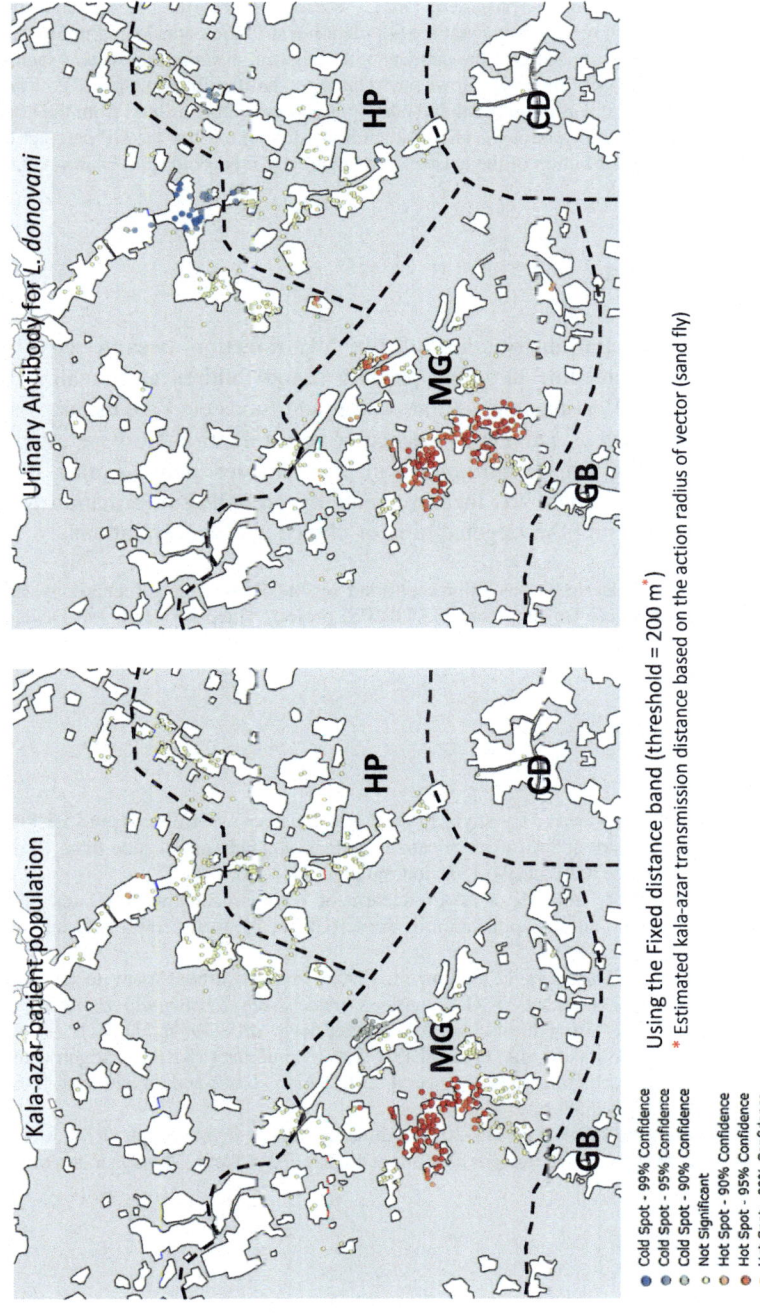

◄ **Fig. 21.10** Result of hot spot analysis (Getis-Ord Gi* statistic). Concerning the spatial distribution of Figs. 21.8 and 21.9. In regards to the patient population distribution considered in Fig. 21.8, the intensive outbreak cluster (hotspot) of the patients with statistical meaningfulness was seen in only the western part of the MG precinct. However, related to the distribution of rKRP42 antibody considered in Fig. 21.9, it was observed that the hotspot was escalating around from the hotspot of the patient outbreak. The hotspot of the rKRP42 antibody did not exist in the HP precinct, instead it is observed that the distribution of the cluster of low antibody titer (cold spot) is around the north side of the precinct

21.4 Discussion

Visualizing the spatial distribution of the VL infection became possible by urine-based mass screening targeting primary schoolchildren and creating a map with a GPS logger. Vast resources (time, labor, and funds etc.) are usually required for VL measures such as active surveillance or vector control. However, the highly precise epidemiology map introduced in this chapter demonstrates that this approach may aid complex VL measurements by providing information for decision making concerning the targeted area of choice and concentrations.

Acknowledgments This study was fully supported by the Science and Technology Research Partnership for Sustainable Development (SATREPS) project, Titled Research and Development of Prevention and Diagnosis for Neglected Tropical Disease, Especially Kala-azar, Japan.

References

1. Rajib C, Dinesh M, Vashkar C, Shyla F, et al. How far are we from visceral leishmaniasis eliminaton in Bangladesh? An assessment of epidemiological surveillance data. PLoS Negl Trop Dis. 2014;8(8):e3020. doi:10.1371/journal.pntd.0003020.
2. Takagi H, Islam MZ, Itoh M, et al. Production of recombinant kinesin-related protein of *Leishmania donovani* and its application in the serodiagnosis of visceral leishmaniasis. Am J Trop Med Hyg. 2007;76:902–5.
3. Islam MZ, Itoh M, Takagi H, et al. Enzyme-linked immunosorbent assay to detect urinary antibody against recombinant rKRP42 antigen made from *Leishmania donovani* for the diagnosis of visceral leishmaniasis. Am J Trop Med Hyg. 2008;79(4):599–604.
4. Islam MZ, Itoh M, Ul Islam MA, et al. ELISA with recombinant rKRP42 antigen using urine samples: a tool for predicting clinical visceral leishmaniasis cases and its outbreak. Am J Trop Med Hyg. 2012;87(4):658–62.
5. BBS. Population and housing census-2011 community report: mymensingh; 2011. http://www.bbs.gov.bd/WebTestApplication/userfiles/Image/PopCen2011/Com_Mymensingh.pdf.

Chapter 22
Geography and Reality of Kala-Azar Endemic in Bangladesh, Analysis Using GIS and Urine-Based Mass Screening

Bumpei Tojo, Makoto Itoh, Shyamal Kumar Paul, Mohammad Sohel Samad, Emi Ogasawara, Fumiaki Nagaoka, Dinesh Mondal, Rashidul Haque and Eisei Noiri

Abstract In this chapter, a map of the endemic area of visceral leishmaniasis (VL), also known kala-azar, was made using epidemiological data of the Upazila Health Complex (UHC). The information from the UHC is very useful in understanding and quantifying the changing geographical distribution of VL endemic areas. A more detailed geographical distribution of the VL endemic areas are surveyed using epidemiological data of a public hospital, the Surya Kanto VL Research Center (SKKRC) in Mymensingh. We examined these 2 sources of geographical information and discovered some interesting endemic hotspots. We also carried out a urine specimen-based VL mass screening in these hotspots. Herein, we suggest that mass screening in conjunction with epidemiological data from hospitals is effective at population monitoring for the prevention of VL resurgence in existing endemic areas, or for the discovery of new outbreak areas.

Keywords Upazila health complex · SKKRC · GIS · Endemic mapping · Mass screening

22.1 Introduction

Phlebotomus argentipes, a species of sand fly 2–3 mm in length is the vector for transmission of visceral leishmaniasis (VL) in Bangladesh. Because of its short possible flight distance, it is said that the habitation radius of an individual sand fly

---

B. Tojo (✉)
Department of Hemodialysis and Apheresis, The University of Tokyo Hospital,
7-3-1, Hongo, Bunkyo-ku, Tokyo 113-8655, Japan
e-mail: bptojo-tky@umin.ac.jp

© Springer International Publishing 2016
E. Noiri and T.K. Jha (eds.), *Kala Azar in South Asia*,
DOI 10.1007/978-3-319-47101-3_22

is only several hundred meters. The geographical distribution of the VL endemic generally has a space accumulation characteristic following the habitat of the sand fly. The early detection of infected people (including asymptomatic individuals) in outbreak locations and their appropriate treatment, along with early warning of the expansion of infected areas through continuous monitoring, have become the basic strategy to combat VL. In 2015, the annual number of cases of VL in Bangladesh was 576. This number is less than 1/10 of the value reported 10 years ago. The focus of disease control of VL has moved towards maintaining this low national endemic level.

Although geographical information on the VL endemic is extremely important for controlling the disease, details of disease distribution have not been thoroughly researched in Bangladesh. The major source of information on the VL endemic situation is epidemiological data from various public hospitals. Generally, patients with VL are poor and from rural areas. Therefore, regional information on VL is mostly handled through the Upazila Health Complex (UHC). The UHC is established in each upazila, of which there are approximately 492 countrywide, and which is the smallest unit of the public hospital system in Bangladesh. Thus, a geographical grasp of the VL endemic distribution can be accurately determined if the patient address is recorded exactly when a patient is registered in the UHC.

22.2 General Background

According to the recent report by the Centers for Disease Control (CDC), there are 16 VL endemic upazila from 2008 to 2013 where the incidence rate was greater than 1 per 10,000 people. Fulbaria and Trishal are the highest endemic upazilas where the incidence rates are 4 times (18.25 and 17.82, respectively, per 10,000 population) higher than any other endemic upazila [1]. These upazilas form the largest part of the VL endemic region in Bangladesh (Fig. 22.1). In this chapter, we focus on this endemic region as a research area. The total incidence rate for the country ranges from more than 4000 cases from 1999 to 2009, to below 4000 in 2010. In 2011, 1501 VL cases were recorded at the UHC from 5 upazila (Fulbaria, Trishal, Gaffargaon, Madhupur, and Ghatail) in this endemic region. It was about 59.3 % of the whole VL incidence (2533 cases) in this year.

The UHC is the lowest level of public hospital with 31–50 beds where medical practitioners are assigned. The UHC is based in one place at each upazila. A total of 89 % (436) of upazila were covered by the UHC in 2014.

The Surya Kanto hospital VL Research Center (SKKRC) is located in the center of the VL endemic areas. SKKRC is a part of the outbuilding (infectious disease ward) of Mymensingh Medical College Hospital. Through the support of the JICA/SATREPS project (Research and Development of Prevention and Diagnosis for Neglected Tropical Disease, Especially VL, 2011–15), Icddr,b, and DNDi, it was renovated as a research center for VL. Diagnosis and treatment of VL patients were started in 2013 at SKKRC. Because the highest numbers of diagnoses are

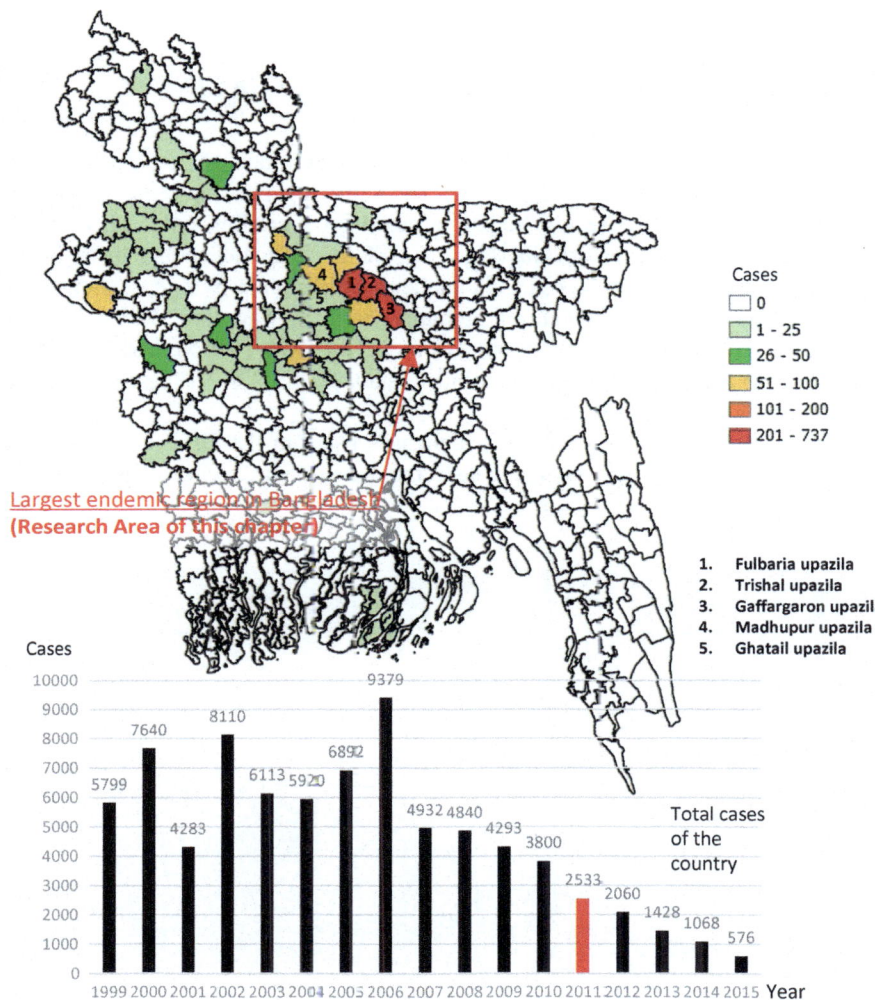

Fig. 22.1 Country trend of Kala-azar cases and location of kala-azar endemic upazila in 2011 (*Source* CDC). There are 16 VL endemic upazila from 2008 to 2013 where the incidence rate was greater than 1 per 10,000 people. Fulbaria and Trishal are the highest endemic upazilas. These upazilas form the largest part of the VL endemic region in Bangladesh

made and treatments are received here, patients gather widely from around country to SKKRC. The epidemiological data that have been accumulated since opening the SKKRC are managed through technical cooperation of the Tokyo University by a Web-based case registration system (INDICE: WHO International Clinical Trials Registry Platform: JPRN-UMIN000011426).

22.3 Methods

22.3.1 Epidemiological Data from UHC and SKKRC

In the UHC, results of diagnostic testing and treatment plans for each patient that visits the center recorded in a log book. The same log book is used for several diseases, although there are log books prepared for recording only VL cases in some special hyperendemic upazila (Fig. 22.2). The VL incidence data listed in the log book were obtained from 4 upazila that showed particularly high incidences of VL. Based on the log book data of Fulbaria (from 2011 to June, 2015), Trishal (from 2011 to May, 2014), Gaffargaon (from 2012 to May, 2015), and Madhupur (from 2011 to March, 2015), a geographical database for 2436 VL cases was made as per the following procedures.

In most cases, the address of the patient was listed in the UHC log book, and included the name of the village where the patient was living. Therefore the

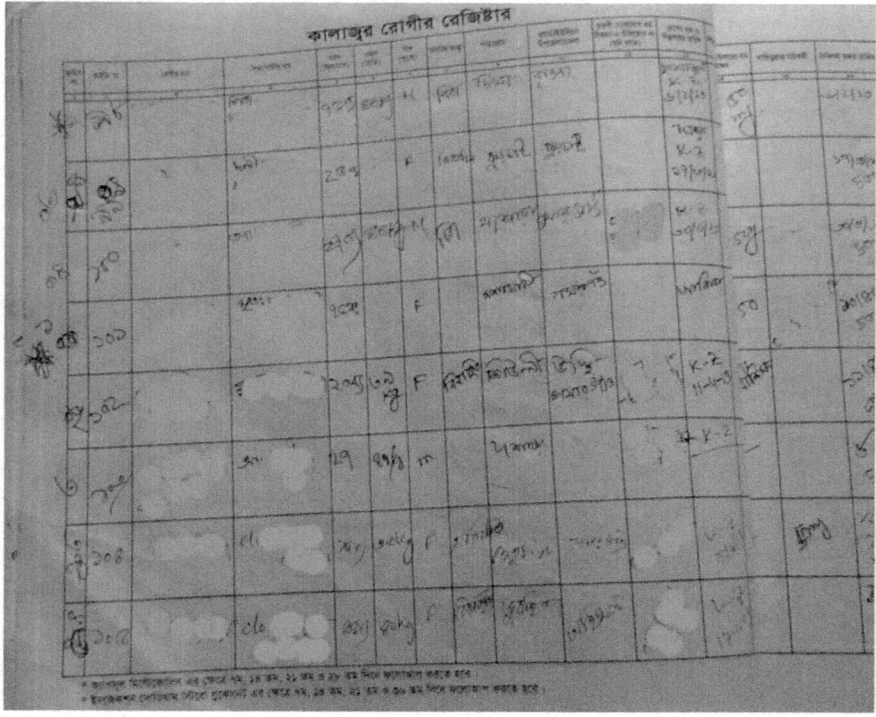

Fig. 22.2 Log book for kala-azar recording at Gaffargaon UHC. In the UHC, results of diagnostic testing and treatment plans for each patient that visits the center recorded in a log book. The same log book is used for several diseases, although there are log books prepared for recording only VL cases in some special hyperendemic upazila

individual patient incidence can be associated with the geographical location of the village if the positional data of each village is available. The village name itself can be found in the population census (2011); however, the population census doesn't include data on the location of each village (as geographical coordinates, or a point on the map). In addition, in the population census, the name of a village listed in patient log book is often not found. Instead, a topographical map data of 1:50,000 scale made by the Survey of Bangladesh (SOB) was digitized and a total of 933 spots of village coordinate data (in Fulbaria, Trishal, Gaffargaon, Madhupur) was prepared (Fig. 22.3).

A total of 668 villages were confirmed in these 4 upazila by the population census [2, 3]. But the Madhpur upazila was separated from the Dhanbari upazila in the census. When 135 villages included in Dhanbari are deducted by the SOB data, the number of the remaining villages becomes 798. When adjustments are made for differences between SOB and census data, the number drops to 130 (see annexed table of Fig. 22.3). Using the epidemiological data of SKKRC, positional information of 410 patients were collected using a Handy GPS receiver (eTrex20, Garmin). Throughout this mapping work, ESRI ArcGIS 10.2.2 software was used.

22.3.2 Urine-Based Mass Screening for VL Targeting School Children

Mass screening in this chapter is based on a detection technology of the VL protozoan (*L. donovani*) antibody from a urine specimen using ELISA testing. Because the urine collection is noninvasive and convenient in comparison to collecting blood, it is easy to get the cooperation of voluntary test subjects, and there is very little labor required for specimen collection. A mass screening targeting 200–300 schoolchildren can be finished by several staff in several hours. While the test can detect antibodies with a high sensitivity, it may detect antibodies from past infections and it is therefore difficult to distinguish between past and recent Leishmaniasis events. However, the mass screening in this study was limited to schoolchildren of about 5–10 years old to ensure only relatively recent endemics are reflected in the screening results.

The procedure for producing the antigen and the method of ELISA are described elsewhere [4–6]. Briefly, flat-bottomed 96-well microtiter plates coated with a recombinant antigen, rKRP42, were used. The wells were loaded with 100 μL urine and incubated at 37 °C for 1 h. After the plates were washed, peroxidase-conjugated goat anti-human IgG (Tago, Camarillo, CA) diluted 1:4000 was added and incubated at 37 °C for 1 h. After the plates were washed again, ABTS substrate (KPL Inc., Gaithersburg, MD) was used for coloration. The optical density (OD) was measured at 415 nm, and an OD of 492 nm was used as a reference. Each sample was assayed in duplicate. Antibody levels were expressed arbitrarily as units that were estimated from the standard curve constructed for each plate with serially

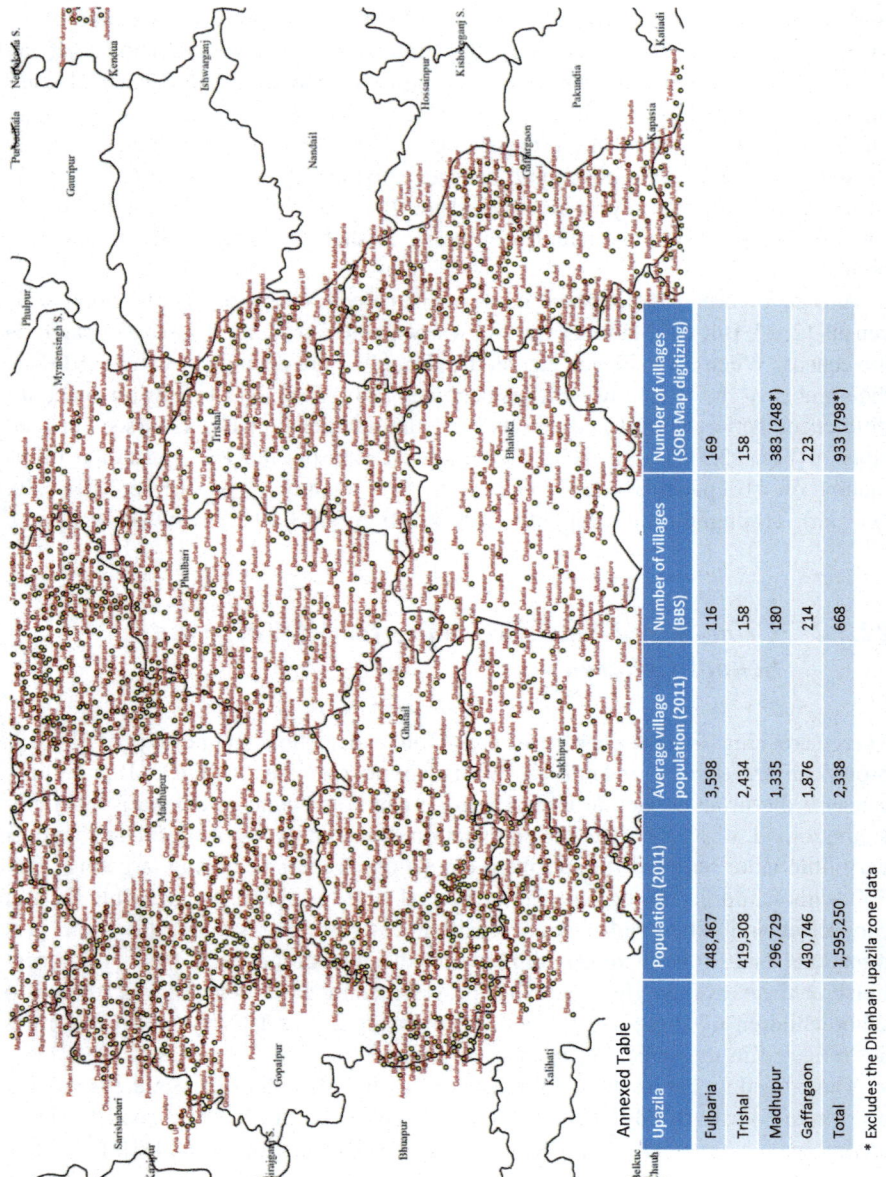

Fig. 22.3 Result of digitizing the SOB Map and comparison between census (2011). The village name itself can be found in the population census however, the population census doesn't include data on the location of each village. In the population census, the name of a village listed in patient log book is often not found. Instead, a topographical map data of 1:50,000 scale made by the SOB was digitized and a total of 933 spots of village coordinate data was prepared

Annexed Table

Upazila	Population (2011)	Average village population (2011)	Number of villages (BBS)	Number of villages (SOB Map digitizing)
Fulbaria	448,467	3,598	116	169
Trishal	419,308	2,434	158	158
Madhupur	296,729	1,335	180	383 (248*)
Gaffargaon	430,746	1,876	214	223
Total	1,595,250	2,338	668	933 (798*)

* Excludes the Dhanbari upazila zone data

diluted positive sera. The cutoff value was 57.9 U. The cut-off point for anti-rKRP42 IgG was first calculated as the mean plus 3 standard deviations (SD) of log (unit + 1) values of the nonendemic healthy controls, and [antilog of (the mean + 3 SDs) − 1] was regarded as the cut-off unit.

Mass screening was carried out for 1574 schoolchildren (primary school) from 8 schools in the beginning of August, 2015. Each schoolchild's personal background (full name, sex, school year, parents' full name, VL medical history, and history of treatment) was recorded in addition to the collection of urine specimens. A standard function (questionnaire form making) of Adobe Acrobat 11 standard software was used, and the necessary number of electronic questionnaire forms was prepared. This electronic questionnaire form was transferred to a tablet device, and the responses to questions for each schoolchild in the screening was recorded on this form. Microsoft Excel 2013 and IBM SPSS 22 were used for summarizing and statistical analysis of the questionnaire data and urine antibody measurements.

22.4 Results

22.4.1 Geography of VL Endemic in Research Area

After we compared the names of the villages of the 2436 VL patients with the names of the villages in the GIS data, 2114 cases (87 %) were collated. There were 2091 cases from the 4 upazila for which the mapping was intended, and the remaining 23 cases were patients visiting from outside the 4 upazila.

According to the population census of 2011, the population size of each village in the 4 upazila was between 1335–3598 (average 2338) (Fig. 22.3 annexed table). Yearly cases of more than 5 per village were approximately equal to an incidence rate (IR) of more than 14–37 (21.4 average) per 10,000 population (derived from the average village population from the census), and this rate is near the average IR (17.82–18.25) of Fulbaria and Trishal in 2008–2013 [1]. In these 4 upazila, the VL outbreak during 2011–2015 was primarily observed in 2011 because 51.6 % of cases were recorded during this year. In this epidemic year, 81.9 % of cases occurred at villages where ≥5 cases were observed. Sixty-seven of these hyperendemic villages (≥5 cases) existed in 2011 (Fig. 22.4).

As a result of mapping the patient distribution (seen in Fig. 22.4), it becomes clear that the hyperendemic villages are located side by side and they form clusters of hyperendemic zones. Large hyperendemic zones are observed in 6 places, 4 of the 6 are in the Fulbaria upazila and 2 of the 6 are in the Trishal upazila. In 2012, the yearly total cases decreased by 1/3 and the area of the hyperendemic zones also decreased. In 2014, the yearly total cases became nearly 1/6 and the hyperendemic zone almost disappeared. It is clear at a glance that endemic-free zones (area formed by villages where no incidence occurred) greatly spread until 2014. There are 649 (69.6 %) villages where no VL cases occurred during this 5 year period.

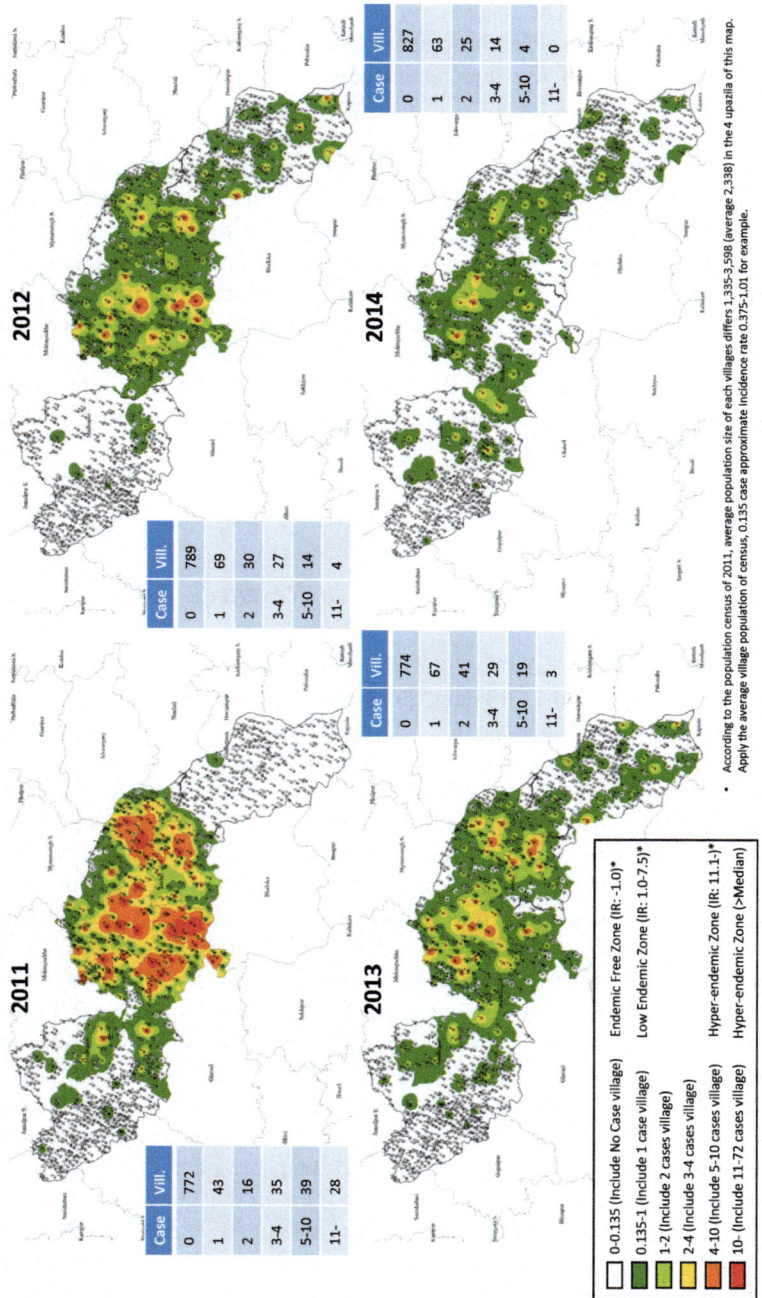

Fig. 22.4 Geographical distribution of UHC patient per year (2011–14). In these 4 upazila, the VL outbreak during 2011–2015 was primarily observed in 2011 because 51.6 % of cases were recorded during this year. In this epidemic year, 81.9 % of cases occurred at villages where ≥ 5 cases were observed. Sixty-seven of these hyperendemic villages (≥ 5 cases) existed in 2011

22.4.2 Detailed Geographical Distribution of VL Patients in the Research Area

From 2013 to the end of 2015, a total of 723 patients with VL were treated at SKKRC. Among these patients, 56.2 % came from endemic upazila such as Fulbaria, Trishal, Gaffargaon, Ghatail, Madhupur, Muktaggacha, and Jamalpur. Figure 22.5 is the geographical distribution of the patients treated at SKKRC. When plotting the patient locations on the map the clustering setting threshold was set at 500 m.

The characteristics of the patient distribution is as follows: first, most patients were distributed within approximately 50 km of SKKRC; and second, known endemic upazila are distributed over the lower left side of the circle surrounding SKKRC, and the upper right side of the circle is devoid of endemic upazila. However, patients were distributed relatively equally in the lower left area. This was in contrast with the large difference in incidence between upazila in the report of the CDC. By the data of the CDC (2011), if the patient numbers of the UHC of Jamalpur and Ghatail was assumed as 1, Madhupur and Muktagacha were equivalent to approximately 4 times, and more than 16 times in case of Fulbaria and Trishal. As a result, in upazila where the UHC-based incidence is low, the ratio of the SKKRC patients to the total incidence of upazila becomes large.

In Fig. 22.6, the geographical distribution of the UHC patients (Fig. 22.4) was recounted in the period of 2013–2015 ($n = 652$) and was put into the geographical distribution of the SKKRC patients ($n = 303$) of the same period. All incidence data ($n = 303$) of SKKRC were gathered into 177 clusters, and 66.1 % of these clusters were spatially sporadic (incidence per cluster $= 1$) patient occurrences. The spots where patients concentrated to some extent (incidence per cluster ≥ 3) was 17 % (30 places). The positions of such large clusters were scattered throughout Trishal, Ghatail, Madhupur, and the Muktagaccha upazila. Fulbaria upazila was the highest of the UHC patient distribution, but most of the large clusters of SKKRC patients were not distributed here.

22.4.3 Case Study of Mass Screening Targeting Unknown Endemic Villages

Based on the geographical tendency of new VL endemics seen in SKKRC patient data, fact-finding of VL endemics in these unknown endemic villages was performed. Specifically, a total of 8 primary schools were chosen, and urine-based mass screening was performed ($n = 1574$). Seven schools were located in 5 endemic villages where UHC or SKKRC patient clusters were observed; 2 of the 7

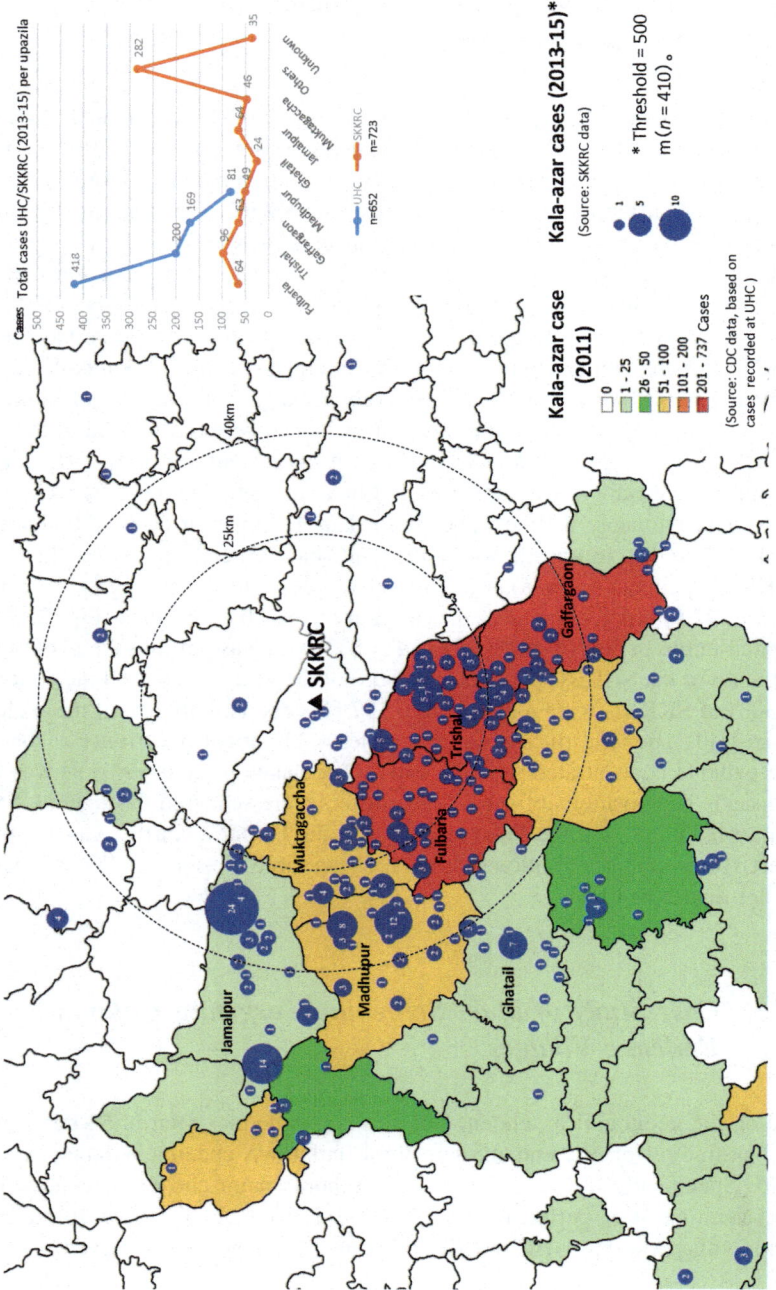

Fig. 22.5 Geographical distribution of SKKRC patients (2013–15). When plotting the patient locations on this map the clustering setting threshold was set at 500 m. From 2013 to the end of 2015, a total of 723 patients with VL were treated at SKKRC. Among these patients, 56.2 % came from endemic upazila such as Fulbaria, Trishal, Gaffargaon, Ghatail, Madhupur, Muktaggacha, and Jamalpur

schools were located in known endemic villages (MG, BR) in the Trishal upazila. Only the NH school was located in the city area of Mymensingh. The MH school was selected as a representative of a very low incidence or endemic free area. See Fig. 22.6 for the position of all 8 schools.

Figure 22.7 is a result of mass screening, the total positive rate was 14.7 %. Only 18 schoolchildren had an antibody titer of more than 1000 U. It is rare that a person with a titer lower than 1000 U develops VL immediately. Therefore, for these 211 positive schoolchildren, mild or asymptomatic infections are suspected.

The hospital based incidence (2013–2015) for each village is quite different, but becomes approximately flat (0.3–0.4 %) when the population of the village is taken into account (see Fig. 22.7 annexed table). However, in most elementary schools, the patients with VL among the schoolchildren's family/precinct population was much larger than the hospital-based number of patients in that village/village population (especially at MG, where they differed by 0.3–5.2 %). The survey results counted the number of patients over a long period, but in contrast, hospital-based incidence is recorded only after 2013. Therefore, the number of patients before 2013 will be reflected in the gap between the (1) and (2) percentages of the annexed table of Fig. 22.7. The number of neglected cases (data loss at hospitals or patients who were treated at any place except the public hospital) also may be reflected in this gap.

In JY village, the number of interfamilial patients as reported by the schoolchildren was 38.8 % of the numbers of the real patients in all the schoolchildren's families, this number was clarified by field survey data (Table 22.1). The 8/15 patients listed in the questionnaire data could not be found, however, 11 interfamilial patients of the schoolchildren who were not listed in the questionnaire data were newly found by this field survey.

The highest urine antibody-positive rate was 23.2 % for JY. Other than for MG, GJ+MK 19–21 %, BR 12.9 %, and JG+ST 8.5 %, all positive rates were relatively high. The mean of the antibody titer each school was in the following order: JY > GJ+MK > BR > JG+ST > MG > NH. It is well known that a high-level VL endemic has continued in the MG and BR precinct for a long time. In MG in particular, 71 interfamilial patients (5.2 % of the population) were listed in the questionnaire data. In other words, transmission of the protozoan frequently occurred between inhabitants of MG and BR. This might be reflected in the high percentage of rKRP42 positive results.

However, JY and MK+GJ were the neglected (or new) endemic villages that can be found out because of the SKKRC epidemiological data. However, judging from the urine antibodies of the schoolchildren, transmission of the protozoan between inhabitants in JY and MK+GJ precinct is at the same level as that seen in MG and BR. JY in particular is the precinct where the mean urine antibody titer is the highest. Transmission there may be taking place at a higher frequency than that seen at MG and BR. It will be important to pay attention to future patient outbreaks

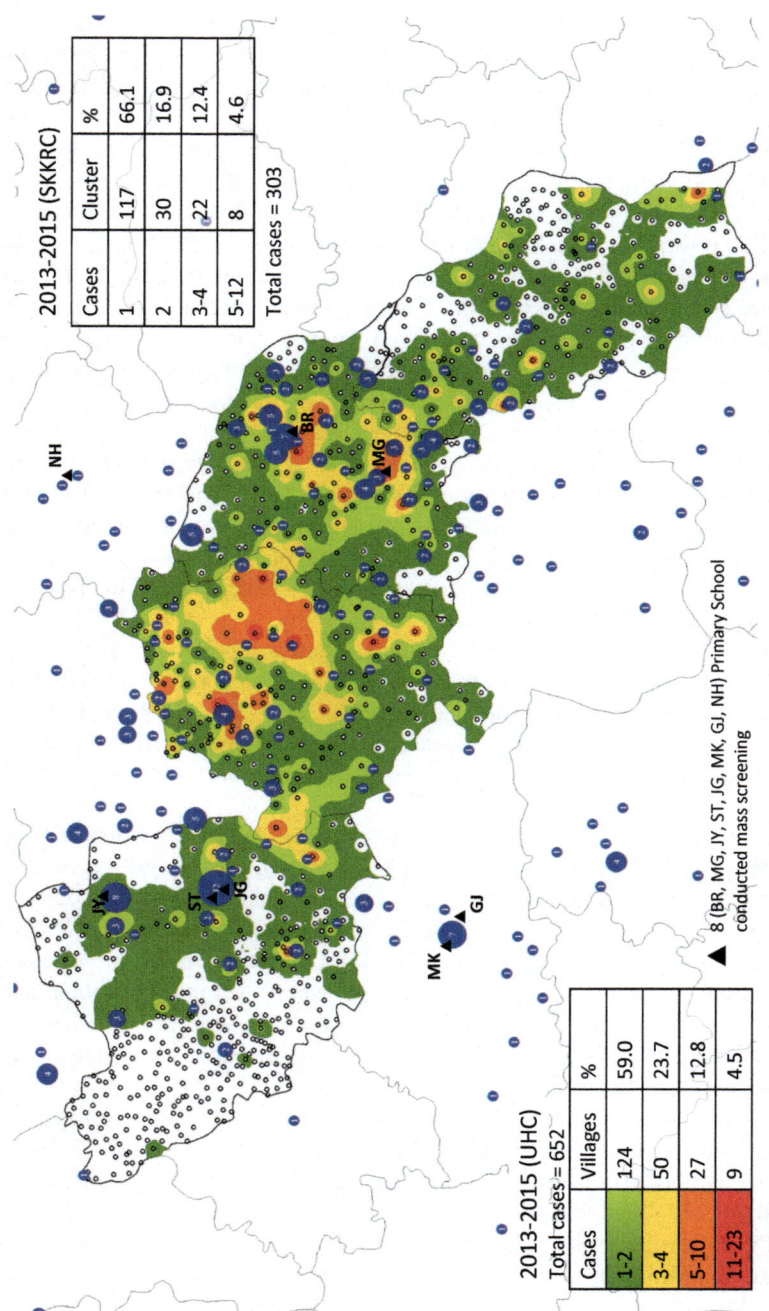

Fig. 22.6 Geographical distribution of the UHC and SKKRC patients (2013–2015). A total of 8 primary schools were chosen, and urine-based mass screening was performed (n = 1574). Seven schools were located in 5 endemic villages where UHC or SKKRC patient clusters were observed; 2 of the 7 schools were located in known endemic villages (MG, BR) in the Trishal upazila

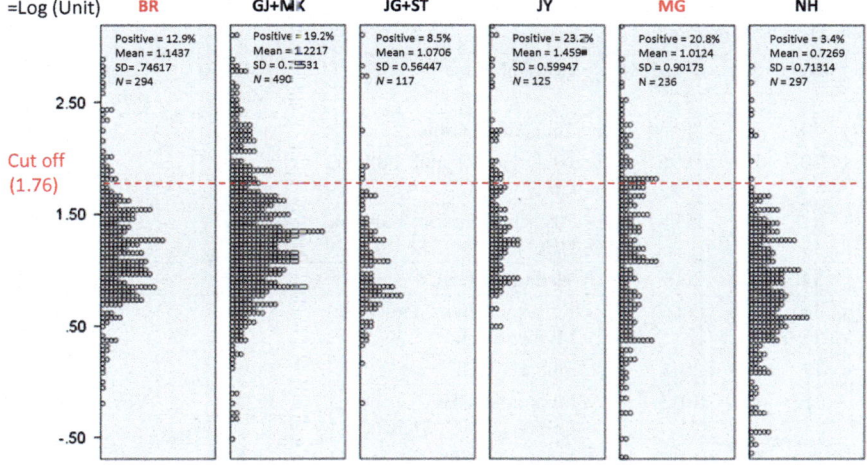

Fig. 22.7 Results of mass screening. The total positive rate was 14.7 %. Only 18 schoolchildren had an antibody titer of more than 1000 U. It is rare that a person with a titer lower than 1000 U develops VL immediately. Therefore, for these 211 positive schoolchildren, mild or asymptomatic infections are suspected

in the JY and MK+GJ precincts. In the MK+GJ precinct, there were fewer patients in comparison with the high urine antibody positive rate. It is possible that there are many reservoirs of VL transmission (incomplete treatment or asymptomatic patients) who have not yet been found in this precinct.

22.5 Discussion

There are many differences between each UHC in the quality of their patient record systems for VL. The record system is relatively good in hyperendemic upazila such as Fulbaria and Trishal. However, there are many differences in the precision of the records between other upazila. The geographical distribution of VL endemic areas was mapped from information of 4 upazila with considerable precision. As shown in Table 22.1, some endemic villages may be overlooked by the data loss at UHC.

Table 22.1 Real numbers of the patients[*] in all schoolchildren families according the field survey in the JY school precinct[**]

No.	Age	Sex	Treatment year	Hospital name	SKKRC data	Questionnaire data
1	70	F	–	Thanarbaid Ginic		
2	72	F	1996	Jalchatra Hospital, Fulbaria UHC		
3		M	1998	Pirgacha Mission, Fulbaria UHC		
4	54	F	2007	Madhupur UHC		
5	30	F	2009	Kaliakor Govt. Medical in Mymensirigh		
6	7	F	2011	Fulbaria UHC		
7	35	M	2013	Madhupur UHC, Fulbaria UHC, SKKRC		Yes
S	38	F	2013	Madhupur UHC, Fulbaria UHC		
9	30	M	2014	Madhupur UHC, Fulbaria UHC, SKKRC	Yes	
10	20	M	2014	SKKRC	Yes	Yes
11	15	M	2014	SKKRC	Yes	Yes
12	35	M	2014	SKKRC	Yes	Yes
13	6	M	2014	Madhupur UHC, Fulbaria UHC, SKKRC	Yes	Yes
14	30	M	2014	Fulbaria UHC		
15	40	M	2015	Mymensingh Medical Hospital		
16	21	M	2015	Fulbaria UHC		Yes
17	23	M	2015	Madhupur UHC, Fulbaria UHC, SKKRC	Yes	
18	40	M	2013–2015	Madhupur UHC, Fulbaria UHC, SKKRC		Yes

In JY village, the number of interfamilial patients as reported by the schoolchildren was 38.8 % of the numbers of the real patients in all the schoolchildren's families. The 8/15 patients listed in the questionnaire data could not be found, however, 11 interfamilial patients of the schoolchildren who were not listed in the questionnaire data were newly found

Source Field Survey Data (2016.1.21–27)

[*]Patient ratio per population is 2.3 % (18/766)

[**]Total population of JY precinct (number of all family member of schoolchildren} is 766 according to the summary of the questionnaire (mass screening)

This kind of geographical-epidemiological information will be important social information for infrastructures when thinking about long-term VL endemic control in this country. The urgent task at this time is increasing the quality of the information recorded at the time of patient registration in every UHC. At the same time, if urine-based mass screening systemically links the registration of the patient with VL to a hospital such as the UHC and SKKRC, an excellent monitoring system for VL will be implemented. As soon as a patient outbreak is confirmed at these hospitals, quantification of the degree of transmission between the inhabitants in the patient outbreak location will be provided by this system. It is hoped implementation of this system will provide unprecedented VL endemic control.

Acknowledgments This study was fully supported by the Science and Technology Research Partnership for Sustainable Development (SATREPS) project, Titled Research and Development of Prevention and Diagnosis for Neglected Tropical Disease, Especially VL, Japan.

References

1. Rajib C, Dinesh M, Vashkar C, Shyla F, et al. How far are we from Visceral Leishmaniasis eliminaton in Bangladesh? An assessment of epidemiological surveillance data. PLoS Negl Trop Dis. 2014;8(8):e3020. doi:10.1371/journal.pntd.0003020.
2. BBS. Population & Housing Census-2011 Community Report: Mymensingh. http://www.bbs.gov.bd/WebTestApplication/userfiles/Image/PopCen2011/Com_Mymensingh.pdf.
3. BBS. Population & Housing Census-2011 Community Report: Tangail. http://www.bbs.gov.bd/WebTestApplication/userfiles/Image/PopCen2011/C_Tangail.pdf.
4. Takagi H, Islam MZ, Itoh M, et al. Production of recombinant kinesin-related protein of Leishmania donovani and its application in the serodiagnosis of visceral leishmaniasis. Am J Trop Med Hyg. 2007;76:902–5.
5. Islam MZ, Itoh M, Takagi H. et al. Enzyme-linked immunosorbent assay to detect urinary antibody against recombinant rKRP42 antigen made from Leishmania donovani for the diagnosis of visceral leishmaniasis. Am J Trop Med Hyg. 2008;79(4):599–604.
6. Islam MZ, Itoh M, Ul Islam MA, et al. ELISA with recombinant rKRP42 antigen using urine samples: a tool for predicting clinical visceral leishmaniasis cases and its outbreak. Am J Trop Med Hyg. 2012;87(4):658–62.

Chapter 23
Animal Models of Visceral Leishmaniasis and Applicability to Disease Control

Yasutaka Osada, Satoko Omachi, Chizu Sanjoba and Yoshitsugu Matsumoto

Abstract Visceral leishmaniasis (VL), also called kala-azar, is the most severe form of leishmaniasis and is often fatal if not treated. The disease is caused by the *Leishmania donovani* complex, which includes *L. donovani* and *L. infantum/chagasi*. Because of a lack of or limited availability of vaccines or chemotherapeutics for VL, extensive preclinical studies using various animal models have been undertaken to explore therapeutic strategies to overcome the disease. Mice, hamsters, dogs, and non-human primates have been studied, each of which has advantages and disadvantages depending on the purpose. Herein, we review the experimental models of VL, with an emphasis on murine models for *L. donovani* infection.

Keywords Animal model · Kala-azar · *Leishmania donovani* · Visceral leishmaniasis · Mouse model · Progressive

23.1 Visceral Leishmaniasis

Many aspects of visceral leishmaniasis (VL) are described in other chapters in this book, so it will be mentioned here only briefly. VL, also called kala-azar, is the most severe form of leishmaniasis caused mainly by *L. donovani* and *L. infantum/chagasi*. Its distribution is widespread in the Indian subcontinent, Africa, Europe, and Latin America. Of the 0.2–0.4 million new cases of VL every year, 90 % occur in Bangladesh, India, Brazil, Ethiopia, and Sudan [1]. There are 2 types of VL based on transmission characteristics: anthroponotic VL caused by *L. donovani* [2], and anthropozoonotic VL caused by *L. infantum* [3]. VL is clinically characterized by prolonged fever, weight loss, anemia, and hepatosplenomegaly, with an

Y. Osada (✉)
Laboratory of Molecular Immunology, Department of Animal Resource Sciences, Graduate School of Agricultural and Life Sciences, The University of Tokyo, 1-1-1 Yayoi, Bunkyo-ku, Tokyo 113-8657, Japan
e-mail: aosada@mail.ecc.u-tokyo.ac.jp

© Springer International Publishing 2016
E. Noiri and T.K. Jha (eds.), *Kala Azar in South Asia*,
DOI 10.1007/978-3-319-47101-3_23

incubation period varying between 2 and 6 months [4]. VL has a high mortality rate in untreated cases. Animal model testing is necessary to study the pathology of VL and to evaluate anti-leishmanial drugs and vaccines.

Animal models are expected to mimic the pathological features and immunological responses observed in humans when exposed to a variety of *Leishmania* species with different pathogenic characteristics. Several experimental models of VL have been developed, but none of these entirely reproduce the disease state in humans [5]. For in vivo testing of new compounds, several animal species have served as experimental hosts for VL. Important among them are mice, Syrian golden hamsters, dogs, and non-human primates [6]. A suitable laboratory host for infection of the *L. donovani* complex is very important for conducting research on various aspects including host-parasite interactions, pathogenesis, prophylaxis, and evaluation of anti-leishmanial compounds.

23.2 Mouse Model for VL

Due to multiple advantages like ease of housing, presence of established genetic modification methods, and the best availability of immunological reagents, mice are the most user-friendly animal models for infectious diseases, including VL. Well known for the Th1/Th2 paradigm in *L. major* infection [7–12], inbred mice, including BALB/c, C57BL/6, and CBA, have been used for understanding responsible genes for susceptibility/resistance against *L. donovani* infection [13–15]. For example, identification of the *Slc1 la1* gene aided in the understanding of susceptibility at early stages of infection in BALB/c mice, which reflects the strength of the innate immune response in controlling early parasite growth independently of acquired immune mechanism [13, 14]. Genetically resistant mouse strains (e.g., CBA) possess a functional *Slc1 la1* gene, which confers innate resistance to early *Leishmania* parasite growth. In contrast, susceptible mice strains (e.g., C57BL/6 and BALB/c) possess a nonfunctional *Slc1 la1* gene and early parasite growth in the liver cannot be controlled [16]. In addition, mice with artificial gene modifications have served as good tools to understand the role of a particular gene or its product in immunity against *L. donovani* infection, such as IFN-γ and TNF-α for protection, and IL-10 for exacerbation [17]. In fact, protective immunity in murine VL primarily depends on the IL-12-driven Th1 response leading to an increase of IL-2 and IFN-γ production resulting in death of the parasite [17–21].

When various inbred strains of mice are compared, BALB/c mice seem to be the most susceptible to *L. donovani* infection [22]. However, even the most susceptible strain develops acquired immunity to control hepatic parasite growth at later stages of infection [23, 24], while showing relatively persistent infection in the spleen [18]. There is an early increased parasite burden in the liver, but clearance of hepatic parasites occurs at 4–8 weeks after infection due to efficient granuloma formation resulting from interaction of *Leishmania*-specific T cells with parasitized resident macrophages [24]. Hence, disease progression in mice can be predicted from the degree of maturation of hepatic granulomas in challenged animals correlating with cell-mediated immunity [19, 25]. In contrast to the liver response, the spleen acts as a site of parasite persistence resulting in chronic infection in mice. Failure of parasitized splenic marginal zone macrophages to produce IL-12, breakdown of splenic architecture, and lack of granulomatous reaction leads to a high parasite burden in the spleen [26]. For an experimental challenge of mice with *L. donovani/L. infantum*, amastigotes isolated from the spleens of infected hamsters are commonly used as infective materials [27–30].

A mouse model of persistent infection with *L. donovani* facilitates evaluation of certain drugs and immunotherapeutics that may require a period of time to act to control parasite burden. We recently developed a mouse model of *L. donovani* infection that shows parasite persistence in both the spleen and liver. The *L. donovani* D10 strain, which was originally isolated from a Nepalese patient with VL [31], was inoculated intraperitoneally with 1×10^8 promastigotes. At 2, 4, 8, and 12 weeks postinfection BALB/c mice were euthanized, and parasite burdens in the spleen and liver were quantitated as Leishman Donovan units (LDU) in organ stamp smears [32]. The LDU increased progressively in the spleen and liver over this time period (Fig. 23.1a, b). The spleen LDU was 805 ± 121 and the liver LDU was $3,139 \pm 489$ at 12 weeks after infection. *L. donovani*-infected BALB/c mice had enlarged spleens and livers compared with those of the uninfected (Fig. 23.1c). Splenomegaly and hepatomegaly in BALB/c mice caused by *L. donovani* infection were progressive over the 12 weeks of infection (Fig. 23.1d, e). In mice infected for 12 weeks, the spleen weights were approximately 8 times, and the liver weights enlarged approximately 1.4 times, higher compared with those of uninfected mice.

Infection with *L. donovani* D10 brings different outcomes in BALB/c, C57BL/6, and C3H/HeN mice. In contrast to the continuous increase of parasite burden seen in BALB/c mice, C57BL/6 mice showed resistance to the parasite (Fig. 23.2a, b). Splenomegaly and hepatomegaly in BALB/c mice caused by *L. donovani* infection were progressive over the 24 weeks of infection (Fig. 23.2c, d). In C57BL/6 mice, the spleen and liver LDU at 12 weeks was much lower than that of BALB/c mice (Fig. 23.2a, b). Moreover, the liver LDU declined between 12 and 24 weeks, and no parasites were detected in the liver at 24 weeks after infection. Although parasites in the spleen were detectable at 24 weeks in C57BL/6, the parasite burdens did not show any increase over time as seen in BALB/c mice. The spleen and liver weights also showed different kinetics over 24 weeks of *L. donovani* infection between BALB/c mice and C57BL/6 mice. In conjunction with parasite burdens, the development of pathologic features was also limited in C57BL/6 J mice. For the spleen, BALB/c and C57BL/6 mice manifested enlargement of the tissue at 12 weeks. In contrast to BALB/c mice, however, C57BL/6 mice did not show progressive splenomegaly at 24 weeks of infection (Fig. 23.2c). For the liver, tissue enlargement was not found in C57BL/6 mice during the experimental period (Fig. 23.2d).

In contrast to the susceptible BALB/c mice and C57BL/6 mice with self-limiting disease, in C3H/HeN mice there were no parasites in either the spleen or liver at 12 weeks of *L. donovani* infection. In parallel to the absence of parasites, neither

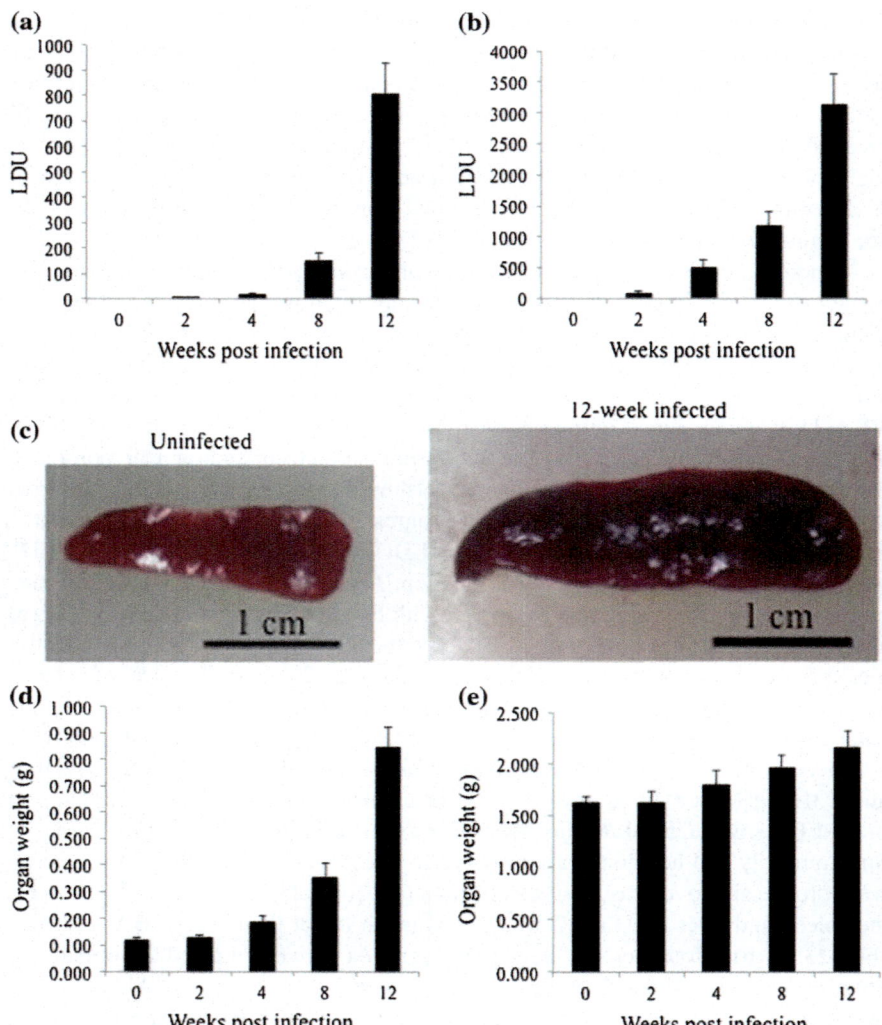

Fig. 23.1 Continuous increase of parasite burdens in the spleen (**a**) and liver (**b**) of *L. donovani*-infected BALB/c mice. Progressive hepato-splenomegaly in *L. donovani*-infected BALB/c mice. Representative images of the spleens from uninfected mice and 12-week infected mice (**c**). Means and standard deviations (SDs) of the spleen (**d**) and liver (**e**) weights over the course of infection in BALB/c mice infected with *L. donovani* are shown. Mean and standard deviations (SDs) at each time point are shown. $P < 0.01$ by non-parametric one-way analysis of variance (ANOVA). LDU: Leishman Donovan units

the spleen nor liver of infected C3H/HeN mice was significantly larger than those of uninfected mice. Resistance/susceptibility to *L. donovani* infection in those inbred mice was associated with different Th1/Th2 balances. When productions of IFN-γ

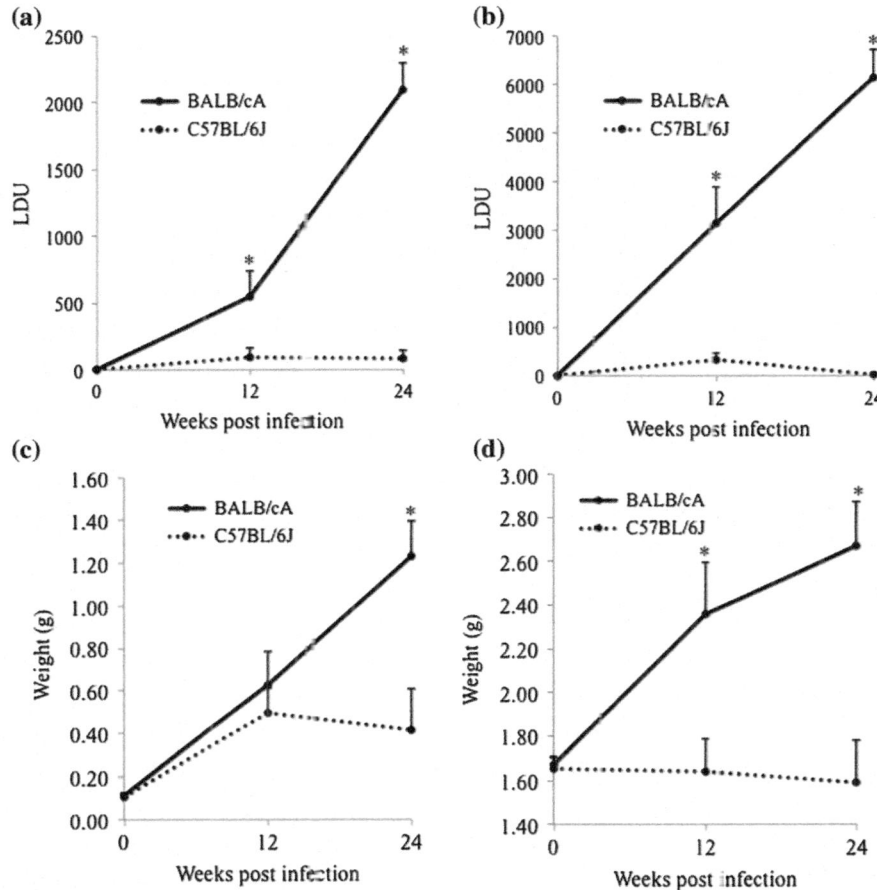

Fig. 23.2 The spleen (**a**) and liver (**b**) parasite loads in *L. donovani*-infected BALB/c and C57BL/6 mice. Enlargement of spleen (**c**) and liver (**d**) weight in *L. donovani*-infected BALB/c and C57BL/6 mice. Means and standard deviations (SDs) in each group are shown. *$P < 0.01$. by Student's *t*-test. LDU: Leishman Donovan units

and IL-4 by splenocytes upon antigen recall were analyzed in *L. donovani*-infected mice at 12 weeks after infection, C3H/HeN (5.34 ± 1.18 ng/mL) and C57BL/6 (3.93 ± 2.18 ng/mL) mice showed higher IFN-γ levels than did BALB/c (1.93 ± 0.26 ng/mL) mice in response to *Leishmania* antigen (Fig. 23.3a). In *Leishmania* antigen-induced IL-4 production, BALB/c (47.31 ± 18.84 pg/mL) mice produced significantly higher IL-4 levels than did C3H/HeN (11.55 ± 5.65 pg/mL) and C57BL/6 (3.06 ± 2.46 pg/mL) mice (Fig. 23.3b). Together, these results suggest that our mouse model of *L. donovani* infection is advantageous in evaluating the long-term efficacy (up to 24 weeks of infection) of novel interventions as well as immunological studies on protective mechanisms.

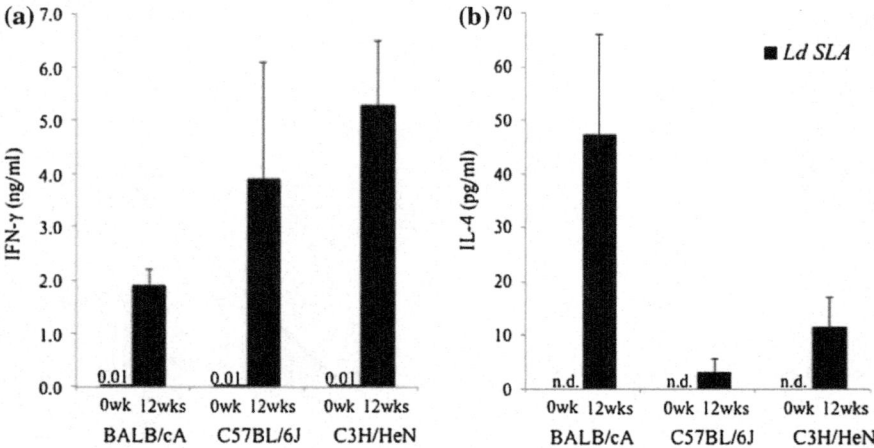

Fig. 23.3 *Leishmania* antigen-induced IFN-γ (**a**) and IL-4 (**b**) production by spleen cells from BALB/c, C57BL/6, and C3H/HeN mice at 12 weeks after experimental *L. donovani* infection and non-infection. Means and standard deviations (SDs) in each group are shown. n.d., not detected

23.3 Hamster Model for VL

Compared to investigations in mice, experimental studies in *L. infantum* and *L. donovani*-infected Syrian golden hamsters (*Mesocricetus auratus*) often reveal several clinical signs of progressive VL (hypergammaglobulinemia, hepatosplenomegaly, anemia, cachexia, and immunosuppression) that closely mimic active canine and human disease [5, 33–36]. Surprisingly, there are significant amounts of Th1 cytokines (IFN-γ, IL-2, and TNF-α) in the spleen, but there is little or no IL-4 [35, 37]. However, deactivating Th2 cytokines (TGF-β and IL-10) may act on infected macrophages as well as anti-*Leishmania* antibodies (which have no protective role in leishmaniasis) that opsonize amastigotes to allow the parasites to multiply and induce IL-10 production in macrophages [35, 37]. These high activation and deactivation processes are likely to occur mainly in the spleen and liver [38]. Syrian hamsters exhibit reduced expression of the gene encoding inducible nitric oxide synthase (iNOS) in response to IFN-γ, and this is thought to lead to low nitric oxide (NO) generation, subsequently defaulting in parasite killing [17, 35, 38]. Increased secretion of TNF-α during later stages of the disease probably results in profound weight loss in infected hamsters [39]. Despite serving as a good model for VL pathologic features, hamsters are not widely used because a lack of available reagents such as antibodies to cell markers and cytokines. Only recently RT-PCR primers for some cytokines have been developed [35, 37], and when these primers were used, no quantitative change in the expression of different cytokine RNA was observed during VL in hamsters, suggesting the necessity to develop a more sensitive method for evaluation, such as quantitative PCR.

23.4 Dog Model and Non-human Primate Model

Dogs have been used as an experimental model for *Leishmania* infections from the beginning of the 20th century and experimental infections have also been achieved with *Leishmania* species for which the dog is not a natural reservoir, e.g., *L. donovani* from India [40]. Infection of dogs with *L. infantum/chagasi* is an important laboratory model because it reproduces a natural infection similar to human infections [41]. Studies on experimentally infected dogs have demonstrated that 3 years after infection, asymptomatic or resistant dogs responded to *L. infantum* antigen both in lymphocyte proliferation assays in vitro and delayed-type hypersensitivity reactions, whereas no serum antibodies to parasite antigens were detected [42]. In contrast, symptomatic or susceptible animals failed to respond to the parasite antigen in cell-mediated assays both in vitro and in vivo and showed considerably higher serum antibodies to leishmanial antigens, which are not immuno-protective [42]. In addition, peripheral mononuclear cells from asymptomatic dogs produced significantly higher levels of IL-2 and TNF-α than did those from symptomatic and control uninfected dogs [43]. Similar results were observed with a group of mixed-breed dogs with natural *Leishmania* infections, also grouped as asymptomatic or symptomatic on the basis of clinical signs of canine visceral leishmaniasis [42].

Some of the observations made in rodent models might not be similar or relevant to findings for human hosts due to the distance in phylogeny. The development of a non-human primate model of leishmaniasis, which largely mimics the human situation, is described for studies of different aspects of the disease that would not be possible in humans for ethical reasons. This would also complement studies in other model systems. However, for financial and ethical reasons, the use of primates in biomedical research is limited. Monkeys are normally the final experimental animals to be used in studies of the safety and efficacy of drugs and vaccines developed in other laboratory animals. Earlier efforts in establishing VL in New and Old World monkeys demonstrated that owl monkeys [44] and squirrel monkeys [45] developed an acute and fulminating, but short-lived, infection. Anti-leishmanial screening was performed in owl and squirrel monkeys. Old World monkeys developed low and/or inconsistent infections [33]. A single intravenous inoculation of hamster-spleen-derived *L. donovani* amastigotes into Indian langurs produced consistent and progressive infection, all the clinicoimmunopathological features as observed in VL, and death between 110–150 days after infection [46, 47].

23.5 Conclusions

VL has yet to be controlled in many countries in spite of its severity, and a good animal model of VL will facilitate accelerating research and drug development for the disease. According to previous studies, the most widely studied animal model of

VL is *L. donovani* infection in BALB/c mice. There is an early increase in parasite burden, but particularly over the course of 4–8 weeks the *Leishmania* parasites are eliminated in the liver and the infection is controlled [16, 21, 32]. In contrast, the clinic-pathological features of the hamster model of VL closely mimic active VL [5, 33–35, 48, 49]. However, the widespread use of hamsters is limited due to a lack of available reagents such as antibodies to cell markers and cytokines. Therefore, a mouse model like ours that combines the strength of both models, i.e., user friendliness of mice and the clinical reproducibility of hamsters, will contribute to understanding the pathogenesis, to improving immunological studies, and to developing new candidate anti-leishmanial drugs and vaccines.

References

1. Alvar J, et al. Leishmaniasis worldwide and global estimates of its incidence. PLoS ONE. 2012;7(5):e35671.
2. Joshi A, et al. Can visceral leishmaniasis be eliminated from Asia? J Vector Borne Dis. 2008;45(2):105–11.
3. Mondal S, Bhattacharya P, Ali N. Current diagnosis and treatment of visceral leishmaniasis. Expert Rev Anti Infect Ther. 2010;8(8):919–44.
4. Chappuis F, et al. Visceral leishmaniasis: what are the needs for diagnosis, treatment and control? Nat Rev Microbiol. 2007;5(11):873–82.
5. Handman E. Leishmaniasis: current status of vaccine development. Clin Microbiol Rev. 2001;14(2):229–43.
6. Gupta S. Visceral leishmaniasis: experimental models for drug discovery. Indian J Med Res. 2011;133:27–39.
7. Chatelain R, Varkila K, Coffman RL. IL-4 induces a Th2 response in Leishmania major-infected mice. J Immunol. 1992;148(4):1182–7.
8. Heinzel FP, Sadick MD, Holaday BJ, Coffman RL, Locksley RM. Reciprocal expression of interferon gamma or interleukin 4 during the resolution or progression of murine leishmaniasis. Evidence for expansion of distinct helper T cell subsets. J Exp Med. 1989;169(1):59–72.
9. Heinzel FP, Sadick MD, Mutha SS, Locksley RM. Production of interferon gamma, interleukin 2, interleukin 4, and interleukin 10 by CD4 + lymphocytes in vivo during healing and progressive murine leishmaniasis. Proc Natl Acad Sci USA. 1991;88(16):7011–5.
10. Locksley RM, Heinzel FP, Sadick MD, Holaday BJ, Gardner KD. Murine cutaneous leishmaniasis: susceptibility correlates with differential expansion of helper T-cell subsets. Ann Inst Pasteur Immunol 1987;138(5):744–749.
11. Sadick MD, Heinzel FP, Shigekane VM, Fisher WL, Locksley RM. Cellular and humoral immunity to Leishmania major in genetically susceptible mice after in vivo depletion of L3T4 + T cells. J Immunol. 1987;139(4):1303–9.
12. Sadick MD, et al. Cure of murine leishmaniasis with anti-interleukin 4 monoclonal antibody. Evidence for a T cell-dependent, interferon gamma-independent mechanism. J Exp Med. 1990;171(1):115–27.
13. Bradley DJ. Letter: genetic control of natural resistance to Leishmania donovani. Nature. 1974;250(464):353–4.
14. Blackwell JM. Genetic susceptibility to leishmanial infections: studies in mice and man. Parasitology. 1996;112(Suppl):S67–74.
15. Liew FY, O'Donnell CA. Immunology of leishmaniasis. Adv Parasitol. 1993;32:161–259.

16. Kaye PM, et al. The immunopathology of experimental visceral leishmaniasis. Immunol Rev. 2004;201:239–53.
17. Wilson ME, Jeronimo SM, Pearson RD. Immunopathogenesis of infection with the visceralizing Leishmania species. Microb Pathog. 2005;38(4):147–60.
18. Engwerda CR, Kaye PM. Organ-specific immune responses associated with infectious disease. Immunol Today. 2000;21(2):73–8.
19. Murray HW, et al. Acquired resistance and granuloma formation in experimental visceral leishmaniasis. Differential T cell and lymphokine roles in initial versus established immunity. J Immunol. 1992;148(6):1858–63.
20. Murray HW, Nathan CF. Macrophage microbicidal mechanisms in vivo: reactive nitrogen versus oxygen intermediates in the killing of intracellular visceral Leishmania donovani. J Exp Med. 1999;189(4):741–6.
21. Squires KE, et al. Experimental visceral leishmaniasis: role of endogenous IFN-gamma in host defense and tissue granulomatous response. J Immunol. 1989;143(12):4244–9.
22. Engwerda CR, et al. A role for tumor necrosis factor-alpha in remodeling the splenic marginal zone during Leishmania donovani infection. Am J Pathol. 2002;161(2):429–37.
23. Lipoldova M, Demant P. Genetic susceptibility to infectious disease: lessons from mouse models of leishmaniasis. Nat Rev Genet. 2006;7(4):294–305.
24. Stanley AC, Engwerda CR. Balancing immunity and pathology in visceral leishmaniasis. Immunol Cell Biol. 2007;85(2):138–47.
25. Carrion J, et al. Immunohistological features of visceral leishmaniasis in BALB/c mice. Parasite Immunol. 2006;28(5):173–83.
26. Nieto A, et al. Mechanisms of resistance and susceptibility to experimental visceral leishmaniosis: BALB/c mouse versus Syrian hamster model. Vet Res. 2011;42:39.
27. Miralles GD, Stoeckle MY, McDermott DF, Finkelman FD, Murray HW. Th1 and Th2 cell-associated cytokines in experimental visceral leishmaniasis. Infect Immun. 1994;62(3):1058–63.
28. Satoskar AR, et al. IL-12 gene-deficient C57BL/6 mice are susceptible to Leishmania donovani but have diminished hepatic immunopathology. Eur J Immunol. 2000;30(3):834–9.
29. Murray HW. Accelerated control of visceral Leishmania donovani infection in interleukin-6-deficient mice. Infect Immun. 2008;76(9):4088–91.
30. Maroof A, et al. Therapeutic vaccination with recombinant adenovirus reduces splenic parasite burden in experimental visceral leishmaniasis. J Infect Dis. 2012;205(5):853–63.
31. Pandey K, et al. Characterization of Leishmania isolates from Nepalese patients with visceral leishmaniasis. Parasitol Res. 2007;100(6):1361–9.
32. Murray HW. Cell-mediated immune response in experimental visceral leishmaniasis. II. Oxygen-dependent killing of intracellular Leishmania donovani amastigotes. J Immunol. 1982;129(1):351–7.
33. Hommel M, Jaffe CL, Travi B, Milon G. Experimental models for leishmaniasis and for testing anti-leishmanial vaccines. Ann Trop Med Parasitol 89 Suppl 1995;1:55–73.
34. Requena JM, Soto M, Doria MD, Alonso C. Immune and clinical parameters associated with Leishmania infantum infection in the golden hamster model. Vet Immunol Immunopathol. 2000;76(3–4):269–81.
35. Melby PC, Chandrasekar B, Zhao W, Coe JE. The hamster as a model of human visceral leishmaniasis: progressive disease and impaired generation of nitric oxide in the face of a prominent Th1-like cytokine response. J Immunol. 2001;166(3):1912–20.
36. Dea-Ayuela MA, Rama-Iniguez S, Alunda JM, Bolas-Fernandez F. Setting new immuno-biological parameters in the hamster model of visceral leishmaniasis for in vivo testing of antileishmanial compounds. Vet Res Commun. 2007;31(6):703–17.
37. Melby PC, Tryon VV, Chandrasekar B, Freeman GL. Cloning of Syrian hamster (Mesocricetus auratus) cytokine cDNAs and analysis of cytokine mRNA expression in experimental visceral leishmaniasis. Infect Immun. 1998;66(5):2135–42.
38. Goto H, Prianti M. Immunoactivation and immunopathogeny during active visceral leishmaniasis. Rev Inst Med Trop Sao Paulo. 2009;51(5):241–6.

39. Tracey KJ, et al. Cachectin/tumor necrosis factor induces cachexia, anemia, and inflammation. J Exp Med. 1988;167(3):1211–27.
40. Chapman WL Jr, Hanson WL, Waits VB, Kinnamon KE. Antileishmanial activity of selected compounds in dogs experimentally infected with Leishmania donovani. Rev Inst Med Trop Sao Paulo. 1979;21(4):189–93.
41. Rioux JA, Golvan YJ, Croset H, Houin R. [Leishmanioses in the Mediterranean "Midi": results of an ecologic survey] (Translated from fre). Bull Soc Pathol Exot Filiales. 1969;62 (2):332–3 (in fre).
42. Pinelli E, et al. Cellular and humoral immune responses in dogs experimentally and naturally infected with Leishmania infantum. Infect Immun. 1994;62(1):229–35.
43. Pinelli E, et al. Infection of a canine macrophage cell line with leishmania infantum: determination of nitric oxide production and anti-leishmanial activity. Vet Parasitol. 2000;92 (3):181–9.
44. Chapman WL Jr, Hanson WL, Hendricks LD. Toxicity and efficacy of the antileishmanial drug meglumine antimoniate in the owl monkey (Aotus trivirgatus). J Parasitol. 1983;69 (6):1176–7.
45. Chapman WL Jr, Hanson WL. Visceral leishmaniasis in the squirrel monkey (Saimiri sciurea). J Parasitol. 1981;67(5):740–1.
46. Anuradha et al. The Indian langur: preliminary report of a new nonhuman primate host for visceral leishmaniasis. Bull World Health Organ 1992;70(1):63–72.
47. Dube A, et al. Leishmania donovani: cellular and humoral immune responses in Indian langur monkeys, Presbytis entellus. Acta Trop. 1999;73(1):37–48.
48. Gifawesen C, Farrell JP. Comparison of T-cell responses in self-limiting versus progressive visceral Leishmania donovani infections in golden hamsters. Infect Immun. 1989;57 (10):3091–6.
49. Pearson RD, Roberts D. Host immunoglobulin on spleen-derived Leishmania donovani amastigotes. Am J Trop Med Hyg. 1990;43(3):263–5.

Chapter 24
Pharmacovigilance on Therapeutic Protocols for Visceral Leishmaniasis

Eisei Noiri, Bumpei Tojo, Yoshifumi Hamasaki, Masao Iwagami, Takeshi Sugaya, Michiyo Harada, Progga Nath, Ariful Basher, Dinesh Mondal, Rashidul Haque, Fasifur Rahman and Shyamal Paul

Abstract Because the primary goal in visceral leishmaniasis (VL) treatment is a cure of this deadly parasitic disease, post-marketing drug safety surveillance is often overlooked. Because overworked patients will not routinely return to the clinic, chronic sequelae of treated VL are not well understood. For example, high doses of drugs that are renally excreted can potentially induce kidney fibrosis. The team of SATREPS (Science and Technology Research Partnership for Sustainable Development) established a patient registration system for VL cases treated in the Surya Kanta Kala-azar Research Center (SKKRC), monitored those subjects periodically by sending field staff on site, and performed surveillance by using a simple urine biomarker kit to detect kidney injury. Herein, this chapter examines the prevalence of people with high urinary levels of L-FABP (FABP1) after VL treatment, which may indicate decreased kidney function. The risk varied according to the treatment protocols and presumably patients' treatment histories.

Keywords Drug monitoring · Amphotericin B · Liposomal amphotericin B · Paromomycin · Miltefosine · Inulin clearance · L-FABP · AKI biomarker

24.1 Introduction

Leishmania parasites belong to the eukaryote family and are phylogenetically closer to humans compared with other human-invading organisms such as bacteria and viruses. Therapeutic approaches to parasites are often harsh to the patient because the working concentration which is toxic to the parasite is nearly toxic to the host. The pharmacological approach to viruses targets the specific virus DNA synthesis and virus

E. Noiri (✉)
Department of Hemodialysis and Apheresis, The University of Tokyo Hospital, Tokyo, Japan; Department of Nephrology and Endocrinology, The University of Tokyo Hospital, Tokyo, Japan
e-mail: noiri-tky@umin.ac.jp

enzymatic activities. However, the approach to parasites differs from that for viruses, and this can be seen from amphotericin B (AMPH-B) [1], a leishmaniasis treatment. AMPH-B binds to ergosterol, a component of fungal and parasitic cell membranes, and causes increased membrane permeability that leads to fatal injury to *Leishmania donovani*. But at a higher concentration, AMPH-B can bind to cholesterol, a key component of mammalian cell membranes and cause similar harmful increases in permeability. For this reason, researchers have been seeking a parasite-specific mechanism that is not shared with mammals. One successful achievement was ivermectin to treat onchocerciasis, also known as river blindness, induced by a filarial nematode; cumulative doses decrease parasite fertility to achieve disease remission. Based on this discovery and achievement, Willian C. Campbell and Satoshi Omura received the 2015 Nobel Prize in Physiology or Medicine.

In addition to the aforementioned acute toxicity and side effects often seen during parasite treatments, the effects of long-term drug therapy should also be considered. After regulatory approval of the PPAR-gamma agonist, rosiglitazone, for type 2 diabetes by the FDA, the post-marketing surveillance reported a significant increase in cardiovascular events [2]. However, if patients are suffering from a life-threatening disease such as cancer or a severe infectious disease such as visceral leishmaniasis (VL), the initial goal is in curing the disease and saving lives. But once such goals are achieved, then the long-term effects of the treatment need to be reevaluated.

There were virtually no reports of drug safety monitoring after the treatment of neglected tropical diseases (NTDs), though a tremendous number of people are treated by the earlier-mentioned harmful drugs. This is because these patients have the lowest income and education within the population and often cannot visit clinics for health checkups. After completion of the initial therapy, they must return to work immediately, often at an unstable job with a low wage, and their address and mobile phone number are often changing. In this study, a surveillance team repeatedly visited patients after discharge from the Surya Kanta Kala-azar Research Center (SKKRC), using information about the initial patient location in the field. Herein, we show for the first time a drug-monitoring report after VL treatment.

24.2 Methods

We enrolled patients who were given diagnoses of VL in SKKRC in a patient registration system (UMIN000011426). Laboratory testing data ordered by physicians in SKKRC were recorded and urine was stored. Urine collection was also done at discharge from hospital. In this study, the surveillance team visited each patient on site and re-examined urine using an immunochromatography (ICT) method of L-FABP (Renischem L-FABP POC®; CMIC Holdings Co., Ltd., Tokyo, Japan) when it had been more than 3 months after completion of VL therapy. Patients who completed VL treatment within 3 months were not included because the current chronic kidney injury (CKD) definition requires decreased kidney function for 3 months. Subjects who showed urinary L-FABP positive, according to a cut-off level of 12.5 ng/mL in ICT, were called for further study, including serum creatinine measurement, ultrasound, and inulin clearance [3] to determine their actual renal condition. Serum creatinine, urine L-FABP (Nordia L-FABP®; Sekisui Medical Co., Ltd., Tokyo, Japan), and urine creatinine levels were measured using Biolis 24i Premium biochemistry analyzer (Tokyo Boeki Medisys Inc., Tokyo, Japan). Urinary protein levels were determined using a dip-test (N-Multistix SG-L, Siemens). We also checked inulin (Inulead inj.® Fujiyakuhin, Japan) clearance (Cin) for participants who agreed.

To better elucidate the characteristics of patients with positive L-FABP ICT, we made a comparison group: patients who were negative or not examined for L-FABP ICT during the follow up period but were positive at discharge from SKKRC. For these people, we measured levels of serum creatinine, urine L-FABP, urinary protein, and Cin. As another comparison, we also measured serum creatinine levels and urine L-FABP levels among healthy people who agreed to contribute to the survey in Mymensingh.

Finally, we conducted a receiver operation characteristic (ROC) curve analysis for the potentiality of urine L-FABP to discriminate the subjects with decreased Cin (<60 mL/min/1.73m^2 and <75 mL/min/1.73m^2).

The study was carried out according to the Declaration of Helsinki. The study protocol was approved by The University of Tokyo Institutional Review Board, and the Ethical Review Committee and Research Review Committee of both ICDDR,B and Mymensingh Medical College. Informed consent was obtained from each participant or the participant's family.

24.3 Results

As shown in Fig. 24.1, 648 patients were treated for VL between December 2013 and December 2015. The surveillance team identified 197 patients (30.4 %) on site. Of these, 45 received single dose AmBisome® 10 mg/kg for primary VL. One hundred eleven patients received multiple doses of AmBisome® 5 mg/kg for treatment failure or relapse. Five patients with PKDL were treated by miltefosine, while 20 PKDL patients received single dose AmBisome® and paromomycin for 10 days as an alternative to miltefosine.

Of the 197 patients, we found 20 people (10.2 %) who tested positive for L-FABP ICT. Figure 24.2 demonstrates a typical L-FABP ICT test and diagnostic sheet. Table 24.1 summarizes the background data of the 197 subjects, 20 L-FABP positive subjects, and 177 negative subjects. People who were L-FAPB ICT positive were more likely to be older men. There was no primary case treated by single dose AmBisome® in the L-FAPB ICT positive group.

Among 20 subjects who tested L-FABP positive by ICT, 13 agreed to revisit SKKRC for further health checkups of renal function. Twelve of 13 agreed to conduct an inulin Cin test to determine renal function further (Table 24.2). As a comparison, 9 of 35 patients who tested negative or were not examined by L-FABP ICT during the follow up period but tested positive at discharge from SKKRC

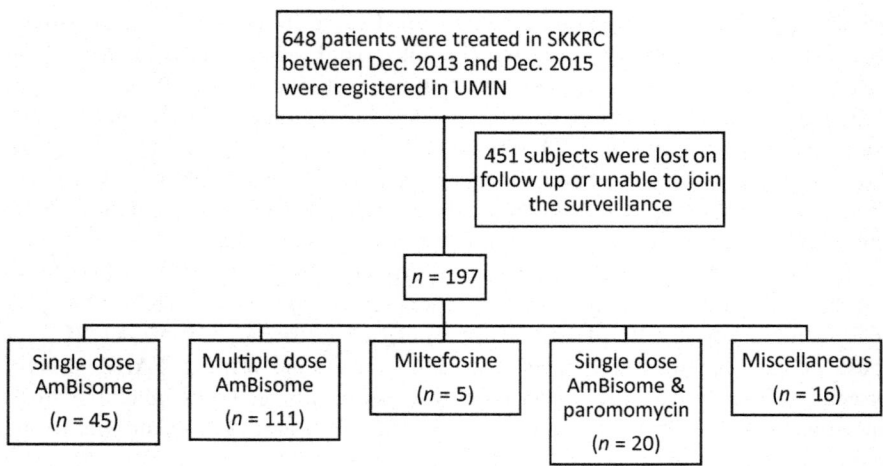

Fig. 24.1 Treatment breakdown of participants to the follow up surveillance

accepted health checkups at SKKRC (Table 24.3). Serum creatinine levels were 1.09 ± 0.37 and 0.86 ± 0.18 mg/dL in subjects with positive L-FABP ICT in urine ($n = 13$) and those with negative results or who were not examined by L-FABP ICT during the follow up period but positive at discharge ($n = 9$), respectively (t-test p-value of 0.05). We also conducted Cin for 16 subjects, of which 12 were L-FABP ICT positive and 4 were recruited from those with negative or not examined by L-FABP ICT but positive at discharge. Figure 24.3a, b demonstrate the correlations of Cin with estimated creatinine clearance by Cockcroft-Gault formula [4] and estimated glomerular filtration rate (GFR) by modification of diet in renal disease (MDRD) formula [5], respectively. The Cin examination, an alternative indicator of glomerular filtration rate, performed at SKKRC had an excellent resolution to reprise the accuracy of Cockcroft-Gault and MDRD formula. The correlation of Cin with urine L-FABP was $r^2 = 0.56$; lower than the Cockcroft-Gault and MDRD formula. Figure 24.4 shows the potentiality of urine L-FABP to discriminate subjects with lower Cin. The area under the curve (AUC) of ROC curve to detect Cin less than 60 mL/min/1.73m^2 was 0.92, with a sensitivity of 1.0 and a specificity of 0.83 using a cut-off level of 9.8 ng/mL (Fig. 24.4a). The AUC-ROC curve to detect Cin less than 75 mL/min/1.73m^2 was 0.95, with a sensitivity of 0.86 and specificity of 0.88 using the cut-off level of 8.0 ng/mL (Fig. 24.4b).

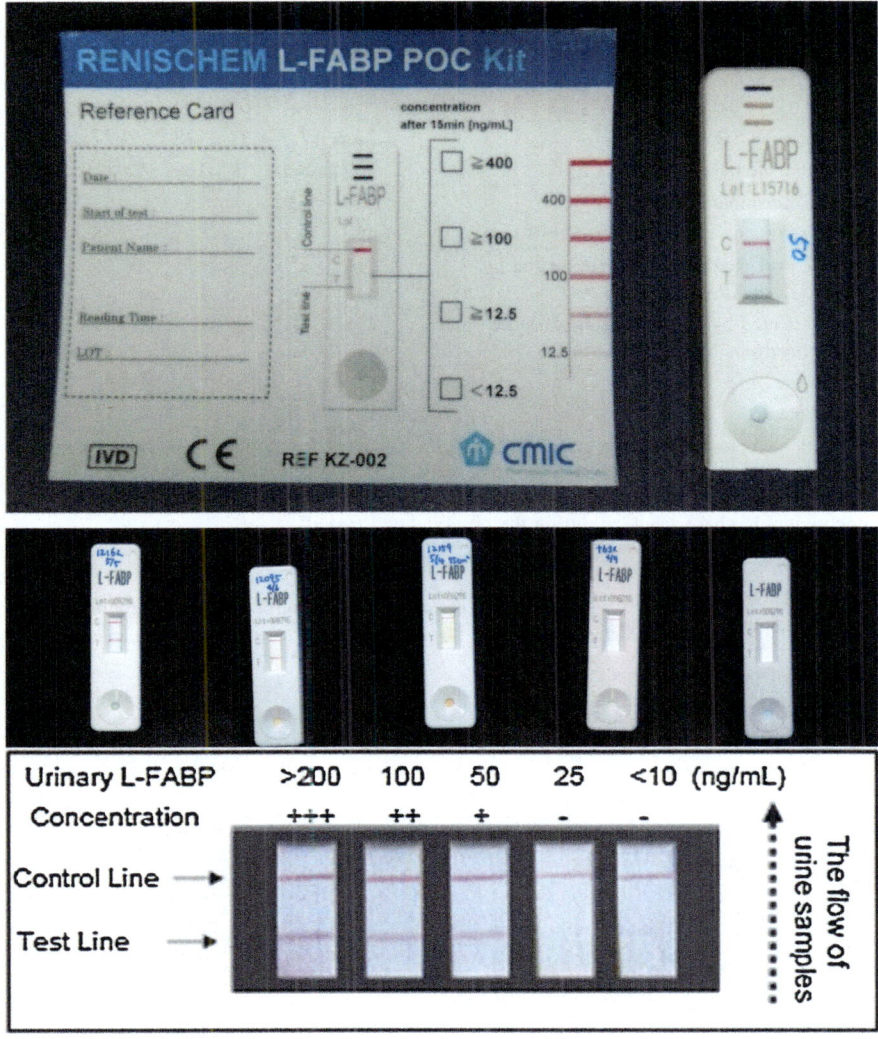

Fig. 24.2 L-FABP immunochromatography (ICT). The cursory cut-off is 12.5 ng/mL. The color-matching sheet can provide more detail

24.4 Discussion

We showed that the L-FABP ICT used during the follow up period after VL treatment is promising to identify people with decreased kidney function after VL treatment. VL is one of the well-known diseases in NTDs and is life threatening. However, the disease does not develop rapidly but progresses slowly. Almost all patients work for a low wage in agriculture, and they do not visit the clinic because

Table 24.1 Background of all subjects and L-FABP ICT positive subjects

	All ($n = 197$)	L-FABP positive ($n = 20$)	L-FABP negative ($n = 177$)
Age (mean ± SD)	27 ± 17	37 ± 18	27 ± 17
Sex (male : female)	129 : 68	19 : 1	110 : 67
Visceral Leishmaniasis (VL)			
Primary: treated by single dose AmBisome®	45	0	45
Relapse and treatment failure: treated by multiple dose AmBisome®	111	13	98
PKDL: treated by single dose AmBisome® and paromomycin or Miltefosine	20	5	15
Miltefosine	5	1	4
Miscellaneous	16	1	15

It should be noted that single dose AmBisome® treatment for primary visceral leishmaniasis (VL) did not cause any urine L-FABP positive subjects. Within those PKDL and L-FABP positive, only 1 was treated by miltefosine. In SKKRC, only AmBisome® was used as a liposomal amphotericin B pharmaceutical during the observational period

Table 24.2 Serum creatinine (SCr), proteinuria, and inulin clearance (Cin) data of L-FABP ICT positive subjects after treatment of visceral leishmaniasis (VL) at SKKRC

ID	SCr (mg/dL)	Proteinuria	Cin (mL/min/1.73m^2)
1	0.84	–	93.6
2	0.69	–	91.3
3	0.87	–	88.4
4	0.7	–	97.6
5	0.99	–	60.4
6	1.09	–	47.0
7	0.78	–	92.0
8	1.7	+	37.1
9	1.13	+	69.5
10	1.4	–	46.5
11	1.1	+	71.2
12	1.89	–	
13	0.93	–	101.2

All subjects were male. One subject ID12 could not collect urine at the end point of the Cin exam

they do not want to miss out on day labors and income. They only visit the clinic when their health has greatly deteriorated. Furthermore, if they feel better again, they often discontinue treatment early. Because of this, and together with concerns about the tolerance and cardiotoxicity of conventional therapies such as SSG [6], new therapeutic regimens were introduced. Clinics now use a several different treatment protocols after distinguishing the clinical situation of VL such as if it is a primary case, a treatment failure case from the initial treatment, a relapse case, or PKDL. AmBisome® [7] and paromomycin [8] are the most commonly prescribed

Table 24.3 Serum creatinine (SCr), proteinuria, and inulin clearance (Cin) data of L-FABP ICT negative subjects during the follow up period but positive at discharge after treatment of visceral leishmaniasis (VL) at SKKRC

ID	SCr (mg/dL)	Proteinuria	Cin (mL/min/1.73m^2)
14	0.99	−	87.2
15	0.87	−	
16	0.66	−	
17	1.11	−	60.1
18	0.83	+	
19	0.87	−	87.2
20	0.87	−	88.4
21	0.99	±	
22	0.53	±	

All subjects were male. Four subjects accepted inulin clearance exam

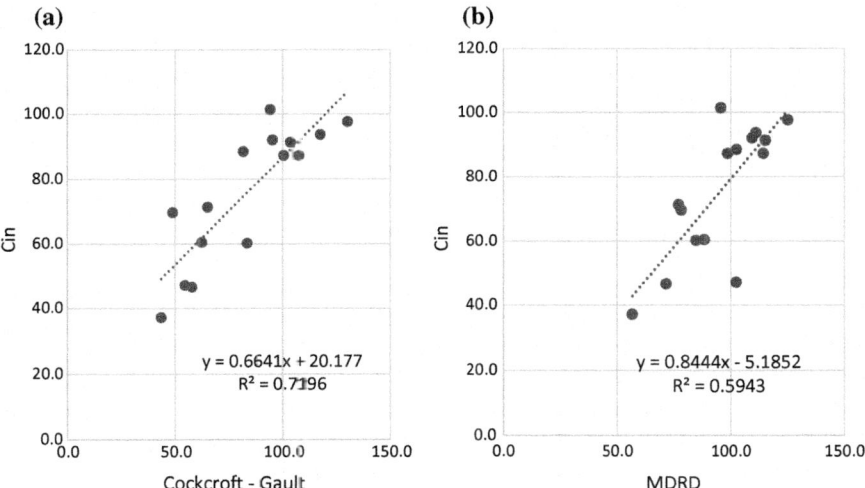

Fig. 24.3 The correlations between Cockcroft-Gault formula and Cin, MDRD formula and Cin. Cin denotes inulin clearance (mL/min/1.73m^2)

therapeutic agents for all of those cases, partially because of a low availability of miltefosine at this moment due to business decisions of pharmaceutical companies. Paromomycin, an aminoglycoside antibiotic, also has the potential to cause drug-induced acute kidney injury (AKI), though it was reportedly low risk. AKI has a significant impact on the outcomes of critically ill patients and the progression to CKD [9]. The potential for drug-induced renal injury was discussed in Chap. 12 in which a small number of patients showed increased urinary levels of L-FABP at discharge. Urine tests at discharge are often declined by patients in spite of encouragement from the medical staff, and is far from a routine examination. Therefore, urine surveillance was performed on subjects treated in SKKRC for VL

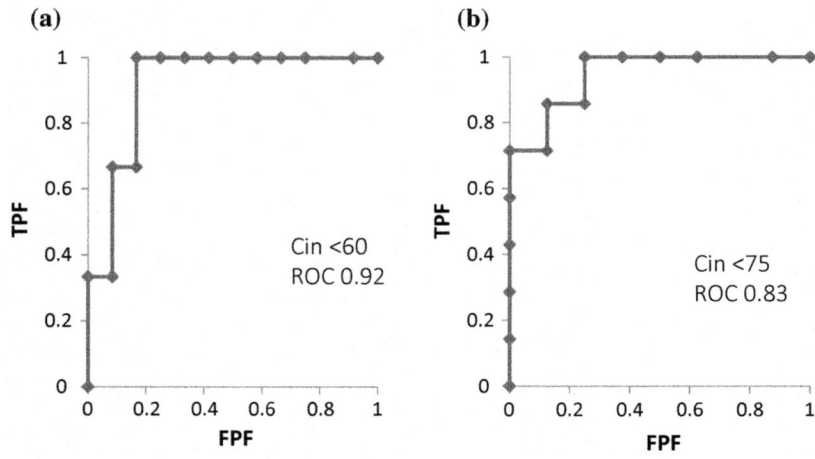

Fig. 24.4 Receiver operating characteristic (ROC) curve and analysis for urinary L-FABP at different Cin. L-FABP denotes L-type fatty acid binding protein. Cin denotes inulin clearance (mL/min). TPF denotes true positive fraction equivalent to sensitivity. FPF denoting false positive fraction is equivalent to $1 - \frac{Specificity(\%)}{100}$

in this study. Because the method has to be simple and reasonably sensitive, urine L-FABP ICT was developed and used in this study.

Urine L-FABP levels of endemic healthy subjects are 0.84 ± 1.68 ng/mL ($n = 59$: male 39, female 20; age 26.1 ± 13.3 years). The number who showed positive lines of L-FABP ICT among endemic healthy subjects was 1 in 59 (≈ 1.7 %), whereas those who were treated in SKKRC for VL showed ≈ 10.2 % positivity. The prevalence of CKD is increasing after VL treatment under the current most popular protocol. This trend persists even accounting for the poor health of subjects living in poverty under NTDs compared with the endemic healthy subjects. Urine L-FABP ICT is a useful tool to monitor the occurrence of silent kidney disease. As detailed in Chap. 12, the promoter of L-FABP has hypoxia responsive elements and hepatocyte nuclear factor that are sensitive to hypoxic stimuli and promote the expression of L-FABP in proximal tubules [10]. Thus, the increase of urine L-FABP forecasts the reduction of peritubular capillary blood flow as an ominous forewarning of CKD or evidence of CKD itself [11].

This report further determined the characteristics of urine L-FABP in comparison with Cin and the gold standard of kidney filtering capacity, GFR. The accuracy of the Cin test was confirmed in Fig. 24.3a, b, which show excellent significant correlations with both Cockcroft-Gault and the MDRD formula. Therefore, urinary L-FABP levels can function as a biomarker in a similar manner as lower Cin values. The AUC of the ROC curve to detect Cin less than 60 mL/min/1.73m^2 was 0.92, with a sensitivity of 1.0 and specificity of 0.83, and for less than 75 mL/min/1.73m^2 was 0.95, with a sensitivity of 0.86 and specificity of 0.88. In practical situations, the urine L-FABP ICT examination is the first line test at the clinic level in endemic areas. If the test result is positive, a portion of urine should be sent to a reference center like SKKRC to be examined using a biochemistry analyzer. This is an example of the lean screening process for drug-induced long-term side effects in NTDs. It is of note that patients in this study were enrolled after completing treatment of VL at SKKRC more than 3 months and less than 2 years ago.

In this study, the treatment protocol of single dose AmBisome® was safer compared with other protocols in terms of drug-induced kidney injury. But if patients were given divided doses for several days to eliminate VL parasites from the body, it increased the risk of CKD. The combination of AmBisome® with paromomycin has to be monitored more carefully, though the number of observations is not sufficient at this moment. One more important point for risk evaluation will be the basal condition of patients. Because patients who visited SKKRC for treatment failure or relapse have at least 1 exposure history to pharmaceuticals for VL (often AmBisome® or equivalent), the treatment received in SKKRC was a second exposure and therefore they were more susceptible to drug-induced organ injury, including of that to the kidney.

It is certainly necessary to perform further study with increased numbers of subjects and multiple medical sites, as well as initiate similar surveillance in different areas, like East-Africa, to further investigate this treatment protocol. However, this is the first report to focus on the drug-induced, long-term side effects specifically after the treatment of VL. From the point of a product benefit-risk balance, the initial prescription decision for VL should be further investigated to clarify the best dose and combination protocols. Urine L-FABP ICT can be a useful tool to monitor a patient's health condition. Future study will allow for adjustments to the protocol to reassess the effectiveness and improve the overall benefit-risk balance of NTDs.

Acknowledgments The authors deeply appreciate the team members (Mustafiz, Rabbani, Pranjal, Subir, Lija, Mumu, Asimu) for devoting such hard work towards helping neglected people suffering with NTDs while living in a neglected region. The authors also appreciate the kind contributions and suggestions from the executives of the Myemnsingh Medical College and Hospital. A part of this study was covered by Japan Agency for Medical Research and Development (EN).

References

1. Zgheib NK, Capitano B, Branch RA. Amphotericin B. In: Debroe ME, Porter GA, Bennett WM, Deray G, editors. Clinical nephrotoxins: renal injury from drugs and chemicals. US, New York: Springer; 2008. p. 323–52.
2. FDA. FDA drug safety communication: FDA eliminates the risk evaluation and mitigation strategy (REMS) for rosiglitazone-containing diabetes medicines. 2015. http://www.fda.gov/Drugs/DrugSafety/ucm476466.htm. Accessed 29 Feb 2016.
3. Smith HW. The reliability of inulin as a measure of glomerular filtration. In: Smith HW, editor. The kidney: structure and function in health and disease. New York: Oxford University Press; 1951. p. 231–8.
4. Cockcroft DW, Gault HM. Prediction of creatinine clearance from serum creatinine. Nephron. 1976;16:31–41.
5. Levey AS, Bosch JP, Lewis JB, Greene T, Rogers N, Roth D. A more accurate method to estimate glomerular filtration rate from serum creatinine: a new prediction equation. Modification of Diet in Renal Disease Study Group. Ann Intern Med. 1999;130:461–70.
6. Rijal S, Chappuis F, Singh R, Boelaert M, Loutan L, Koirala S. Sodium stibogluconate cardiotoxicity and safety of genetics. Trans R Soc Trop Med Hyg. 2003;97:597–8.
7. Sundar S, Chakravarty J, Agarwal D, Rai M, Murray HW. Single-dose liposomal amphotericin B for visceral leishmaniasis in India. N Engl J Med. 2010;362:504–12.
8. Sundar S, Jha TK, Thakur CP, Sinha PK, Bhattacharya SK. Injectable paromomycin for visceral leishmaniasis in India. N Engl J Med. 2007;356:2571–81.
9. Hsu C-Y. Yes, AKI truly leads to CKD. J Am Soc Nephrol. 2012;23:967–9.
10. Noiri E, Doi K, Negishi K, Tanaka T, Hamasaki Y, Fujita T, Portilla D, Sugaya T. Urinary fatty acid-binding protein 1: an early predictive biomarker of kidney injury. Am J Phyiol Renal Physiol. 2009;296:F669–679.
11. Kamijo A, Sugaya T, Hikawa A, Yamanouchi M, Hirata Y, Ishimitsu T, Numabe A, Takagi M, Hayakawa H, Tabei F, Sugimoto T, Mise N, Kimura K. Clinical evaluation of urinary excretion of liver-type fatty acid-binding protein as a marker for the monitoring of chronic kidney disease: a multicenter trial. J Lab Clin Med. 2005;145:125–33.

Index

A
Acute kidney injury (AKI), 142, 143, 145–147, 152, 154, 304
AID/HIV, 13, 33, 37, 47, 48, 55, 61, 70, 72, 75, 83, 85
Amastigote, 3–5, 12, 47, 72, 89, 90, 92, 94, 110, 117, 127, 152, 162, 189, 289, 292, 293
Amphotericin B, 8, 13, 14, 33, 44, 47, 53, 55, 56, 58–61, 69, 70, 72, 74, 75, 80, 81, 83, 84, 109, 249, 298, 302
Anthroponotic, 7, 11, 14, 15, 54, 72, 202, 205, 212, 215, 225, 287
Anthropozoonotic, 287
Antigen detection, 11, 127, 130, 132, 137, 139, 164
Aspiration biopsy, 127, 129, 174
Awareness, 11, 16, 241

B
Bangladesh, 7, 8, 16, 23, 24, 28, 33–35 38, 40, 47, 48, 53, 54, 69, 125, 126, 130, 136, 149, 150, 153, 168, 174, 175, 188, 190, 201, 202, 204, 206, 208, 213, 216, 217, 223, 225–230, 232–241, 251, 254, 257, 263, 271, 273, 275, 287
BCG, 86

C
Chronic kidney diseases (CKD), 147, 299, 304, 305
Combination therapy, 13, 33, 45, 46–53, 57–59, 61, 84
Compliance, 46, 48, 49, 53, 54, 59–61, 70, 72, 83, 87, 136, 139, 153, 164, 169
Cutaneous leishmaniasis, 24, 37, 60, 69, 86, 89, 117, 119, 167, 191, 215, 225

D
Diagnostic criteria, 109, 125–132, 139, 141, 161, 162, 164, 168, 169, 174, 177, 179, 181, 187, 240, 274, 300
Diagnostic test, 7, 16, 126, 127, 129
Dichlorodiphenyl-trichloroethane (DDT), 34, 212, 216, 217, 227, 240, 241
Differential diagnosis (PKDL), 9, 11, 126
Direct agglutination test (DAT), 11, 125, 128, 161, 162, 164, 205
Drug resistance, 15, 33, 42, 53, 54, 57–60, 135

E
Ecological scales, 207, 208, 211
ELISA (enzyme-linked immunosorbent assay), 11, 136–139, 142, 155, 156–158, 162, 164, 168–170, 205, 258, 261, 262, 275
Epidemiology, 8, 14, 24, 173–175, 181, 201, 205, 211, 223, 225, 241, 270

F
Fatty-acid-binding protein 1 (FABP1, L-FABP), 141–150, 152–157
Flood, 226, 229, 235, 237, 238, 241, 263, 264
Formol gel test, 127
Fulminating leishmaniasis, 85, 293

G
Geography, 277
Generic product, 250, 251, 253–255
Gp63, 5
Granulocyte-macrophage colony-stimulating factor, 47, 86

H
Histopathology (PKDL), 12

I

Indoor residual spraying (IRS), 16, 205, 212, 239
IL-10, 3, 6, 14, 47, 67, 86, 91, 189–192, 288, 292
IL-18, 84–87
Immunochemotherapy, 46, 85, 86
Immunochromatography, the (Dip Stick), 126, 143, 155, 156, 299, 301
India, 7, 8, 12, 15–17
Indian subcontinent, 8–10, 12, 16, 33–35, 38, 45, 48, 57, 80, 87, 125–131, 161, 202, 205, 208, 211, 212, 217, 223, 225, 228, 234, 235, 238–240, 250, 287
immunopathology, 104

K

Kala-azar, 8, 12, 13, 17
Kinetoplastoid DNA (kDNA), 167, 168, 173, 174, 176

L

LAMP, 12, 136, 137, 167–170, 205
Latex agglutination test (KAtex), 130, 138, 168
Latex agglutination test (KAtex), 130, 135–139
Leishman-Donovan bodies, 289
Leishmania, 7, 11, 12, 14, 15
Leishmania donovani s.l., 3–5, 67, 90, 103, 126, 161, 167, 168, 174, 208, 212, 225, 232, 287, 298
Leishmanin skin test, 13
Life cycle, 3, 5, 200, 207
Lipophosphoglycan, 5
Liposomal amphotericin B, 23, 28, 29, 44, 45, 53, 56, 60, 69, 70, 81–83, 249, 302

M

Macular, 10, 12, 67
Malathion, 217
Miltefosine, 8, 13, 14, 28, 29, 33, 40, 45–48, 53, 57, 58, 60–62, 70–73, 75, 80, 82, 83, 87, 103, 299, 302, 304
MRP8, 103
MRP14, 103
Mucosal, 6, 9, 11, 13, 86

N

Nepal, 7, 8, 16
Nodular, 11, 15, 67

P

P. argentipes, 15, 201, 202, 204–208, 212, 215–217
Papular, 67
Parasite detection, 127, 142
Paromomycin, 8, 14, 33, 45, 47, 53, 56, 58, 60, 68, 69, 80, 82, 83, 87, 300, 302, 304, 305
PCR, 12, 24, 67, 127, 135–137, 139, 167–169, 173–177, 179, 181, 205, 208, 292
Pentamidine, 43, 55, 68, 102, 109, 117
Pentavalent antimony, 68, 81
PKDL, 12, 23, 27–29, 67–72, 74, 75, 86, 168, 169, 202, 225, 227, 302, 304
Policy makers, 249
Polymorphism, 14, 179, 181
Post kala-azar dermal leishmaniasis (PKDL), 7–17
Post kala-azar dermal leishmaniasis, 12
Promastigote, 3, 4, 15, 90, 91, 104, 105, 108, 109, 117, 137, 161, 162, 175, 202, 289

Q

Quality assurance, 119, 125, 128, 131, 132, 176, 179, 284

R

Rapid diagnostic test (RDT), 126–129, 131, 132
Receiver-operating-characteristic (ROC) curve, 145, 146, 156, 158, 299, 302, 304, 305
Regional strategy, 58
Renal toxicity, 57, 147, 152
RK39, 11, 24, 125, 126, 128, 129, 131, 142, 148, 152, 162, 168, 187, 190, 240
RKRP42, 168, 258, 261, 262, 267, 268, 275, 281
RNaseP or ribonuclease P RNA, 176

S

Sandfly, 15, 168, 225
Schoolchildren, 260–264, 266–268, 270, 275, 277, 281, 283, 284
Sea-level rise, 235
Sepsis-induced AKI, 146
Serological tests, 125, 127
Serological Tests, 128–130, 136, 174
Seroprevalence, 129
Sodium antimony gluconate (SAG), 40
Sodium stibogluconate (SSG), 8, 26, 33, 39, 41, 42, 44, 54, 60, 68, 80, 83, 126, 150
Sodium stibogluconate (SSG), 135, 138, 139
Sri Lanka, 92, 206

T

Temperature, 12, 108, 131, 137, 142, 150, 161, 162, 164, 165, 167, 169, 176, 200, 201, 207, 224, 233–235, 258

Th-1, 46, 47, 86, 90, 91, 92, 93, 189, 288, 290, 292
Th-2, 6, 86, 90, 91, 189, 288, 290, 292
Treatment, 7, 8, 12–17, 23, 24, 28, 29, 33, 39–41, 43–46, 48, 53–55, 57–62, 67, 68, 70, 72, 75, 80, 81, 83, 86, 87, 89, 118, 119, 125, 126, 129, 130, 132, 135–139, 147, 149, 150, 152, 153, 158, 163, 170, 173, 174, 177, 181, 189, 190, 213, 225, 227, 238–241, 249, 250, 254, 260, 271, 272, 274, 277, 283, 297–300, 302, 303, 305

U
Urine storage, 150

V
Vaccine, 13, 47, 71, 75, 85, 89–94, 102, 287, 288, 293, 294
Visceral leishmaniasis, 7, 8
VL Care, 125, 126, 130

Z
zymodeme MON-37, 108

Printed by Printforce, the Netherlands